TROPICAL PASTURE SCIENCE

TROPICAL PASTURE SCIENCE

BY

P. C. WHITEMAN
University of Queensland, Australia

WITH CONTRIBUTIONS

BY

S. A. WARING
University of Queensland, Australia

E. S. WALLIS
University of Queensland, Australia

R. C. BRUCE
Queensland Department of Primary Industries,
Australia

OXFORD UNIVERSITY PRESS
1980

Oxford University Press, Walton Street, Oxford OX2 6DP

OXFORD LONDON GLASGOW
NEW YORK TORONTO MELBOURNE WELLINGTON
KUALA LUMPUR SINGAPORE HONG KONG TOKYO
DELHI BOMBAY CALCUTTA MADRAS KARACHI
NAIROBI DAR ES SALAAM CAPE TOWN

British Library Cataloguing in Publication Data

Whiteman, Peter Carlile

Tropical pasture science
1. Pastures — Tropics
I. Title
633'.2'00913   SB193.3.T/   79–41743
ISBN 0 19 859470 4
ISBN 0 19 859471 2 Pbk

Set by Hope Services, Abingdon
Printed in Great Britain by
Lowe & Brydone, Thetford, Norfolk

Preface

The course manual on which this book is based was an edited compendium of lecture and practical-course notes published expressly for the Australian–Asian Universities Co-operation Scheme Short Courses and the short courses sponsored by the FAO.

Since the original manual was published further short courses have been held in Indonesia, Thailand, East Africa, and Sri Lanka, and experience from these courses has demonstrated certain deficiencies in the material. As a result this book, while having its origins in the course manual, is an entirely rewritten text with, hopefully, a wider appeal to general undergraduate and postgraduate teaching in agricultural science. A companion text *Manual of practical experiments in tropical pasture science* has been published by the Department of Agriculture, University of Queensland.

The present book was sponsored by the Food and Agriculture Organization of the United Nations and was strongly encouraged by Dr Fernando Riveros, tropical pasture specialist in the Crop and Grassland Production Service, FAO, Rome. Major credit for the publication of this text is due to the Australian Freedom from Hunger Campaign who provided funds for preparation and typing.

Our sincere thanks are due to Jenny Chandra, Heather Taylor, Hazel Harte, Pauline O'Leary, and Mary Ellen Tooma for their excellent typing, correction of drafts, and preparation of figures.

Brisbane, Australia   PCW
March 1979

Contents

Acknowledgements

The authors thank the following publishers for permission to use tables and figures: American Society of Agronomy; Australian National University Press; Blackwell Scientific Publications Limited; Cambridge University Press; Commonwealth Agricultural Bureau; C.S.I.R.O. Publications — Australian Journals of — Agricultural Research, Biological Sciences, Botany, Experimental Agriculture and Animal Husbandry, and Rural Research; Doxiadis Associates International Co. Ltd, Consultants; W. H. Freeman and Company, Scientific American Inc.; Iowa State University Press; John Hopkins University Press; McGraw-Hill Book Company; Macmillan Journals Ltd; Sydney University Press; Tropical Grassland Society of Australia; University of Queensland Press; John Wiley and Sons, Wiley — Interscience Publications.

Introduction

Livestock production in the tropics can be increased through increasing the output per animal and the productivity per unit area of land. A major factor in increasing livestock productivity will be the improvement of animal nutrition and feed supplies, especially in the case of ruminant animals. Improved animal disease and parasite control, breeding, and management will also be important, but initially a major emphasis must be placed on providing better nutrition.

The importance of animal products in human nutrition has been recognized, and particular emphasis has been placed on the role of animal proteins in relation to the quality of the human diet. It is only recently that the essential importance in human diets of structural lipids derived from animal products has been recognized (Crawford 1973). The ruminant animal is particularly valuable because of its ability to convert forages, roughages, and by-products which cannot be used directly in human nutrition.

The tropics and subtropics (30 °S to 30 °N) already contain a significant proportion of the world's total ruminant animal population, as follows: cattle 55 per cent, sheep 36 per cent, goats 67 per cent, buffaloes 80 per cent, camels 86 per cent (McDowell 1972). However, in many less-developed countries productivity is extremely low, being limited largely by nutrition, but also by low reproductive rates, poor disease control, and traditional systems of management. In some areas where improved disease control and management schemes have been instituted, increasing animal populations have led to overgrazing of rangelands. This has often been exacerbated by expansion of arable agriculture into traditional grazing lands further reducing animal forage resources (Pearse 1970). Thus livestock development schemes which do not first improve the fodder resources may cause further problems owing to increased animal numbers without concomitant increases in forage resources.

The improvement of ruminant livestock production in the tropics can be approached in two ways, either through increasing the productivity of already utilized resources or through the introduction and development of new or little-used resources. In most areas of high population density in the tropics, as in south-east Asia, the traditional agricultural patterns have placed little emphasis on ruminant animals except as draught animals or scavengers of crop residues and waste ground. Cattle and buffalo were part of the rice production system but not generally considered as productive units in their own right. Only in regions where climate and soils are unsuited to intensive arable agriculture are livestock grazed as the major commercial activity. But even in these areas there has been no tradition of sowing forage crops or pastures solely for the use of grazing animals. Such a concept has been alien to most farmers and herdsmen in the tropics, although fodder crops are grown for draught animals in some countries, e.g. India and Egypt. In other areas of the tropics, particularly in East and West Africa, herding of animals has, of course, been a major activity. But again, even where herdsmen have been relatively settled or only semi-nomadic, the concept of sown pastures was not developed. Thus throughout the tropics generally there has been little development of sown pastures and hence little selection of species for pasture use.

The main feeding systems for ruminants within intensively cropped areas are based on combinations of the following: (i) crop residue and stubbles, (ii) grazing or cutting of fodder along roadsides and field margins, (iii) communal grazing areas often along forest margins, and (iv) by-products such as rice bran which are commonly used as supplements. In most regions dependant on these feed resources the quantity is likely to be limiting while the quality is variable. The special problems of supporting a large ruminant population in India in competition with an intensive agriculture have been discussed so often as not to require repetition here.

Animal production from within arable cropping systems can be greatly improved through the integration of forage production within the rotational systems; through the better utilization of crop residues, by-products, and wastes in combination with supplements; and through the integration of sown pastures with tropical plantation crops. The higher fertility cropping areas must also be incorporated into a stratified animal-production system, receiving animals from the more extensive breeding areas to be grown and fattened on improved pastures or forages.

While it seems obvious that there are many avenues for improving ruminant production in existing farming systems, there is also a major potential for increasing the areas available for grazing animals. Throughout the tropics most of the areas of fertile soils, cleared of forest, under reasonable rainfall conditions or irrigated, are already densely settled and devoted to cropping. These areas are usually in the major river valleys, or in fertile well-watered regions such as the islands of Java and Bali in Indonesia. Within these areas opportunities for large-scale pasture development for cattle grazing will be limited.

However, in many countries considered to be densely populated there are large areas of *marginal* land which could be developed with sown pastures. These areas may be defined as marginal because of insufficient rainfall to support regular cropping, low nutrient status, or soils difficult to cultivate by traditional methods. In the higher rainfall tropics large areas of marginal land are under rainforest. Any proposals for large scale clearing for crop or pasture development should be carefully examined in relation to the relative values as a timber resource, in watershed management, and in wild life conservation. Other areas in the wet tropics, which may have been under rainforest, are presently covered in natural grassland. Very commonly the major component of this grassland is *Imperata cylindrica* as in Indonesia and the Phillipines. While these grasslands support only low levels of production, they offer a great potential for improvement through correct use of fertilizers and the introduction of adapted pasture grass and legume species. As an example, these areas in Indonesia represent one of the major unused resources with an estimated total area of 4 000 000 ha (Whiteman 1973). Development of sown pastures in these grasslands might also be seen as the initial stage in improving soil fertility allowing subsequently for the integration of arable cropping into the livestock production system.

Throughout the tropical world there is major scope for increasing livestock fodder resources, but there are also major limitations to development. A survey of the south-east Asian region by the Asian Development Bank (1968) found that 'In all the countries of the region there is a desperate need to augment quickly the national manpower pool of persons professionally trained in agriculture and related sciences. The shortage of trained manpower creates deficiencies that pervade the entire spectrum of effort to promote agricultural growth'. This applies particularly in the fields of tropical pasture agronomy, tropical animal husbandry, and veterinary science. For large-scale developments the lack of property managers experienced in herd management, the use of machinery, and in pasture-seed production creates further problems. Availability of capital is also a major limitation even though international agencies such as the International Bank for Reconstruction and Development (World Bank) and the Asian Development Bank are providing increasing funds for livestock development programmes. Finally the importance of social and land tenure problems must be recognized in all livestock development projects.

REFERENCES

Asian Development Bank (1968). *Asian agricultural survey*, Vol. 1, Sect. 3, Chap. 3, pp. 75–90, Constraints on agricultural development. Manila.

Crawford, M. A. (1973). A re-evaluation of the nutrient role of animal products and new systems of livestock management from wildlife. *Prod. 3rd World Conf. Animal Production*, Melbourne. Vol. 2, Paper No. 4.

McDowell, R. E. (1972). *Improvement of livestock production in warm climates*. W. H. Freeman, San Francisco.

Pearse, C. K. (1970). Range deterioration in the Middle East. *Proc. 11th Int. Grass. Cong.*, Australia. Sect. 1, p. 26.

Whiteman, P. C. (1973). Tropical pasture development potential for livestock production in Indonesia. *Proc. 3rd World Conf. Animal Production*, Melbourne. Vol. 1, Sect. 3, p. 10.

1

Climatic factors affecting pasture growth and yield

The yield of a pasture is affected by a range of factors influencing the growth of the individual plants in the pasture sward, but any measure of overall pasture productivity must include the output of animal products derived from the pasture. Thus not only are factors affecting pasture yield important, but those factors affecting the nutritive value or quality of the pasture must also be considered.

The main factors affecting pasture growth and yield can be grouped into four broad categories:

(i) Climatic factors: radiation, day length, temperature, humidity, wind, precipitation.
(ii) Soil factors: chemical fertility, physical properties, soil moisture characteristics, topography.
(iii) Pasture species: genetic potential for yield and nutritive value, adaptation to the environment, plant competition, acceptability to the grazing animal, and long-term persistence.
(iv) Pasture management: type of animal grazed, stocking rate and stocking system used, fertilizer strategies, weed control, and other cultural practices.

The interactions of the factors are discussed in the chapters which follow, leading finally to some assessments of the options available for pasture development within the constraints imposed.

1.1 CHARACTERISTICS OF TROPICAL CLIMATES

The tropics can be defined in geographical terms as the region between the Tropics of Cancer and Capricorn (23.5 N and 23.5 °S latitude respectively). Latitudinal boundaries so defined are unlikely to reflect identifiable vegetation or land-use boundaries. Various definitions have been proposed to delineate the tropical climate boundaries so that these better reflect biological boundaries.

Although solar insolation is the primary factor, most definitions depend on temperature, either the mean annual temperature or the mean temperature of the coldest month. On the basis of mean annual temperature different authors have proposed boundaries at 20, 21.1, and 23 °C, while on the basis of mean temperature of the coldest month 18, 20, and 21 °C have been proposed (Blumenstock 1957).

In terms of effects on vegetation, minimum temperatures in the coldest month will have a greater effect on plant growth and species distribution than mean temperatures. Blumenstock (1957) compared boundaries based on all the above temperature limits and suggests that Köppen's (1936) boundary of a mean of 18 °C in the coldest month was most appropriate (Fig. 1.1(a)).

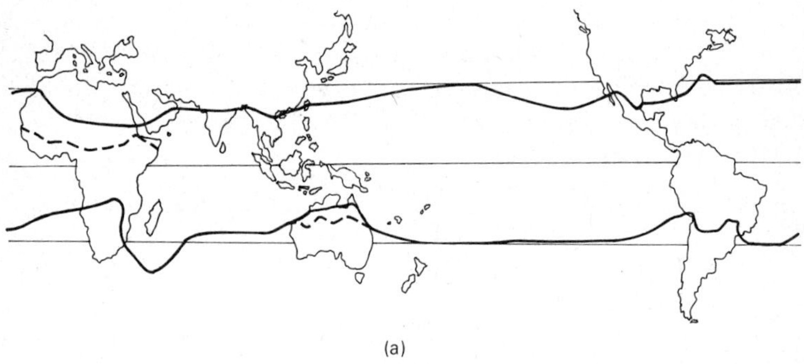

(a)

Fig. 1.1(a). Boundary of the tropical climatic realm (solid line) and boundary between the dry and humid tropics (dashed line). (From Blumenstock 1957.)

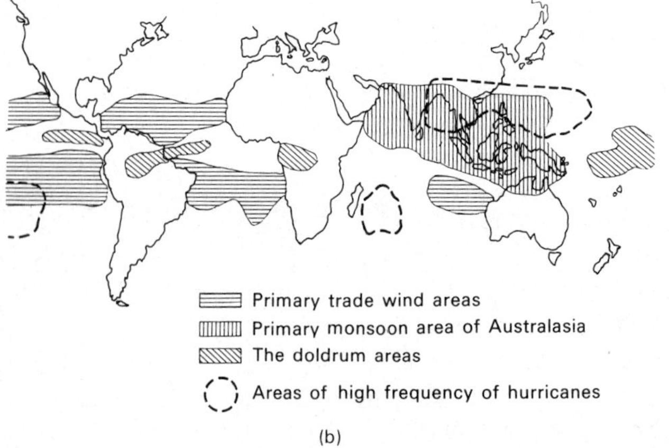

(b)

Fig. 1.1(b). Main circulation regions in the tropics. (Adapted from Blumenstock 1957.)

As a first approximation, separation of tropical climates from extra-tropical climates appears reasonable on the basis of thermal boundaries. Temperature characteristics of tropical climates are readily recognized in human comfort and animal adaptations, while a generalized division of plant species into tropical and temperate forms is widely recognized. Adaptation of some plant families to the higher temperature and radiation regimes of the tropics is manifested in fundamental differences is biochemical and photosynthetic pathways. These adaptations are discussed in Section 1.2.

Within tropical climates there is a further major division into humid and dry tropical climates. Many factors affecting evaporative demand and soil moisture storage and availability should be taken into account in determining the moisture regime. However Köppen's (1936) simple division of humid from monsoon and periodically dry climates is based on a minimum of 60 mm precipitation in the driest month. This has been shown to provide a useful boundary realistically coinciding with major changes in vegetation communities (Blumenstock 1957). The non-humid tropical climates can be subdivided on the basis of length, and time of year of the dry season, and a generalized version of the major climatic types is shown in Table 1.1.

In the tropical region defined by thermal limits, the further sub-divisions on the basis of rainfall regimes are influenced by the atmo-spheric circulation patterns. Blumenstock (1957) recognizes five major circulation regions (Fig. 1.1(b)) as follows:

(i) *Middle latitude westerlies*

This region is defined in terms of the frequency of extra-tropical cyclonic influences which bring a variety of weather types including cold and warm fronts, general rains, and cold front storms. In contrast to other tropical realms these regions can experience occasional cold waves of varying severity. These could influence species diversity and land-use patterns.

(ii) *Primary trade wind area*

These realms embrace those areas in which winds from the north-east in the Northern Hemisphere and south-east in the Southern Hemisphere occur at least 60 per cent of the time in every month of the year. Trade winds carry waves or eddies of unsettled weather, considerable cloudi-ness, and often intense steady widespread rain. Because of the regular unidirectional flow pronounced rain shadow effects are common on the leeward side of islands with high mountains, e.g. Hawaii.

Table 1.1. Generalized climatic types (based on Köppen; Vink 1975)

Zone A: Tropical rainy climates (average monthly temperature never below 18°C)
   Af: always humid, at least 60 mm precipitation in the driest month.
   Am: monsoon climate with moderately dry season.
   Aw: savannah climates with dry winter.
   As: savannah climates with dry summer.

Zone B: Dry climates (determined by relation between annual precipitation and
        annual temperature)
   BS: steppe climates.
   BSh: hot B climate.
   BSK: cool B climate.
   BSK[1]: cold B climate.
   BW: desert climate.

Zone C: Warm – temperate rainy climates (between 18 and 3°C isotherms in
        coldest month)

Zone D: Subarctic climate with cold winter (between −3°C in coldest month and
        10°C in warmest month)

Zone E: Polar climates (warmest month between 10 and 0°C or even lower)

A climates

A graphical representation of boundaries of Af, Am, and Aw climates (after
Köppen) is shown below:

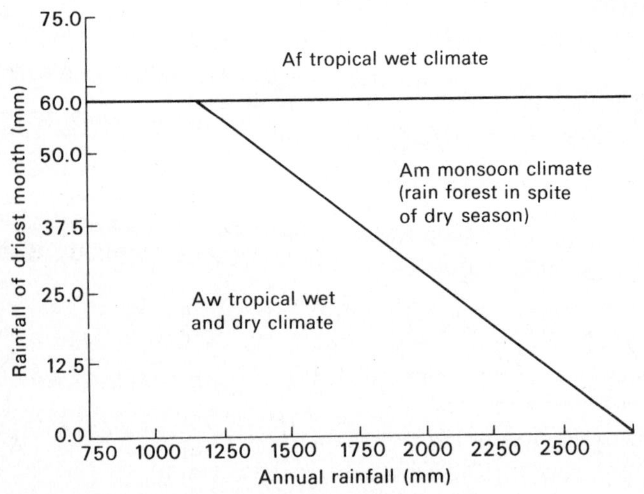

(iii) *Primary monsoon area*

The main characteristics of the monsoon regions are the seasonal reversal of winds and the marked seasonality of cloudiness and rainfall. The monsoon winds often contain waves and eddies, bringing an alternation during the rainy season of spells of heavy prolonged rain, intermittent rains, and periods of clear weather. Very heavy rains yielding up to 250 mm per day may persist for over five days in some hilly regions.

(iv) *Doldrums region*

This includes not only ocean areas but large areas of land in the Amazon Basin and Equatorial Africa. They identify areas within which horizontal temperature gradients are small, rainfall is well distributed, and humidity and cloudiness are higher throughout the year than in other lowland tropical regions.

(v) *Hurricane or typhoon region*

This can be divided into regions of high and low hurricane (or cyclone) frequency (Fig. 1.2). The high winds and high intensity rainfalls generated by these events represent a significant hazard in land use. The high intensity rainfalls can have pronounced effects on runoff and consequent erosion.

Between these major regions defined by Blumenstock (1957) are transition areas where major influences interact to affect the climate. Within this generalized definition of the major climatic characteristics of the tropics we must recognize that the climate of any local region will be modified by particular conditions of radiation, temperature, wind, humidity, and rainfall distribution. Furthermore these factors are even further modified at the microclimate level where the major effects on plant growth are exerted. The effects of these particular environmental parameters are discussed below.

1.2 THE RADIATION REGIME

Agriculture has been defined as 'an exploitation of solar energy made possible by an adequate supply of nutrients and water to maintain plant growth' (Monteith 1958). The radiation regime is a basic determinant of plant growth through the direct input of energy into the photosynthetic system, into the transpiration process, and as a determinant of leaf temperature. The illumination regime also determines the

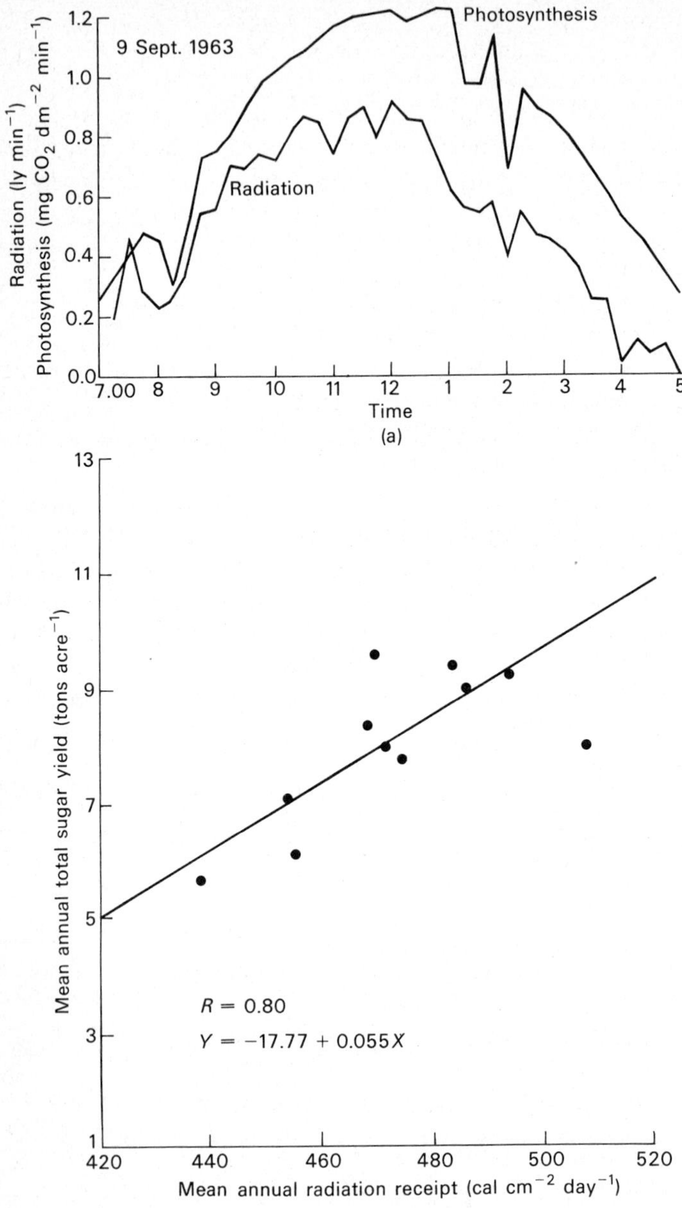
9 Sept. 1963
Photosynthesis
Radiation
Radiation (ly min^{-1})
Photosynthesis (mg CO$_2$ dm^{-2} min^{-1})
1.2
1.0
0.8
0.6
0.4
0.2
0.0
7.00 8 9 10 11 12 1 2 3 4 5
Time
(a)
Mean annual total sugar yield (tons acre^{-1})
13
11
9
7
5
3
1
$R = 0.80$
$Y = -17.77 + 0.055X$
420 440 460 480 500 520
Mean annual radiation receipt (cal cm^{-2} day^{-1})
(b)

photoperiod (or daylength) which has major effects on growth and reproduction of pasture species.

Solar radiation

At the equator the amount of radiation received at the outer atmosphere varies only about 13 per cent throughout the year (Blumenstock 1957). Seasonal variation in the radiation receipt increases greatly with increasing latitude. The total energy received at the outer atmosphere declined with increasing latitude as follows:

Latitude	0	10	23½	40	60	80
Energy receipt (kcal cm^{-2} a^{-1})	312	306	287	246	178	134

The amount of radiation received at the earth's surface depends upon the degree of atmospheric filtration determined by (i) the latitude and therefore the angle of the sun to the atmosphere; (ii) the altitude of the measuring station, and (iii) the cloudiness and atmospheric turbidity caused by water vapour, dust, smoke, and other aerosols.

Radiation received at the ground varies from about 0.2 to about 0.8 of that received at the outer atmosphere. Within the tropics, where variation due to latitude is relatively small compared with higher latitudes, the major determinant of radiation receipt is cloudiness. The close relationship between cloudiness and radiation received at the ground is demonstrated by the data in Table 1.2.

Table 1.2. Relationship between average cloudiness and radiation receipt at ground level (after Blumenstock 1957)

Area	Mean cloudiness values (tenths)		Insolation at ground (g cal day^{-1})
(i) Nairobi, Kenya	'Summer'	0.6	590
	'Winter'	0.8	350
(ii) Katanga, Congo	'Summer'	0.4	525
	'Winter'	0.7	395

In the humid tropics radiation receipt can be markedly reduced by cloud cover. The regions of highest radiation levels are in the arid zones

Fig. 1.2(a). The relationship between changing radiation levels and photosynthesis in a cotton leaf throughout the day. (b) Effect of mean daily radiation receipt on yield of sugar per crop of sugar-cane in Hawaii. (From Chang 1968.)

on the tropical margin, the major desert areas of Sahara, Saudi Arabia, and Central Australia, as shown in Table 1.3.

Table 1.3. Average annual radiation receipts at ground level in selected regions

Region	Radiation receipt $(kcal\ cm^{-2}\ a^{-1})$
Sahara – Sudan	200–20
Australia – North and centre	180–200
Australia – South	140–80
S.E. Asia – Indonesia, Malaysia	140–50
Western Europe	80–100
United Kingdom	70–90

During periods of prolonged cloud cover, such as in monsoonal wet seasons, plant growth can be limited by low radiation receipts. This is demonstrated in many areas of the world where the irrigated dry-season rice crop greatly outyields the wet season crop. The rate of photosynthesis is related to the amount of radiation energy received, both at the level of a single leaf (Fig. 1.2(a)) and for a whole crop canopy (Fig. 1.2(b)).

In terms of pasture productivity in the tropics, the relationship between photosynthesis and radiation is of particular importance. It has been found that the tropical grasses have evolved a different biochemical pathway of photosynthesis, apparently better adapted to the higher radiation and temperature conditions of the tropics, and giving the potential for higher growth rates. This new biochemical pathway was elucidated by Hatch and Slack (1966) and is different from the original pathway in temperate species demonstrated by Calvin and Benson (1948). The Calvin pathway is termed the C_3 pathway, and the Hatch and Slack pathway in tropical grasses, the C_4 pathway. These are compared in Fig. 1.3

The C_4 pathway has been identified in nearly 100 genera in at least ten plant families. The largest group of plants having the C_4 pathway are the tropical grasses in the sub-family Panicoideae, while a range of dicotyledonous species, particularly arid zone plants, e.g. some species of *Atriplex*, and *Amaranthus* also have the C_4 pathway.

Associated with the C_4 pathway in the tropical grasses are a number of other features which distinguish them from the temperate grasses and the tropical legumes. These differences have important consequences in pasture productivity. At the anatomical level the leaves of the C_4 grasses are characterized by having two types of chloroplast containing

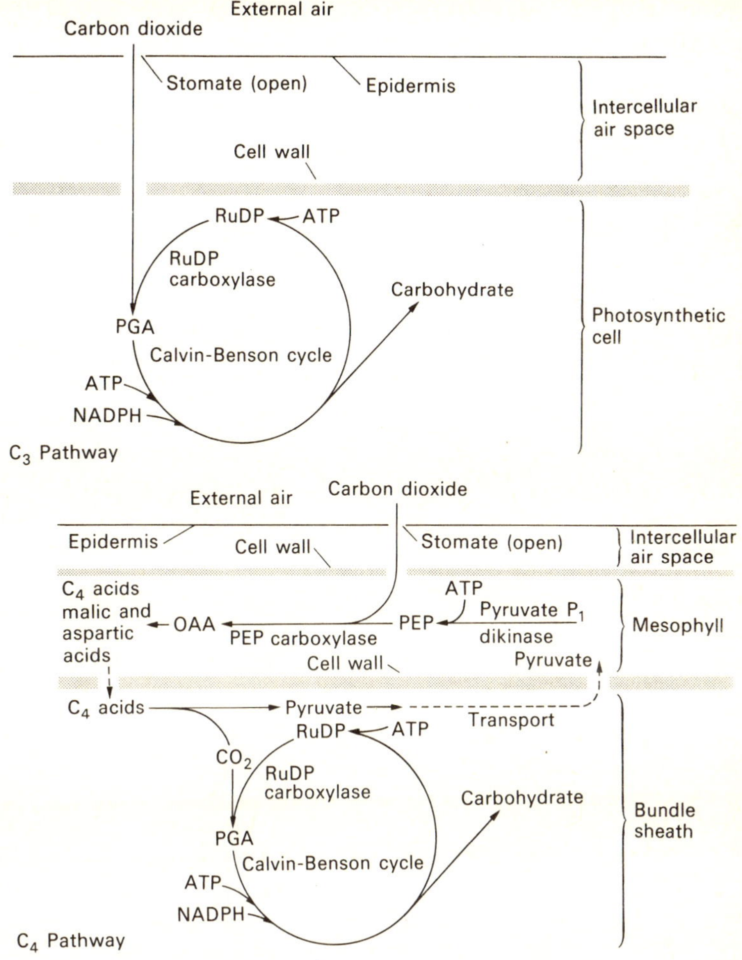

Fig. 1.3. Comparison of the C_3 and C_4 biochemical pathways of photosynthesis (Bjorkman and Berry 1973).

cells, the mesophyll cells and the bundle-sheath cells which surround leaf veins. Chloroplasts in the pallisade cells are much smaller chloroplasts containing very little starch, while the bundle-sheath chloroplasts contain abundant starch grains (Laetsch 1968). These morphological features are interpreted as adaptations for the rapid transport of pre-cursors and end-products of photosynthesis.

Associated with the morphological differences between C_3 and C_4 plants are differences in biochemical and photosynthetic pathways. The most striking features are the higher rates of net photosynthesis in C_4 grasses, and their response to increasing levels of light intensity up to full sunlight (Fig. 1.4). Another feature is the lack of photorespiration in the C_4 plants during photosynthesis in the light. Also C_4 plants in a sealed atmosphere are able to reduce the ambient CO_2 concentration to zero in the light, that is, they have a CO_2 compensation point of zero, compared with C_3 plants (tropical legumes) which have a CO_2 compensation point of about of about 40 p.p.m. CO_2 (Ludlow and Wilson 1972).

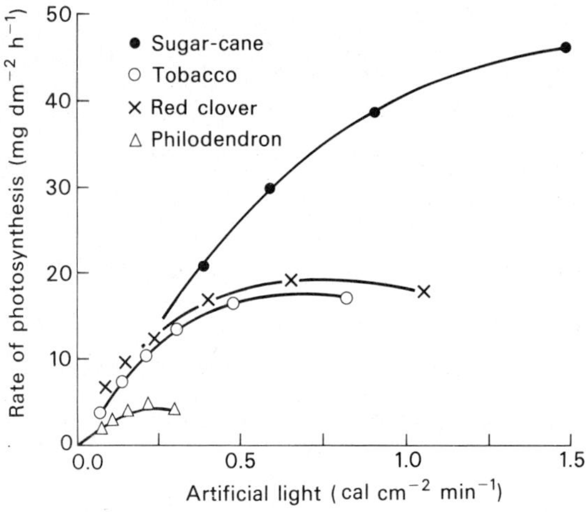

Fig. 1.4. Comparison of net photosynthesis of C_4 (sugar-cane) and C_3 plants as function of radiation flux density of incandescent light (Hesketh and Moss 1963).

The higher rate of net photosynthesis in tropical grasses compared with the legumes is not simply due to the lack of photorespiration, which can account for less than half of the difference (Ludlow and Wilson 1972). A major effect appears to be in large differences in intra-cellular resistances to CO_2 transfer. The main features of comparisons between C_4 and C_3 plants are summarized in Table 1.4.

Table 1.4. Main features of comparison between C_4 and C_3 plants

C_4 Plants	C_3 Plants
Biochemical	
First photosynthetic products are 4-carbon acids — malic and aspartic	First photosynthetic product is a 3-carbon acid–Phosphoglyceric acid
The CO_2 acceptor molecule, phosphoenolpyruvate (PEP) and its associated enzyme (PEP-carboxylase) is more reactive with CO_2 than the CO_2 acceptor system in C_3 plants	The CO_2 acceptor molecule ribulose diphosphate (RuDP) and its associated enzyme RuDP carboxylase is less reactive with CO_2 than the CO_2 acceptor system in C_4 plants
PEP carboxylase is not inhibited by oxygen	RuDP carboxylase is somewhat inhibited by oxygen
Optimum temperature for PEP carboxylase activity is between 30–5 °C	Optimum temperature of RuDP carboxylase activity is between 20–5 °C
Physiological consequences	
Rate of photosynthesis in single leaves is higher than in C_3 plants, with maximum values up to 100–20 mg CO_2 dm^{-2} h^{-1}	Maximum rate of photosynthesis by single leaves rarely exceeds 45 mg CO_2 dm^{-2} h^{-1}
Light saturation of photosynthesis is approached at 11×10^4 lm m^{-2} (approximately full sunlight)	Light saturation of single leaves usually reached in the range 4–5×10^4 lm m^{-2}
No apparent photorespiration	Significant photorespiration (7–15 mg CO_2 dm^{-1} h^{-1})
CO_2 compensation point (Γ) is zero in the light	CO_2 compensation point usually in the range 30–60 $\mu\ell$ ℓ^{-1} CO_2 concentration
Rate of photosynthesis is unaffected by reduced atmospheric oxygen concentrations	Rate of photosynthesis is enhanced (by 20–60%) at low oxygen concentrations
Changes in stomatal resistance (r_s) exert a more sensitive control over the rate of CO_2 uptake, while mesophyll resistance (r_m) is low	Less sensitive to changes in stomatal resistance (r_s). Mesophyll resistance is a much larger component in overall resistances to CO_2 uptake
Ratio of photosynthesis/transpiration (P/E) tends to be lower in C_4 grasses	Ratio P/E is higher than for C_4 grasses
Morphological attributes	
Leaves have two types of chloroplast-containing cells, bundle-sheath cells surrounding the vascular tissue and mesophyll cells surrounding the bundlesheath cells	Only one type of chloroplast-containing cells, chloroenchyma cells, distributed throughout the leaf mesophyll

(continued)

Table 1.4. (*Cont'd*)

C_4 Plants	C_3 Plants
Two types of chloroplasts are developed. The mesophyll cell chloroplasts are small and contain little starch. Bundle-sheath chloroplasts are large and contain abundant starch grains	Only one type of unspecialized chloroplast. Both photosynthetic processes and starch deposition occur simultaneously
Mitochondria of bundle-sheath cells are much larger than in the mesophyll cells	Mitochondria are small

The higher photosynthetic rates in C_4 grasses at both the individual leaf level and on a whole sward basis, lead, of course, to higher crop growth rates even though the dark respiration rate of the C_4 species is higher than the C_3. For example, Ludlow and Wilson (1972) found that the mean value of dark respiration at ambient CO_2 concentration and 30 °C in the tropical grasses was 3.97 mg CO_2 dm^{-2} h^{-1} and for the tropical legumes was 2.83 mg CO_2 dm^{-2} h^{-1}. Photorespiration rate in the tropical legumes is in the range of 7–15 mg CO_2 dm^{-2} h^{-1}.

The other important consequence of the higher efficiency of solar energy utilization by C_4 plants is in the ability to produce dry matter. With very high nitrogen rates (2240 kg N ha^{-1} a^{-1}) and a 90-day cutting cycle, a dry matter yield of 85 860 kg ha^{-1} a^{-1} has been produced from napier grass (*Pennisetum purpureum*) (Vicente-Chandler, Silva, and Figarella 1959). Although this cutting interval yielded pasture of relatively low quality, and the level of nitrogen used was not economic, it does indicate the potential of the tropical grasses to fix CO_2 and energy into dry matter. This value of 85.8 t ha^{-1} compares with a maximum value recorded by Hughes (1970) for a temperate grass (*Lolium multiflorum*) of 21.4 t ha^{-1}.

Thus, where nitrogen and other nutrition is adequate, the potential rate of CO_2 fixation and dry-matter production gives the tropical grasses a marked competitive advantage in utilization of solar energy over the associated legumes. Maintenance of a legume in a tropical pasture sward must then be achieved through a nitrogen limitation to the potential grass productivity and/or by grazing and defoliation management to favour the legume.

Photoperiod

Another important feature of the effects of latitude on the radiation regime is mediated through the seasonal changes in daylength or photoperiod. Although the seasonal amplitudes of daylength are not as great in the low latitude tropics as in higher latitudes, as shown in Table 1.5, these changes have important consequences on the reproductive development of plants. Because photoperiod responses are mediated at low light intensities, the twilight periods are important and are therefore included in Table 1.5.

Table 1.5. The relationship between latitude and maximum and minimum daylengths (after Blumenstock 1957)

Latitude	Day length (hours and mins) (sunrise to sunset)		Day length (hours and mins) (including civil twilight)	
	Max.	Min.	Max.	Min.
0°	12:10	12:10	12:50	12:50
5°	12:30	11:50	13:00	12:30
10°	12:40	11:30	13:10	12:10
15°	13:00	11:10	13:50	12:00
23½°	13:30	10:40	14:20	11:30
40°	15:00	9:20	16:10	10:20
50°	16:20	8:00	17:50	9:20
65°	22:00	3:30	24:00	5:00

In a wide range of pasture plants the change from vegetative growth to reproductive development is induced by changing daylength. This response is perceived in the leaves of the plant, which transmit a floral stimulus to the vegetative meristem to cause a change to reproductive development. This process is termed floral induction. Usually plants must pass through a juvenile phase before becoming sensitive to photoperiod effects. This 'ripeness to flower' stage is often dependant upon achieving a minimum number of leaves before becoming receptive to the photoperiod effects.

The development of the floral stimulus has been shown to be mediated through the *phytochrome* pigment system (Butler, Norris, Siegelman, and Hendricks 1959). This pigment can exist in two forms, a red-light absorbing form (Pr) and a far-red-light absorbing form (Pfr). In darkness many plants experience a rapid phytochrome conversion from the far-red to the red-absorbing form, and this is presumed to be the initial step of the photoperiodic reaction. The Pfr form is thought to be biologically active, suppressing flowering in 'short-day' plants and hastening it in 'long-day' species. Thus 'short-day' plants appear to require a longer dark period to retain the Pr form for a longer period. These reactions can be summarized as follows:

$$\text{Far-red absorbing form of phytochrome } (P_{730}) \;\underset{\text{Red light}}{\overset{\text{Darkness Far-red light}}{\rightleftharpoons}}\; \text{Red absorbing form of phytochrome } (P_{660})$$

Thus the critical component in photoperiod responses under natural conditions is not the length of the light period, but the length of the dark period. On the basis of response to photoperiod conditions, plants can be classified into a number of major groups:

Short-day plants (SDP). These flower in response to a range of relatively short days, i.e. require relatively long dark periods to build up the floral stimulus. Therefore they will not flower given night lengths shorter than some critical minimum period. Under otherwise inductive night lengths, flowering can be prevented by giving a light break of low light intensity in the middle of the dark period.

Long-day plants (LDP). These flower in response to a range of relatively long days, i.e. require relatively short dark periods for floral initiation, and will not flower in night lengths longer than some critical maximum. Flowering may be promoted experimentally by continuous (24 hours) light. Under short-day conditions with non-inductive long nights, flowering may be promoted by giving a night light break.

Intermediate plants (IP). Some species will only flower within certain limits of day length, i.e. they have both a critical maximum and a critical minimum night length.

Indeterminate or day-neutral plants (DNP). Species in this category are not sensitive to photoperiodic control and will flower over a wide range of daylengths.

This classification is complicated by some species which have particular multiple requirements of long days followed by short days (LD–SD), e.g. some *Bryophyllum* and *Cestrum* species. Some winter cereals, such as rye, may have SD–LD requirements. Also some species may be strictly controlled in flowering by photoperiod while others are only quantitatively accelerated or delayed by the photoperiod. Photoperiod responses are also important in a range of other plant development processes including vegetative growth, tuber and bulb formation, dormancy, and seed germination. Therefore photoperiod response is important when considering the adaptation of pasture species to particular tropical regions. Not only will photoperiod determine whether a species will flower and set seed in a particular environment, but also will determine the length of the vegetative growing period which is of particular importance in pasture production. The photoperiod responses of a range of important tropical pasture species are reviewed in Section 3.3.

1.3 THE TEMPERATURE REGIME

The ambient temperature regime of a plant in the field is difficult to define as it is a dynamic factor; in common with other environmental factors it changes continually throughout a 24-hour diurnal cycle, as well as on a longer term seasonal basis. The temperature of a plant is basically determined by the radiation regime and the ambient air temperature modified by the aspect in which it is growing, its position in a canopy determining the amount of shading, its rate of transpiration, and other microclimatic parameters such as wind movement and saturation deficit of the air.

For any species a range of cardinal temperatures can be defined, as represented below:

Lethal minimum	Mimum temperature	Optimum temperature	Maximum temperature	Lethal maximum
	for growth	for growth	for growth	

The cardinal temperatures should not be considered as precisely defined single temperatures, but as a temperature range, as these are modified by factors such as plant age (Alberda 1969), by daylength and light intensity (Beinhart 1962), and by plant adaptation through previous exposure to higher or lower temperatures (Downton and Slatyer 1972).

The upper limit of temperature for plant growth is the *maximum* temperature at which growth ceases. This can be related to inhibition of photosynthesis coupled with an increase in the rate of photorespiration and/or dark respiration, so that net photosynthesis approaches zero (Pearson and Hunt 1972; Ludlow and Wilson 1971). Further increase in temperature to the lethal maximum causes irreversible biochemical damage and denaturation of enzymes leading to death of the plant. At the other end of the scale, cessation of growth at the *minimum* temperature may be related to a decline of the rate of reaction in enzyme systems involved in photosynthesis and to changes in chloroplast structure (West 1970) and to stomatal closure (Pasternak & Wilson 1972). For many species the lethal minimum is achieved at -2 to $0\,^{\circ}$C, where death from freezing injury ensues following ice formation between and within plant cells. This leads to dehydration and disruption of cell membranes (Levitt 1962; Santarius 1969). However, many temperate species can maintain low rates of photosynthesis at temperatures approaching $0\,^{\circ}$C, and survive freezing temperatures below $-20\,^{\circ}$C (Levitt and Dean 1970).

Effects of temperature on growth of tropical pasture species

The rate of growth of pasture plants, expressed as the rate of increase in dry matter, is primarily a function of the rate of net photosynthesis and the rate of increase in leaf area. In many grasses increase in leaf area is related to the rate of tiller development, while in legumes it is related to shoot development and rate of appearance of new leaves. Effects of temperature on growth can be analysed in relation to effects on these processes.

Tropical grasses have a higher maximum net photosynthetic rate (P_N) than tropical legumes or temperate species, as previously discussed (Table 1.4). This maximum P_N is achieved at higher temperatures than for other species. The optimum temperature is either a well-defined specific temperature or confined to a small temperature range, unlike tropical legumes or temperate species which exhibit a rather flat topped response curve to temperature (Cooper and Tainton 1968; Ludlow and Wilson 1971). Thus P_N in tropical grasses is more sensitive to changes in temperature than in other plants. The optimum temperature for P_N is generally above $35\,^{\circ}$C depending on light intensity, species, and plant age, with P_N being reduced to zero at between 5 and $10\,^{\circ}$C and between 52 and $61\,^{\circ}$C at the lower and upper temperature limits respectively (Ivory 1975; Table 1.6).

Comparative values for some tropical legumes are also shown in Table 1.6 (Ludlow and Wilson 1971). The optimum temperature for these species is around $31\,^{\circ}$C, with minimums similar to the grasses between 5 and $8\,^{\circ}$C but with lower maximums of $50\,^{\circ}$C.

In temperate grasses it is generally considered that low temperature favours tiller production, particularly where low night temperatures are associated with high day temperatures (Ivory 1975). In some tropical grass species increasing temperature causes an increase in rate of tiller

Table 1.6. The optimum, maximum, and minimum leaf temperatures for net
    photosynthetic rate for a number of tropical grass and legume species.
    (After Ivory 1975)

| Species | Leaf temperature ($^\circ$C) | | | Reference |
	Minimum	Optimum	Maximum	
(a) Tropical grasses				
Brachiaria ruziziensis	9	38	56	Ludlow and Wilson (1971)
Cenchrus ciliaris cv. Biloela	6	39	61	Ludlow and Wilson (1971)
Cenchrus ciliaris (14 ecotypes)		35–40		Treharne, Pritchard, and Cooper (1971)
Chloris gayana		35		Murata, Iyama, and Honma (1965)
Cynodon dactylon		35		Murata, Iyama, and Honma (1965)
Cynodon dactylon		35		Miller (1960)
Eragrostis curvula		30–5		Cooper and Tainton (1968)
Hyparrhenia hirta		30–5		Cooper and Tainton (1968)
Melinis minutiflora	6	39	58	Ludlow and Wilson (1971)
Panicum maximum cv. Hamil	10	38	58	Ludlow and Wilson (1971)
Paspalum dilatatum		35–40		de Jager (1968)
Paspalum dilatatum		35		Murata *et al.* (1965)
P. notatum		35		Murata *et al.* (1965)
Pennisetum purpureum	7	37	59	Ludlow and Wilson (1971)
Sorghum almum	5	40	52	Ludlow and Wilson (1971)
Sorghum spp		35		Downes (1971)
(b) Tropical legumes				
Calopogonium mucunoides	7	34	51	Ludlow and Wilson (1971)
Glycine wightii cv. Cooper	5	31	50	Ludlow and Wilson (1971)
Macroptilium atropurpureum cv. Siratro	6	30	50	Ludlow and Wilson (1971)
Vigna luteola cv. Dalrymple	8	31	49	Ludlow and Wilson (1971)

production, as in *Paspalum dilatatum* (Mitchell 1956), *Astrebla* spp,
(Jozwik 1970), and *Panicum maximum* var. *tricheglume* and *Pennisetum clandestinum* (Ivory 1975). However, tillering in *Chloris gayana*,

Cenchrus ciliaris, and *P. coloratum* var. *makarikariense* was not responsive to temperature, but leaf area was increased with increasing temperature by large increases in the size of individual leaves (Ivory 1975).

Rate of leaf area development in both grasses and legumes is markedly affected by temperature. The effect was demonstrated in an experiment by Ludlow and Wilson (1970) where the growth of a range of tropical grasses and legumes was compared at constant temperatures of 20 or 30 °C, Table 1.7.

Table 1.7. Effect of constant 20 or 30 °C temperature on growth, leaf area, and mean rate of leaf area development in ten grasses and ten legumes (Ludlow and Wilson 1970)

Species	Plant dry wt. (g)		Leaf area (cm^2)		Rate of leaf area development ($cm^2\ day^{-1}$)	
	20°C	30°C	20°C	30°C	20°C	30°C
Grasses						
Brachiaria ruziziensis	0.16	3.06	25	445	0.7	21.2
Panicum maximum						
(creeping Guinea)	0.83	1.32	50	102	1.4	4.8
P. maximum cv. Hamil	0.12	2.57	16	341	0.4	16.2
P. maximum cv. Petrie	0.21	2.05	31	280	0.9	13.3
P. maximum cv. Common	0.16	2.40	22	356	0.6	16.9
P. coloratum	0.22	1.55	30	174	0.8	8.3
Setaria sphacelata cv. Nandi	0.18	0.56	25	68	0.7	3.2
Cenchrus ciliaris cv. Biloela	0.22	2.75	24	334	0.7	15.9
Chloris gayana cv. Samford	0.20	1.17	26	188	0.7	8.9
Melinis minutiflora	0.11	0.15	11	23	0.3	1.1
Mean	0.20	1.76	26	231	0.7	11.0
Legumes						
Macrotyloma uniflorum						
cv. Leichardt	0.16	1.97	20	443	0.6	21.1
Vigna luteola cv. Dalrymple	0.27	2.16	31	272	0.9	12.9
Centrosema pubeacens	0.05	0.95	7	131	0.2	6.2
Macroptilium atropurpureus						
cv. Siratro	0.27	1.10	28	196	0.8	9.3
Calopogonium mucunoides	0.02	0.98	4	162	0.1	7.7
Pueravia phaseoloides	0.07	0.84	7	177	0.2	8.4
Glycine wightii cv. Tinaroo	0.28	0.72	27	121	0.8	5.8
Desmodium uncinatum						
cv. Silverleaf	0.15	0.66	18	83	0.5	3.9
D. intortum cv. Greenleaf	0.06	0.34	7	57	0.2	2.7
Lotononis bainesii cv. Miles	0.02	0.02	2	4	0.06	0.2
Mean	0.15	0.97	15	165	0.4	7.8

At 20 °C plants were harvested after 35 days while at 30 °C were harvested after 21 days. Differences in plant weight within temperatures at this stage were closely related to seed weights. At 20 °C rate of leaf development in the grasses was 6 per cent of the rate at 30 °C, and in the legumes 5 per cent of the rate at 30 °C.

It is more difficult to assess the relative importance of temperature effects on P_N and rate of leaf appearance on whole plant growth under sward conditions. In a sward P_N becomes modified by lower light conditions in the canopy, and by temperature effects on the respiratory rate of non-photosynthetic tissue in the light period, and total respiratory rate of the plant at night. Nevertheless, a number of field experiments with maize (Wilson 1966) and *Cenchrus ciliaris* (Treharne *et al.* 1971) suggest that in the tropical grasses, temperature will affect growth rate predominantly through its effects on P_N. Furthermore the optimum temperatures measured for P_N in both tropical grasses and tropical legumes are similar to the optimum temperatures for total plant growth (Ludlow and Wilson 1971; Whiteman 1968; Ivory 1975).

Most temperate grasses and legumes have optimum temperature for

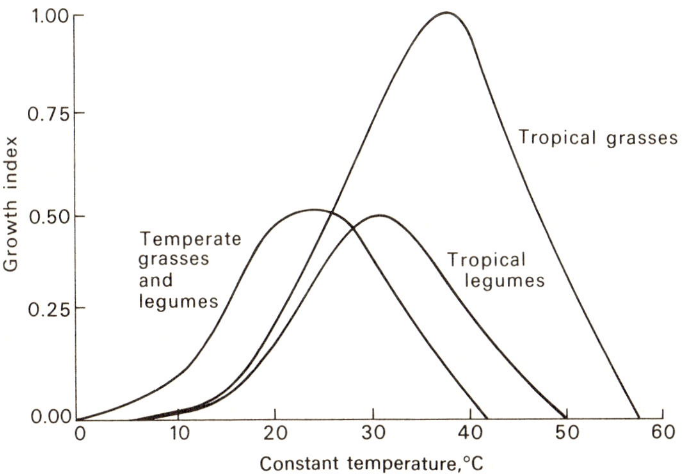

Fig. 1.5. Generalized representation of dry matter production responses of temperate and tropical pasture species to temperature. The growth index represents dry matter production relative to maximum dry matter production of tropical grasses at their optimum temperature.

whole plant growth between 10 and 25 °C, and although growth rate is reduced rapidly below 10 °C there is still some growth at 5 °C and the plant remains healthy (Mitchell and Lucamo 1962; Cooper and Mc-William 1966). In contrast, tropical grasses and legumes make little growth below 15 to 17 °C and reach maximum growth rates at about 30 °C for legumes and between 35 and 40 °C for the grasses. These general temperature responses are shown in Fig. 1.5.

The responses of tropical species to extreme conditions of temperature are discussed in the context of species adaptation to stress in Section 3.3.

1.4 THE MOISTURE REGIME

The tropical realm covers the full range of moisture regimes from the arid/semi-arid regions in central northern Africa and Saudi Arabia and north-west Australia, through the seasonal wet/dry climates of India and northern Australia, to the humid wet tropics centred more-or-less about the equator (Fig. 1.1(a)). On the basis of Köppen's classification a little less than half the total land area in the tropics falls within the truly humid tropics (Blumenstock 1957). Even within the humid tropics periods of moisture stress of varying severity may occur, with important effects on pasture productivity.

Anlaysis of the rainfall to determine potential pasture productivity in a particular region requires more than an estimate of *annual total rainfall*. This gives only an idea of the 'relative wetness' of an area and is not a very useful agronomic index. In terms of plant growth *rainfall distribution* on a daily, weekly, or monthly basis is required to predict patterns of pasture growth. *Rainfall intensity* data is also important allowing estimations of runoff losses, erosion potential, and frequency of high intensity storm rains. The final essential components in the overall moisture regime is an estimate of the seasonal patterns of evaporation. From these data, collected over a period of years, the variability of rainfall about the mean can be determined. This is more important than the mean itself, as this allows prediction of probabilities of having a growing season of a given length, the probability of seasonal and long-term droughts, and the definition of the critical periods of forage deficit throughout the year. Thus a detailed analysis of climatic data is an essential first step in any regional pasture development study, as it indicates the seasonal patterns of pasture growth, the potential pasture productivity, and the species likely to be successful in the region.

Water use by pastures

The water balance of a particular site, or catchment or region can be expressed in general terms by the water balance equation:

$$P = E + T + R + I$$

where P = precipitation (including rainfall, snow, ice, fog, and dewfall; E = evaporation from wet plant, soil, and free water surfaces; T = transpiration through the plant of water extracted from the soil; R = runoff of water from the land surface; and I = infiltration of water into the soil, which can be subdivided into three components:

$$I = U + S + A$$

where U = loss of water to ground water by deep percolation; S = lateral seepage through the soil to drainage channels; and A = the increment in soil moisture storage.

In field studies of plant water use, the evaporation and transpiration components are usually combined into a single component *evapotranspiration* (Et). Since the components R, U, and S are not available to plants, the most important components in determining plant growth are Et and A.

Transpiration through a pasture sward takes place from the walls of the mesophyll cells within leaves and the epidermal cells of leaf and stem surfaces. There is a resistance to water vapour transfer between the cell wall surfaces and the atmosphere. Because the total area of cell wall evaporating surface is very large, and since total leaf area above a given area of ground often is greater (leaf area indices around 4 to 5 are common in pasture swards), evapotranspiration rate can exceed evaporation rate from the same ground area of free water. This is particularly so when hot dry air moves over the crop, and transfer of water vapour transpired is enhanced by roughness and turbulence characteristics of the crop surface (van Bavel, Fritschen, and Reeves 1963).

These high rates of Et can only occur when the soil is well supplied with water. The value A in the water balance equation represents the increment in soil water storage, and the amount available to plants is that fraction held between field capacity and wilting point (approximately −1 to −15 bars tension). The amount of water held between these points depends on a number of soil characteristics which are discussed in Chapter 2. At this stage, it is sufficient to note that soils vary widely in their soil moisture storage characteristics.

Beginning with a soil initially fully recharged with water (i.e. at field capacity), as water loss through the sward proceeds, the water in the soil is held with increasing tension and becomes less available to the plant. At high rates of transpiration, rate of water movement through the soil is insufficient to keep the roots supplied, leading to water deficits in the plant and so to stomatal closure and reduced transpiration. Reduction in transpiration is a function also of the evaporative demand, as conditions of high demand lead to more rapid stomatal closure under conditions of restricted soil water availability, as shown in Fig. 1.6 (Denmead and Shaw 1962).

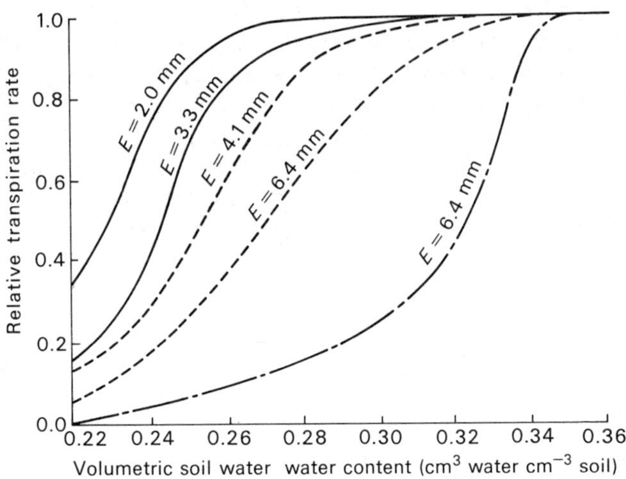

Fig. 1.6. Variation in relative transpiration rate (expressed as the ratio of actual transpiration to the maximum observed values) with soil and water content under different evaporative conditions (expressed as potential transpiration and indicated on diagram in daily potential transpiration in mm of water) (Denmead and Shaw 1962).

Water balance models

While analysis of rainfall distribution data will given an indication of the length of growing season and periods of moisture deficit, this can only be approximate as no account is taken of soil moisture storage, evaporative demand and sward water use. The input data required to construct a water balance model for a particular site will normally include:

(i)   Daily or weekly rainfall – this is the gross amount of precipitation received. To compute the amounts available for infiltration into the soil some estimate of runoff loss should be made depending on rainfall intensity, slope at the site and surface, and infiltration characteristics of the soil. Total rainfall is then reduced by an appropriate factor for rainfall intensity and site characteristics.

(ii)   Mean daily or weekly evaporation – where available, these data can be obtained directly from standard tank evaporimeters. Usually such data are not available and evaporation can be computed from meteorological data (air temperature, relative humidity, wind speed, and solar radiation; which in turn can be estimated from indirect data of latitude, cloudiness, and surface albedo) by the formulae of Penman (see Rose 1966).

(iii)   Soil moisture factors – the total soil water storage can be computed from estimates of field capacity and wilting point (the

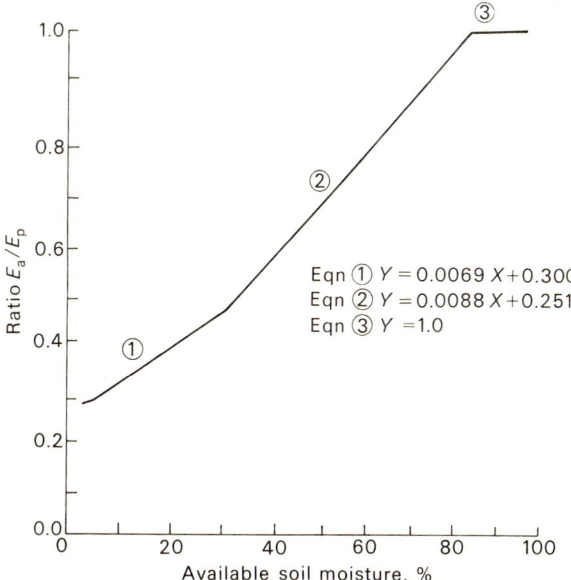

Fig. 1.7. Relationship between E_a/E_p and available soil moisture for a sorghum crop in a semi-arid tropical environment (Clewett 1969).

available moisture range) and soil depth. As water is removed from the soil, rate of actual evapotranspiration declines, as shown in Fig. 1.7, so that estimates of pasture water use must be reduced in relation to the measured or potential evaporative demand. This is usually done by relating the ratio, actual/potential evapotranspiration (E_a/E_p) to the available soil moisture, Fig. 1.7 (Clewett 1969).

From these data a model of the annual changes in water use can be constructed and periods of deficit defined or predicted. On the basis of a similar model Fitzpatrick and Nix (1970) have computed moisture index (MI) values, which vary from 1, when water is non-limiting, to 0, when available soil moisture has been exhausted, for tropical grasses and legume growth in northern Australia. These values were combined with similar index values for the other major environmental limitations to plant growth, a thermal index (TI) and a light index (LI) to define an overall growth index of the form:

GI = LI × TI × MI.

This growth index was used to compare different environments and their effect on the growth of tropical pasture species, as shown for a number of selected sites in Fig. 1.8.

These data show that at Katherine the major limitation is due to moisture. At Cairns and Gayndah both moisture and temperature limit tropical legume growth at different times of the year, while at the sub-tropical site at the highest latitude, Lismore, temperature is the major determinant.

Effects of water stress on plant growth

Throughout the large areas of the tropics, as demonstrated in Fig. 1.8, moisture deficits are the major limitation to pasture production. As available soil moisture declines, internal water deficits develop in the plant, leading to loss of turgor, until the plant reaches a state of permanent wilting when sufficient moisture can no longer be extracted from the soil to maintain turgor. For many plants the permanent wilting point is accepted as being at a soil water potential of –15 bars. As plant turgor declines, progressive stomatal closure causes reduction in the rate of CO_2 uptake. For a range of C_4 tropical grasses Ludlow and Ng (1976) have shown that stomatal closure causing net photosynthesis

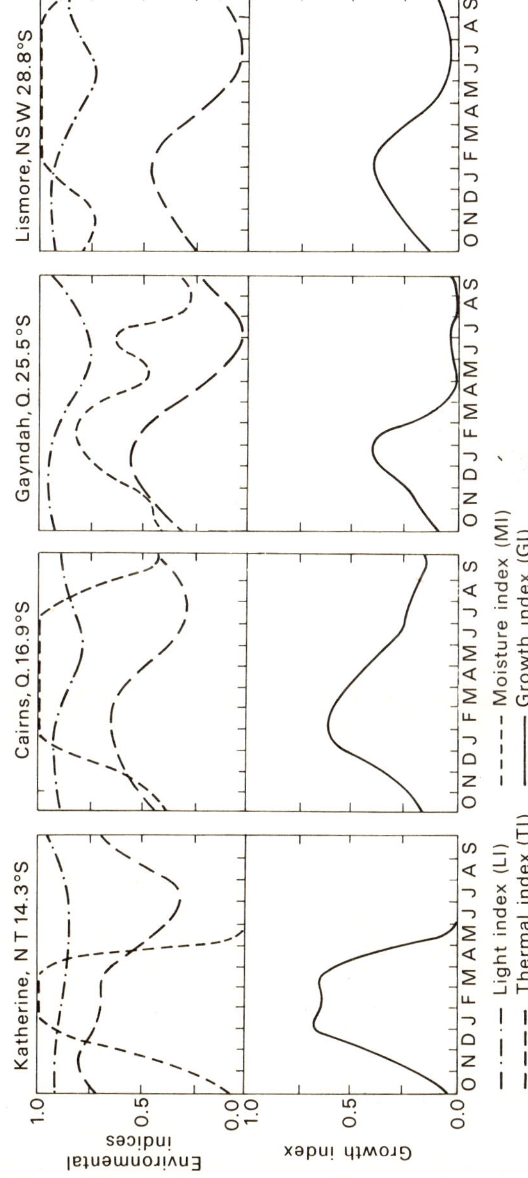

Fig. 1.8. Annual trends in light, thermal, moisture, and growth index values for the tropical legume group at four representative locations (Fitzpatrick and Nix 1970).

to cease occurs at leaf water potentials of –12 bars in controlled environments. However in the field where water deficits usually build up more slowly, and roots exploit a greater soil volume for water, photosynthesis continues at the potential rate at least to leaf water potential below –19 bars. Thus in studies of water stress effects on plant growth controlled environment experiments are not related to real conditions in the field (Ludlow and NG 1976). Nevertheless, at some point in the field, water stress causes net photosynthesis to decline to zero, further leaf appearance and expansion ceases, and with continued water deficit leaves begin to die, from the oldest first to the youngest, and from tip to base (Ludlow 1975).

During periods of water stress, growth processes appear to be suspended, including the ontogenetic aging process (Ludlow 1975). When water stress is relieved, metabolic activity is rapidly resumed at an apparently higher rate than for unstressed plants of the same chronological age. However Ng, Wilson, and Ludlow (1975) suggest that the stressed plants, on the relief of stress, are physiologically younger than unstressed plants by an equivalent to the time spent under stress, and resumed growth rates are similar to growth rates of plants of that age. However, stress does lead to reduced growth by reducing photosynthetic rate and later by reducing the rate of leaf area development, and by death of existing leaf tissue. Prolonged periods of water stress may lead to death of the plant.

Drought resistance in pasture plants

Drought resistance is a complex phenomenon, involving many separate adaptations and biochemical processes, none of which can be easily defined or measured (Gates 1974). Drought resistance involves two components, drought tolerance and drought avoidance. Drought tolerance defines the ability of plant cells to survive periods of water deficit (Levitt, Sullivan, and Krull 1960). This ability is best developed in young meristematic tissue, and many perennial grass species survive protracted drought periods through survival of basal buds in the plant crown. Survival of meristematic tissue can be related to redistribution of water and nutrients to maintain turgor in the meristems (Gwynne 1960).

Drought avoidance involves a wide range of mechanisms designed to reduce water deficits during drought periods or avoid drought periods altogether. A good example of the latter is provided by annual pasture species which grow and set seed over the wet season, and pass through the dry season as seed. Many plants have developed mechanisms to reduce water loss during dry periods by folding or rolling their leaves, shedding of leaves, developing thick cuticles and dense hair covering, or by developing water storage tissue as in the cacti. Other plants have adapted to drought survival by developing deep and extensive root systems to fully exploit the available soil moisture.

It is difficult to predict plant drought resistance in the field from laboratory measurements of plant responses. In seeking pasture species adapted for drought survival, field evaluation must be combined with selection of species from arid and semi-arid areas.

REFERENCES

Alberda, T. (1969). The effects of low temperature on dry matter production, chlorophyll concentration and photosynthesis in maize plants of different age. *Acta bot. neerl.* **18**, 39.

Beinhart, G. (1962). Effects of temperature and light intensity on CO_2 uptake, respiration and growth of white clover. *Pl. Physiol., Lancaster* **37**, 709.

Bjorkman, O. and Berry, J. (1973). High-efficiency photosynthesis. *Scient. Am.* **229**, 80.

Blumenstock, D. I. (1957). Distribution and characteristics of tropical climates. *Proc. 9th Pacific Sci. Cong.*, Vol. 20, p. 3.

Butler, W. I., Norris, K. H., Siegelman, H. W., and Hendricks, S. B. (1959). Detection, assay, and preliminary purification of the pigments controlling photoresponsive development of plants. *Proc. natn. Acad. Sci. U.S.A.* **45**, 1703.

Calvin, M. and Benson, A. A. (1948). The path of carbon in photosynthesis. *Science, N.Y.* **107**, 476.

Chang, J.-H. (1968). *Climate and agriculture. An ecological survey*, pp. 64–5. Aldine, Chicago.

Clewett, G. (1969). Simulation analysis of the grain sorghum growing season with strategic irrigation at Richmond, North Queensland. B.Agric.Sci. Thesis, University of Queensland.

Cooper, J. P. and McWilliam, J. R. (1966). Climatic variation in forage grasses. 2. Germination, flowering and leaf development in Mediterranean populations of *Phalaris tuberosa. J. appl. Ecol.* **3**, 141.

—— and Tainton, N. M. (1968). Light and temperature requirements for the growth of tropical and temperate grasses. *Herb. Abstr.* **38**, 167.

de Jager, J. M. (1968). Variation in photosynthetic activity: infra-red gas analysis. Report of the Welsh Plant Breeding Station, 1967; p. 17.

Denmead, O. T. and Shaw, R. H. (1962). Availability of soil water to plants as affected by soil moisture content and meteorological conditions. *Agron. J.* **54**, 385.

Downes, R. W. (1971). Relationship between evolutionary adaptation and gas exchange characteristics of diverse *Sorghum* taxa. *Aust. J. biol. Sci.* **24**, 843.

Downton, J. and Slatyer, R. O. (1972). Temperature dependence of photosynthesis in cotton. *Pl. Physiol., Lancaster* **50**, 518.

Fitzpatrick, E. A. and Nix, H. A. (1970). The climatic factor in Australian grassland ecology. In *Australian grasslands* (ed. R. Milton Moore) pp. 3–26. ANU Press, Canberra.

Gates, C. T. (1974). Water shortage and agriculture: some responses. *J. Aust. Inst. agric. Sci.* **40**, 121.

Gwynne, M. D. (1960). Drought effects on plants. *New Scient.* **8**, 795.

Hatch, M. D. and Slack, C. R. (1966). Photosynthesis by sugar-cane leaves: a new carboxylation reaction and the pathway of sugar formation. *Biochem. J.* **101**, 103.

Hesketh, J. D. and Moss, D. N. (1963). Variation in response of photosynthesis to light. *Crop Sci.* **3**, 107.

Hughes, R. (1970). Factors involved in animal production from temperate pastures. *Proc. 11th Int. Grassl. Cong.*, pp. A31–8.

Ivory, D. A. (1975). The effects of temperature on the growth of tropical pasture grasses. Ph.D Thesis, University of Queensland.

Jozwik, F. X. (1970). Response of Mitchell grass (*Astrebla* F. Muell.) to photoperiod and temperature. *Aust. J. agric. Res.* **21**, 395.

Köppen, W. (1936). Das geographischen System der Klimate. Hamb. d. Klim., 1(C). Boerntraeger, Berlin.

Laetsch, W. M. (1968). Chloroplast specialization in dicotyledons possessing the C_4-dicarboxylic acid pathway of photosynthetic CO_2 fixation. *Am. J. Bot.* **55**, 875.

Levitt, J. (1962). A sulfhydryl-disulphide hypothesis of frost injury and resistance in plants. *J. theoret. Biol.* **3**, 355.

— and Dean, J. (1970). The role of membrane proteins in freezing injury and resistance. In *The frozen cell* (eds G. W. Walstenholme and M. O'Connor) pp. 149–74. Ciba Foundation Symposium.

—, Sullivan, C. T., and Krull, E. (1960). Some problems in drought resistance. *Bull. Res. Coun. Israel* **8D**, 173.

Ludlow, M. M. (1975). Effect of water stress on the decline of leaf net photosynthesis with age. In *Environmental and biological control of photosynthesis* (ed. R. Marcelle) pp. 123–34. W. Junk, The Hague.

— and Ng, T. T. (1976). Effect of water deficit on carbon-dioxide exchange and leaf elongation rate of *Panicum maximum* var. *trichoglume. Aust. J. Pl. Physiol.* **3**, 401.

— and Wilson, G. L. (1970). Growth of some tropical grasses and legumes at two temperatures. *J. Aust. Inst. agric. Sci.* **36**, 43.

— (1971). Photosynthesis in tropical pasture plants. I. Illuminance, carbon-dioxide concentration, leaf temperature and leaf–air vapour pressure difference. *Aust. J. biol. Sci.* **24**, 449.

— (1972). Photosynthesis of tropical pasture plants. IV. Basis and consequence of differences between grasses and legumes. *Aust. J. biol. Sci.* **25**, 1133.

Miller, V. J. (1960). Temperature effects on the rate of apparent photosynthesis of seaside bent and Bermuda grass. *Proc. Am. Soc. Hort. Sci.* **75**, 700.

Mitchell, K. J. (1956). Growth of pasture species under controlled en-environment. I. Growth at various levels of constant temperature. *N.Z. Jl Sci. Technol. A* **38** 203.

— and Lucanus, R. (1962). Growth of pasture species under controlled environments. 3. Growth at various levels of constant temperature with 8 and 16 hours of uniform light per day. *N.Z. Jl agric. Res.* **5**, 135.

Monteith, J. L. (1958). The heat balance of soil beneath crops. In *Climatology and microclimatology*. UNESCO, Paris.

Ng, T. T., Wilson, J. R., and Ludlow, M. M. (1975). Influence of water stress on water relations and growth of tropical (C_4) grass, *Panicum maximum* var. *trichoglume. Aust. J. Pl. Physiol.* **2**, 581.

Nurata, Y. Iyama, and J. Honma, T. (1965). Studies on photosynthesis of forage crops. 4. Influence of air temperature upon the photosynthesis and respiration of alfalfa and several southern forage crops. *Proc. Crop Sci. Soc. Japan* **34**, 154.

Pasternak, D. and Wilson, G. L. (1972). After effects of night temperatures on stomatal behaviour and photosynthesis of Sorghum. *New Phytol.* **71**, 683.

Pearson, C. J. and Hunt, L. A. (1972). Effects of pretreatment temperature on carbon-dioxide exchange in alfalfa. *Can. J. Bot.* **50**, 1925.

Rose, C. W. (1966). *Agricultural physics*, pp. 78–86. Pergamon Press, Oxford.

Santarius, K. A. (1969). The effect of freezing and desiccation of chloroplasts in the presence of electrolytes. *Planta* **89**, 23.

Treharne, K. J., Pritchard, A. J., and Cooper, J. P. (1971). Variation in photosynthesis and enzyme activity in *Cenchrus ciliaris* L. *J. exp. Bot.* **22**, 227.

van Bavel, C. H. M., Fritschen, L. J., and Reeves, W. E. (1963). Transpiration by seed on grass as an externally controlled process. *Science, N.Y.* **141**, 269.

Vicente-Chandler, J., Silva, S., and Figarella, J. (1959). Effects of nitrogen fertilization and frequency of cutting on the yield and composition of Napier grass in Puerto Rico. *J. Agric. Univ. P. Rico* **43**, 215.

Vink, A. P. A. (1975). *Land use in advancing agriculture*. Springer-Verlag, Berlin.

West, S. H. (1970). Biochemical mechanism of photosynthesis and growth depression of *Digitaria decumbens* when exposed to low temperatures. *Proc. 10th Int. Grass. Cong.*, pp. 514–17.

Whiteman, P. C. (1968). The effect of temperature on the vegetative growth of six tropical legume species. *Aust. J. exp. Agric. Anim. Husb.* **8**, 528.

Wilson, J. W. (1966). Effect of temperature on net assimilation rate. *Ann. Bot.* **30**, 753.

2

Soil factors affecting pasture growth and yield

2.1 SOILS OF THE TROPICS

Soil formation or soil genesis

A useful general text on soil formation or genesis is that by Buol, Hole, and McCracken (1973) and a text emphasizing the tropics is that by Mohr, van Baren, and Van Schuylenborgh (1972). Soil formation in the tropics occurs as a result of similar processes or factors that operate in other regions. The general processes or factors will be discussed first and then special aspects of the tropical situation.

In the study of field soils accurate description is essential so that classification can be made and communication and discussion between investigators is possible. The first step in description is to determine soil morphology, i.e. to identify the soil profile and its horizons on the basis of properties that can be seen or measured directly in the field; in particular colour, texture, structure, horizon thickness, pH etc. Detailed chemical, physical, and mineralogical analysis may be required at a later stage.

The genesis or development of the soil profile can be considered as being due to three major groups of soil forming processes.

(1) *Transformations*
These may be chemical, physical, or biological.

(a) Breakdown
A variety of materials are subject to breakdown, important ones being primary rock minerals, secondary soil minerals, fresh organic matter, and humus. The most significant process is probably rock weathering giving rise to weathering products of very different nature to the parent rock. Thus easily weathered rock minerals such as olivine, augite etc., are almost absent from soils, whereas resistant ones, e.g. quartz, accumulate. A detailed treatment of the weathering process is given by Mohr *et al.* (1972).

(b) Synthesis
Some of the materials released as a result of breakdown are synthesized

into new compounds. Important amongst these are: (i) clay minerals, (ii) hydrated iron and aluminium oxides, (iii) humus. These compounds give soils quite different features to the original parent material. In particular clay minerals and humus are characteristic features of soils but are usually low or absent in parent materials.

Minerals in the clay fraction of soils, formed as a result of breakdown or synthesis, are shown in Table 2.1. This table shows the 13-stage weathering sequence developed by Jackson and Sherman (1953) reflecting the stability of the more important soil minerals.

Table 2.1. Representative minerals and soils associated with weathering soil stages (Jackson and Sherman 1953)†

Weathering stage	Representative minerals	Typical soil groups
Early weathering stages		
1	Gypsum (also halite, sodium nitrate)	Soils dominated by these minerals in the fine silt and
2	Calcite (also dolomite, apatite)	clay fractions are the youthful
3	Olivine–hornblende (also pyroxenes)	soils all over the world, but mainly soils of the desert
4	Biotite (also glauconite, nontronite)	regions where limited water keeps chemical weathering to
5	Albite (also anorthite, microcline, orthoclase)	a minimum
Intermediate weathering stages		
6	Quartz	Soils dominated by these
7	Muscovite (also illite)	minerals in the fine silt and
8	2:1 layer silicates (including vermiculite, expanded hydrous mica)	clay fractions are mainly those of temperate regions developed under grass or trees. Includes
9	Montmorillonite	the major soils of the wheat and corn belts of the world
Advanced weathering stages		
10	Kaolinite	Many intensely weathered
11	Gibbsite	soils of the warm and humid
12	Hematite (also goethite, limonite)	equatorial regions have clay fractions dominated by these
13	Anatase (also rutile, zircon)	minerals, they are frequently characterized by their infertility

†Primary minerals are underscored.

(2) *Translocations*

(a) Within or out of the profile

Movement is usually due to material either dissolving or being suspended in the soil water and moving with it, usually downward or laterally, sometimes upward. The term leaching is often applied if the movement is downward or lateral. The severity of leaching depends mainly on the rainfall and soil permeability to water. If leaching is weak (arid soils) movement of soluble material (soluble salts and $CaCO_3$) occurs into the subsoil. If more severe, e.g. in humid soils, more soluble materials are completely removed and relatively insoluble materials such as clay, iron and aluminium oxides, and organic matter may move into the subsoil where they are deposited forming characteristic B horizons. Lateral leaching is important on sloping sites particularly where an impermeable B horizon occurs. Erosion of finer surface soil also occurs on sloping sites leaving shallow, coarse-textured soils.

(b) Accession to the profile

Soluble salts may be added in rainwater (cyclic salt) or invade a profile if a rise in the water table occurs (e.g. excess watering in irrigation areas). Suspended material may be added as a result of water erosion, e.g. by raindrop splash or deposition from flowing water, or as wind-borne material, e.g. loess.

(3) *Homogenization*

Against the differentiating mechanisms of transformation and translocations homogenizing influences act in a counter direction. Vegetation is most significant retaining ions within the profile by uptake that would otherwise be lost — a soil–plant cycle occurs. The effect is enhanced by water uptake reducing the leaching effect. Other influences are earthworms, termites, and burrowing animals.

Factors of soil formation

The intensity of the processes affecting soil formation is affected by a number of environmental factors or factors of soil formation. Five general factors have been stressed, as indicated in Fig. 2.1, namely time, parent material, climate, relief, and organisms (refer to Jenny (1941) *et seq.* and Buol *et al.* (1973)).

Time

A soil in equilibrium with its environment is termed a mature soil. The time taken to reach equilibrium is obviously going to be quite

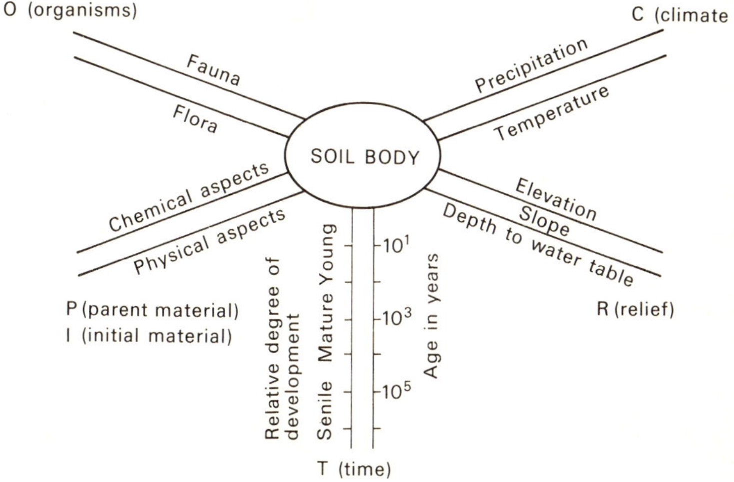

Fig. 2.1. A diagrammatic view of the influence of the factors of soil formation. From Buol *et al.* 1973.

variable depending on the nature of the soil forming conditions. The evidence suggests that it may be less than 50 years for a permeable parent material such as volcanic ash under wet tropical conditions to 30 000 or more years for some soils in Australia (refer to Buol *et al.* 1973).

Young soils, e.g. fluvents, are often characterized by lack of horizon development and mature soils by well-developed horizons, e.g. ultisols. Young soils generally are well supplied with primary minerals which weather and release nutrients such as phosphorus, potassium, calcium, etc., whereas in old soils the nutrients so released may, over time, have been leached from the profile or 'fixed' as secondary compounds of low availability.

Parent material

Parent material has a major influence on soil characteristics, particularly on younger land surfaces. Of particular importance is the quantity of clay-forming minerals since soils high in clay are generally more fertile. Thus parent materials such as basalt usually give rise to soils high in clay whereas if the quantity of resistant minerals such as quartz is high, e.g. with siliceous sandstone, clay content and fertility are low.

The nutrient status of soils is often directly related to that of the soil parent material.

Climate

Rainfall and temperature are usually regarded as the most important elements. Rainfall has very significant effects. Moisture is required in the soil for many important chemical and biological processes and excess will affect leaching through and out of the profile and also erosion of surface material. The different nature of arid and humid soils is well recognized. Thus in arid soils leaching is weak, biological activity is low, calcium carbonate is frequently present, and the pH is usually alkaline. In contrast to this in humid soils, leaching is strong, biological activity is high, calcium carbonate is typically absent but iron and aluminium compounds present, and the pH is usually acid.

Temperature significantly affects soil formation due to its influence on the rate of chemical reaction, so that as long as the moisture status is adequate the rate of soil formation increases with temperature.

Relief

Characteristic soil changes often occur as one moves from upper to lower slope positions. Erosion of finer material from upper positions and depositions in lower positions occurs so that the soil becomes progressively thicker and finer moving down the slope. Greater run-off is usual in upper positions making the soils there locally arid whereas downslopes they are locally humid. Waterlogging may occur and saline soils may form if the groundwater is high in soluble salts (accelerated by clearing of upper slopes). The organic and nitrogen content of the soil usually increases down the slope due to the better moisture conditions and thus fertility is usually higher as long as drainage remains good. Slope sequences, particularly on long slopes, are often due to changes in parent material, as well as the effects described above.

Organisms

Organisms of various types — plants, bacteria, fungi, burrowing animals, earthworms, termites, and Man all play a part in soil formation. Vegetation is of particular importance. Organisms frequently act as a dependant variable, i.e. changes in rainfall, temperature, soil fertility, etc. affect the vegetation and other organisms and this influences soil formation and hence soil features. Less often it is independant the best example here being Man's activities. His detrimental influence is common, e.g. erosion, changes in salinity, loss of organic matter due to cultivation. On the credit site the use of fertilizers and improved pastures, drainage etc. may cause very significant beneficial changes.

Soil formation in the tropics

Using the geographic definition of the tropics, i.e. as that part of the world located between 23.5° north and south of the equator, the main feature of this zone in terms of soil formation is that temperature and rainfall tend, on average, to be higher than elsewhere. However, both temperature and rainfall show a wide range.

The key temperature feature of the tropics, defined by the prefix *iso* in the US Soil Taxonomy (Soil Survey Staff 1975), is that there is 'less than 5 °C difference between the mean summer and mean winter temperatures at 50 cm or to a lithic contact if shallower'. Four soil temperature regimes can be estimated from mean annual temperature and elevation data:

Regime	Mean annual temperature (°C)	Elevation (m)
Isohyperthermic	$\geqslant 22$	0–600
Isothermic	$\geqslant 15-< 22$	600–1800
Isomesic	$\geqslant 8-< 15$	1800–3000
Isofrigid	< 8	> 3000

Annual rainfall in the tropics can vary from zero to 10 000 mm. Four soil moisture regimes are recognized in the tropics in the US Soil Taxonomy:

1. *Udic*

In most years the control section* of the soil is not dry in any part for as long as 90 cumulative days. If precipitation exceeds evapotranspiration in all months, with moisture tension rarely above 1 bar in the control section and water moving downward in all months if not frozen, the regime is called '*perudic*'.

2. *Ustic*

The control section is dry for more than 90 cumulative days but

*Defined as that part of the soil between the following depths: below that depth to which a dry (> 15 bars but not air dry) soil will be moistened by 2.5 cm of water in 24 hours and above that depth to which a dry soil will be moistened by 7.5 cm of water in 48 hours.

less than 180 cumulative days or 90 consecutive days during the year.

3. *Aridic*

The control section is dry for more than 180 cumulative days or moist for less than 90 consecutive days per year.

4. *Aquic*

The soil is saturated with water long enough to cause reduced soil conditions. If the groundwater is always at or very close to the soil surface the term *'peraquic'* is used.

In the lowland tropics the distribution of these moisture regimes is approximately as follows: udic, 29 per cent; ustic, 34 per cent; aridic, 29 per cent; and aquic, 8 per cent.

Despite these variations in the tropical climate the generally hotter and wetter conditions than elsewhere lead to a faster rate of soil formation due to a speeding up of many of the chemical processes. Thus at $25\,^{\circ}$C compared with $10\,^{\circ}$C ionization of water is four times as high and silica is eight times as soluble. Breakdown of primary minerals is rapid and the leaching action of the rainfall removes the soluble products of weathering. As a result of this intense weathering and leaching characteristic secondary minerals such as kaolinite and gibbsite are formed, as indicated in Table 2.1. Intense biological activity is also a feature of such soils with large organic matter inputs and rapid mineralization of organic matter and re-cycling of nutrients.

Under moist tropical conditions the soil forming process termed *laterization* (alternatively desilication, ferritization) is of particular importance. This is defined (refer to Buol *et al.* (1973)) as: 'The chemical migration of silica out of the solum and thus the concentration of sequioxides (goethite, gibbsite etc.) as in oxic horizons (of oxisols) with or without formations of ironstone (laterite: hardened plinthite) and concretions'. The iron-rich material *plinthite*, occurring as dark red mottles and hardening on exposure to the air is a feature of the deep subsoil of many tropical soils. Its particular properties are thought to be due to the operation of a fluctuating water table, after iron accumulation by laterization or other processes. Many other processes of soil formation which are important in temperate regions are also important in the tropics, for example:

Lessivage

The mechanical migration of small mineral particles from the A to B horizons of a soil, producing a relative enrichment in clay in B horizons (argillic horizons).

Podzolization

The chemical migration of aluminium and iron and/or organic matter results in the concentration of silica in the layer eluviated and accumulation of aluminium and iron and/or organic matter in the subsoil.

Land surfaces in the tropics are extremely variable in age ranging from the young surfaces of alluvial or volcanic origin, usually with fertile soils, to the oldest on earth in the ancient crystalline rocks of South America, Africa, and Australia, often with infertile soils due to leaching of nutrients over very long time periods. The soils on these old land surfaces are often considered not to be in equilibrium with the present climate but to have formed under previously wetter conditions (paleosols).

Soil composition and properties

Soils have two groups of material which especially distinguish them from non-soil, namely clay minerals and organic matter. They also determine most of the key physical and chemical properties of soils.

The clay minerals are found mostly in the clay fraction of soils (non-aggregated material < 0.002 mm effective diameter). They can be grouped as follows:

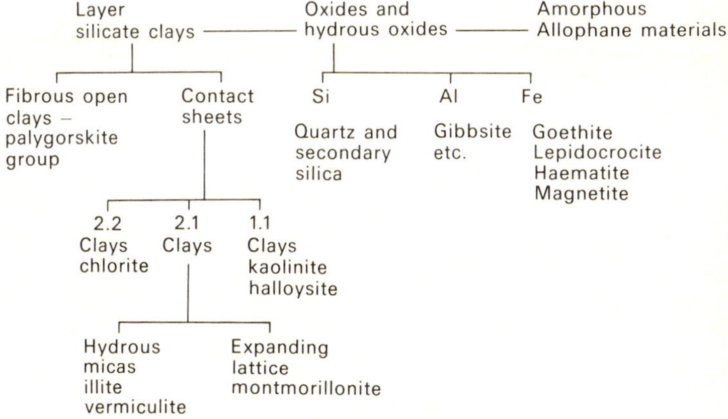

In humid tropical soils the important clay minerals are usually the 1 : 1 layer silicates, hydrated aluminium, and iron oxides particularly gibbsite

($Al_2O_3.3H_2O$) and goethite (KFeO.OH), and allophane (see above and Table 2.1). The layer lattice silicates are sheet structures whereas allophane consists of amorphous mixtures of aluminium and silicon hydroxides.

Organic matter includes a small but important living component consisting mainly of various micro-organisms; plant and animal residues in various stages of decomposition; and the relatively stable end product of decomposition, the soil humus (usually 95 per cent or more of the total organic matter). As with the inorganic fraction it is the fine colloidal organic material (<0.001 mm) that is the most important in its effect on soil properties.

Organic matter levels in soils depend on the quantity of organic material added and the rate of its decomposition and the decomposition of the humus. Under natural conditions an equilibrium level occurs, in which gains and losses are balanced, its magnitude depending on the particular environment and the soil characteristics. Table 2.2 shows annual additions, decomposition rates, and equilibrium values for organic matter (expressed as organic carbon) for some tropical and temperate soils. Conditions favouring higher equilibrium levels are high rainfall, low temperature, soil acidity, poor drainage, and high clay content since plant growth leading to additions is generally favoured more than is decomposition under these conditions. Equilibrium levels in tropical soils are not very different from those of temperate regions (Table 2.2) since in situations where additions are very high (e.g. in udic tropical forests) decomposition rates are very high. Andepts tend to have high equilibrium levels due to the slow decomposition rate, apparently due to the effects of the allophane present (see Table 2.2). Removal of permanent vegetation results in a non-equilibrium with decomposition exceeding additions and a rapid decline in organic matter levels to a new and lower equilibrium value.

Properties of the soil colloids

The colloidal (<0.001 mm diameter) clay and organic matter is sometimes described as the 'active' soil fraction because of its effect on the physical, chemical, and biological phenomena that occur in soil. One reason for this is its very large surface area. Thus a cube of 1 centimetre side has an area of 6 square centimetres. If this cube is divided into smaller cubes of sides 0.00005 mm (i.e. fine colloid) the surface area becomes about 100 hectares. A second reason is that the soil colloids

Table 2.2. Estimates of annual additions, decomposition rates, and equilibrium levels of topsoil organic carbon in some tropical and temperate locations. (From data by Greenland and Nye 1959; Sanchez 1976)

Location	Addition of undecomposed organic matter (t ha^{-1})	Decomposition rate of fresh organic matter into soil organic carbon (%)	Soil organic carbon addition (t ha^{-1})	Soil organic carbon decomposition rate (%)	Soil organic carbon at equilibrium (t ha^{-1})	(%)
Tropical forests						
Ghana (ustic)	5.28	50	2.64	2.5	106	2.4
Zaire (udic)	6.05	47	2.86	5.2	55	1.2
Colombia (udic, andept)	3.85	51	1.97	0.5*	394	9.0
Temperate forests						
California (oak)	0.75	47	0.35	0.4	88	2.0
California (pine)	1.65	52	0.86	1.0	86	1.9
Tropical savannahs						
Ghana (1250 mm rain)	1.43	50	0.71	1.3	55	1.2
Ghana (850 mm rain)	0.44	43	0.19	1.2	16	0.4
Temperate prairie						
Minnesota (870 mm rain)	1.42	37	0.53	0.4	134	3.0

*Low value due to the presence of allophane.

develop both negative and positive charges. Colloidal properties of particular importance are:

1. *Cation and anion adsorption and exchange*

The capacity for ion exchange results from charge development on the soil colloids. Negative charge that is unaffected by soil pH is described as the 'permanent' negative charge. In addition there is a so called 'variable' negative charge which increases as pH increases and positive charge that increases as pH decreases. The charge characteristics of important soil minerals and materials is shown in Table 2.3.

Table 2.3. Charge characteristics of some clay minerals separated from soils of Kenya (milliequivalents per 100g clay). (Mehlich and Theisen (unpublished), reported in Sanchez (1976))

Material	Cation exchange capacity*			Anion exchange capacity*
	Permanent	Variable	Total	
Montmorillonite	112	6	118	1
Vermiculite	85	0	85	0
Illite	11	8	19	3
Halloysite	6	12	18	15
Kaolinite	1	3	4	2
Gibbsite	0	5	5	5
Goethite	0	4	4	4
Allophanic colliod	10	41	51	17
Peat	38	98	136	6

*Measured at pH 8.2. Cation exchange capacity (CEC) measures the negative charge and anion exchange capacity (AEC) the positive charge.

In natural soils consisting of mixtures of various minerals and organic matter negative charge is usually greater than positive, giving a net negative charge. The magnitude depends on the pH and the quantity and nature of the colloid. In tropical soils charge is mostly pH dependant, CEC values are usually low at the pH of the soil (often <10 milliequivalents per 100g soil), the value in surface soils depending on the organic matter present (Fig. 2.2). In subsoils CEC values are usually lower owing to lower organic matter and the charge may become net positive in some acidic subsoils.

The development of charge on the colloids gives the soil the capacity

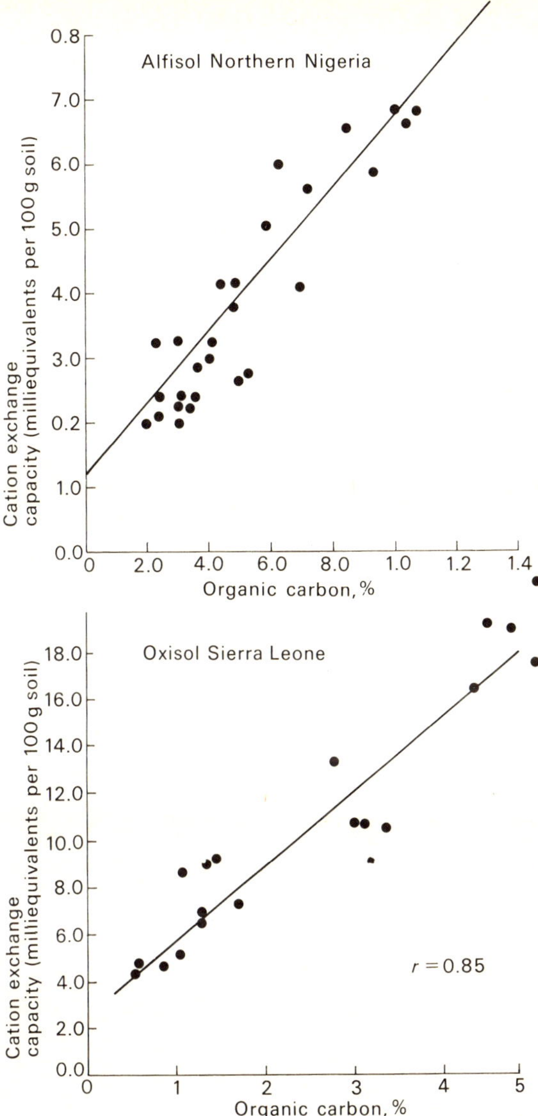

Fig. 2.2.

to hold (adsorb) and exchange both cations and anions. The net negative charge of the colloid is balanced by the adsorption of a diffuse layer of cations, termed the exchangeable cations, the most important being Ca^{2+}, Mg^{2+}, K^+, Na^+, H^+ (and aluminium below pHs of about 5.5, with positive charges of +1–3). An equilibrium exists between these adsorbed ions and ions in solution with 95 per cent or more of the cations in the adsorbed layer, as shown below:

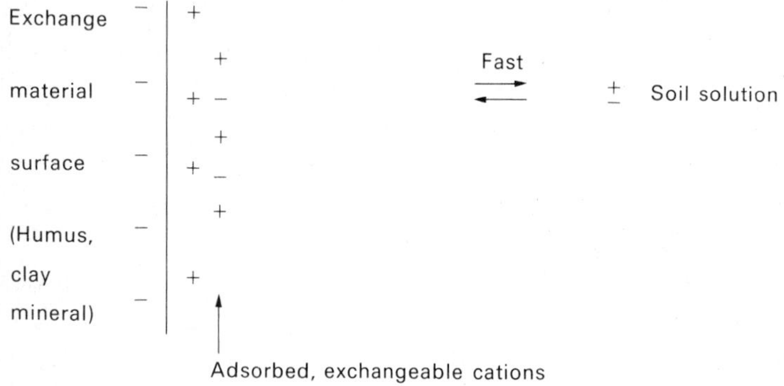

If the equilibrium is disturbed by addition of another cation to the solution, e.g. K^+ from KCl fertilizer, the added K^+ will displace other adsorbed cations resulting in more exchange positions being held by K^+, i.e. cation exchange will have occurred. The importance of cation adsorption and exchange is that nutrient cations, particularly Ca^{2+}, K^+, and NH_4^+ are held against leaching in a form available to plants, since if a nutrient cation is taken up by a plant further cation will be released from the absorbed position to restore the equilibrium. Soluble fertilizer such as KCl will be held against leaching by the cation exchange reaction.

In tropical soils development of significant positive charge occurs, increasing as pH decreases, allowing adsorption and exchanges of negative ions such as Cl^-, NO_3^- and SO_4^{2-}. This may occur even where the net charge is negative owing to the spatial separation of the negative and positive charges. The effect will be stronger in subsoils which develop a net positive charge. Thus the nutrient ions NO_3^- and SO_4^{2-} may be held against leaching in comparable anion exchange reactions.

In addition to the exchange reactions described, nutrient anions such as phosphate and molybdate and cations such as copper may be adsorbed by the operation of different and stronger bonding mechanisms than those described above, reducing leaching substantially but also reducing availability.

2. *Acidity and buffering*

Tropical soils are usually acid and poor plant growth is frequently related to acidity (see later section). The H^+ ions in the soil solution causing the acidity (active acidity) may arise from the exchange complex or from water molecules associated with aluminium ions (potential acidity). H^+ ions may arise from humic colloid or from clay material above a pH of about 6. Below a pH of about 6, H^+ causes some decomposition of the clay releasing aluminium ions which then satisfy the permanent negative charge on the clay. The aluminium ions form co-ordination units with water, these units serving as a source of H^+ ions. Increasing levels of aluminium as the pH drops below 6, are shown in Fig. 2.3. In waterlogged soils containing sulphidic materials (e.g. sulpha-quents) drainage may cause oxidation to sulphuric acid and very low soil pHs (pH 1–2).

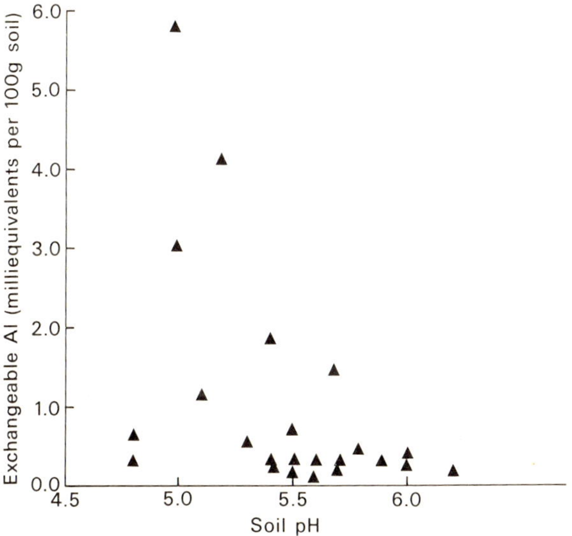

Fig. 2.3.

If alkaline material such as lime is added to an acid soil it will first neutralize the active acidity and then potential acidity is released neutralizing further lime thus tending to resist the expected rise in pH. This resistance to pH change is termed buffering and soils are buffered with respect to the addition of materials that might increase or decrease pH. Thus addition of sulphur acidifies the soil since it oxidizes to sulphuric acid, but the H^+ ions move out of the soil solution onto the exchange complex or become associated with aluminium ions thus reducing the acidifying effect.

The extent of the buffering effect will depend on the quantity and the nature of the soil colloid. Thus sands are poorly buffered and must be fertilized carefully with materials inducing pH changes otherwise detrimental plant effects may occur because of too large a change in soil pH, e.g. from the over-liming of acid soils. On the other hand clays and soils high in organic matter are well buffered. This is usually a desirable agronomic soil feature, but it means that pH may be difficult to shift by the addition of materials such as lime.

Soil classification

For a detailed discussion refer to Buol *et al.* (1973); for tropical soils refer to D'Hoore (1968) and Sanchez (1976). Soil classification is necessary for a number of reasons, refer to Buol *et al.* (1973). It assists communication about soils and helps to extrapolate findings about soils in one region to other regions. Unfortunately there is not as yet any classification system that is universally recognized. Most classification systems combine to varying degrees a genetic approach, i.e. emphasis on the processes or factors thought to have been important in soil development and a factual approach, i.e. emphasis on the actual properties of the soil profile.

In tropical regions three major classification systems with strong genetic bias have been and are still in use. These are the 1938 USDA systems (modified by Thorp and Smith 1949) used primarily in tropical America and Asia, and the French ORSTROM system (Aubert 1968) and Belgian INEAC (Sys, Van Wambeke, and Frankart 1961) system used primarily in Africa. More recently the so called 'old' US system (1938) has been replaced by a greatly changed and much more comprehensive system (Soil Survey Staff 1975). This system is much less genetic placing much greater emphasis on actual soil properties (morphological, chemical etc.) and therefore being desirably less subjective, but often requiring much laboratory work on profile samples before final classification. A soil map of the world (FAO system) is being prepared by the two United Nations agencies, FAO and UNESCO (see Dudal

1968, 1970). The soil units have two levels, roughly equivalent to the suborder and great soil group levels of the US *Soil taxonomy*. It draws strongly on the latter system but the nomenclature has been drawn from a number of classification systems. It is not really a classification but rather a two-level legend of map units and will represent a significant contribution when completed.

In view of its comprehensiveness and probable influence for the future the US *Soil taxonomy* is used in subsequent discussion of tropical soils. The highest category level is order and all ten orders are found in the tropics. The next two category levels are suborder and great group. A simplified outline of the main suborders and great groups found in the tropics is given in Table 2.4. The lower category levels are subgroups, families, and series. Identification of a profile requires a knowledge of the so called diagnostic horizons present and other diagnostic characteristics apart from horizons. The full detail of these and their application during classification are given in *Soil taxonomy* (Soil Survey Staff 1975).

The nomenclature used represents an almost complete break from previous systems. Names of orders can be recognized as such because the name of each order ends in 'sol'. Each name of an order contains a formative element that is carried down as an ending at all lower category levels. Thus for the oxisols 'ox' is the formative element, and all soil names in this order end in 'ox' (see Table 2.4). Names of suborders have two syllables, the first connating the diagnostic properties of the suborder and the second the formative element of the order. Thus oxisols with an aquic soil moisture regime are described as aquox. The name of a great group consists of the name of a suborder and a prefix that consists of one or two formative elements indicative of further diagnostic properties. Thus an aquox with gibbsite dominant in the mineral fraction is described as a gibbsiaquox (refer to Table 2.4).

The major problem in the practical application of the US *Soil taxonomy* is the work often required on the soil before accurate identification is possible. Also the methods used in many laboratories throughout the world often differ from those required in the US system. Thus lack of detailed soil knowledge is a major current drawback in its worldwide application. However, the requirement for objective data does remove a major problem of most systems, i.e. their reliance on subjective decision making. Table 2.5 indicates former great soil groups of the 'old' US system and some other systems included in the

Table 2.4. Simplified definitions of the orders, suborders, and great groups of the US *Soil taxonomy* found in the tropics*

Order	Suborder	Great group

Oxisols   Soils with oxic horizons (< 16 milliequivalents per 100 g clay), consisting of kaolinite, iron oxides, and quartz; low in weatherable minerals. Usually deep, well-drained, red or yellow soils, excellent granular structure, very low chemical fertility, uniform properties with depth

	Orthox	Oxisols with udic soil moisture regimes Haplorthox: simple Eutrorthox: high base saturation Acrorthox: very low base saturation Gibbsiorthox: gibbsite dominant
	Ustox	Oxisols with ustic soil moisture regimes Haplustox: simple Eutrustox: high base saturation Acrustox: very low base saturation
	Torrox	Oxisols with aridic soil moisture regimes
	Humox	Oxisols with high organic matter (usually in highlands) Haplohumox: simple Acrohumox: very low base saturation Gibbsihumox: gibbsite dominant
	Aquox	Oxisols with aquic soil moisture regimes Ochraquox: light-coloured A horizon Umbraquox: dark-coloured A horizon Plinthaquox: with plinthite Gibbsiaquox: gibbsite dominant

Ultisols   Soils with an argillic horizon (20 per cent increase in clay content in the control section) with less than 35 per cent base saturation in the control section. Usually deep, well-drained, red or yellow soils, higher in weatherable minerals than Oxisols, with less desirable physical properties, and relatively low native chemical fertility. Ultisols may have oxic horizons above or below the argillic

	Udults	Ultisols with udic soil moisture regimes Tropudults: simple Paleudults: very deep argillic horizon Rhodudults: dusky red, high in oxides Plinthudults: with plinthite Fragiudults: with fragipans
	Ustults	Ultisols with ustic soil moisture regimes Haplustults: simple Paleustults: deep argillic horizon Rhodustults: dusky red, high in oxides Plinthustults: with plinthite

*From Sanchez (1976). For detailed definitions, nomenclature, and the key to the various categories refer to *Soil taxonomy* (1975).

Order	Suborder	Great group
	Humults	Ultisols high in organic matter (usually in highlands) Tropohumults: simple Palehumults: deep argillic horizon Plinthohumults: with plinthite
	Aquults	Ultisols with aquic soil moisture regimes Tropaquults: simple Paleaquults: deep argillic horizon Plinthaquults: with plinthite Fragaquults: with fragipan Albaquults: light-coloured A horizon
Alfisols		Soils with an argillic horizon with more than 35 per cent base saturation. Similar to Ultisols except for considerably higher native chemical fertility
	Udalfs	Alfisols with udic soil moisture regimes Tropudalfs: simple
	Ustalfs	Alfisols with ustic soil moisture regimes Haplustalfs: simple Paleustalfs: deep argillic horizon Rhodustalfs: dusky red, high in oxides Plinthustalfs: with plinthite Natrustalfs: sodic horizon Durustalfs: with duripan
	Aqualfs	Alfisols with aquic soil moisture regimes Tropaqualfs: simple Plinthaqualfs: with plinthite Natraqualfs: sodic horizon Duraqualfs: with duripan
Aridisols		Soils of aridic moisture regimes, with horizon differentiation
	Argids	Aridisols with argillic horizons Haplargids: simple Paleargids: deep argillic horizon Natrargids: sodic horizon Durargids: with duripan
	Orthids	Aridisols without an argillic horizon Salorthids: salic horizon Paleorthids: deep argillic horizon Calciorthids: calcareous horizon Gypsiorthids: gypsic horizon Camborthid: cambic horizon Durorthids: with duripan
Inceptisols		Young soils with cambic horizon but no other diagnostic horizons
	Andepts	Inceptisols derived from volcanic materials Cryandepts: cold andepts

(continued)

Table 2.4. (cont'd)

Order	Suborder	Great group
		Vitrandepts: high in volcanic glass Dystrandepts: low base saturation Eutrandepts: high base saturation Hydrandepts: well drained with high water content Durandepts: with duripan
	Aquepts	Inceptisols with aquic soil moisture regimes Tropaquepts: simple Andaquepts: affected by volcanic ash Halaquepts: saline Sulfaquepts: acid sulphate soils or cat clays Plinthaquepts: with plinthite
	Tropepts	Other tropical inceptisols Dystropepts: low base status Eutropepts: high base status Ustropepts: with ustic soil moisture regimes Humitropepts: high in organic matter (highlands)
Entisols		Soils of such slight and recent development that only an ochric (yellowish) epipedon or simple man-made horizons have formed
	Aquents	Entisols with aquic moisture regimes Tropaquents: simple Fluvaquents: alluvial Hydraquents: high moisture content, thixotropic Sulphaquents: acid sulphate soils or cat clays
	Fluvents	Entisols of recent alluvial origin Tropofluvents: simple Ustifluvents: with ustic moisture regimes Torrifluvents: with aridic moisture regimes
	Psamments	Sandy entisols Tropopsamments: simple Ustipsamment: with ustic soil moisture regimes Torripsamments; with aridic soil moisture regimes Quartzipsamments: with mainly quartz sand
	Orthents	Other tropical entisols Troporthents: simple Ustorthents: with ustic soil moisture regimes Torriorthents: with aridic soil moisture regimes
Vertisols		Heavy, cracking clayey soils with more than 35 per cent clay and > 50 per cent of 2:1 minerals in clay fractions. Usually shrink and swell with changes in moisture contents, have gilgai microrelief and slickensides on peds
	Usterts	Vertisols with ustic soil moisture regimes Chromusterts: light coloured Pellusterts: dark coloured

Order	Suborder	Great group
	Uderts	Vertisols with udic soil moisture regimes
	Torrerts	Vertisols with aridic soil moisture regimes
Mollisols		Soils with a mollic epipedon (i.e. high in organic matter, soft when dry, and > 50 per cent base saturation)
	Rendolls	Mollisols over limestone, formerly called Rendzinas
	Ustolls	Mollisols with ustic moisture regimes Haplustolls: typical Argiustolls: argillic horizon Paleustolls: deep argillic horizon Calciustolls: calcitic horizon Durustolls: with duripan Natrustolls: sodic horizon
	Aquolls	Mollisols with aquic moisture regimes Haplaquolls: typical Argiaquolls: argillic horizon Calciaquolls: calcic horizon Duraquolls: with duripan Natraquolls: sodic horizon
Spodosols		Soils with a spodic horizon (of iron and organic matter accumulation), usually developed on sandy materials. Equivalent to Podzols in all other classification systems
	Aquods:	Spodosols with aquic moisture regimes Tropaquods: typical Duraquods: with hardpan
	Humods	Spodosols high in organic matter Tropohumods: typical
	Orthods	Other tropical Spodosols Troporthods: typical
Histosols		Organic soils with a histic epipedon (> 20 per cent organic matter). All have a tropo-great group
	Fibrists	Over two-thirds fibre
	Folists	Not saturated with water (foggy tops)
	Hemists	Between one-third and two-thirds fibre
	Saprists	Less than one-half fibre

orders of the US *Soil taxonomy*, and Table 2.6 gives approximate correlations between the FAO legend, the US *Soil taxonomy* and the French classification.

Table 2.5. *Orders of the US* Soil taxonomy *in relation to great soil groups of the previous USDA schemes and other classification systems*

Order	Former great soil groups included
Entisols	Azonal soils, some low humic gley, lithosols, regosols
Vertisols	Grumusols, tropical dark clays, regur, black cotton soils, dark magnesium clays
Inceptisols	Andosols, hydrol humic latosols, soil brun acidic, some brown forest, low humic gley, humic gley
Aridisols	Desert, reddish desert, sierozem, solonchak, some brown and reddish-brown soils, associated solonetz
Mollisols	Chestnut, chernozem, brunizem, rendzina, some brown forest, brown, associated humic gley and solonetz
Spodosols	Podzols, brown podzolic, ground-water podzols
Alfisols	Grey-brown podzolic, grey wooded, noncalcic brown, degraded chernozem, associated planosols and half bog, some terra roxa estruturada and eutric red–yellow podzolics, some katosols and lateritic soils
Ultisols	Red–yellow podzolic, reddish-brown lateritic, humic latosols, associated planosols, and some half bogs, latosols, lateritic soils, terra roxa, and ground-water laterites
Oxisols	Low humic latosols, humic ferruginous latosols, aluminous ferruginous latosols, some latosols, lateritic soils, terra roxa legitima, ground-water laterites
Histosols	Bog soils, organic soils, peat, muck

Tropical soil orders and suborders

A summary of the profile features of tropical soils at the various category levels has been presented in Table 2.4. Useful texts covering the formation, morphology, distribution, and fertility of tropical soils are Buringh (1970), Aubert and Tavernier (1972), Mohr *et al.* (1972), and Sanchez (1976).

The distribution of tropical soils at suborder level is shown in Fig. 2.4(a) and (b). Calculations based on this map show the approximate extent of the major soil suborders in the tropics (Table 2.7). Oxisols are shown to occupy the largest area (22 per cent), followed by aridisols (18 per cent), alfisols (16 per cent), ultisols (11 per cent), inceptisols (8 per cent), entisols (8 per cent), vertisols (2 per cent), and mollisols (1 per cent). Spodosols and histosols although not of sufficient area to be recognized as separate mapping units are locally important in many areas. The mountain areas group (12 per cent) is not differentiated,

Table 2.6. Approximate correlation between the FAO, US *Soil taxonomy* and the
French Soil Classification Systems with special reference for the
Tropics (from Aubert and Tavernier 1972)

FAO legend	US *Soil taxonomy*	French classification
Fluvisols	Fluvents	Sols minéraux bruts et sols peu évolués d'apport alluvial et colluvial
Regosols	Psamments	Sols minéraux bruts et sols peu évclués d'apport éolien
Arenosols		
Ferralic	Oxic Quartzipsamments	Sols ferralitiques moyennement or fortement désaturés, à texture sableuse
Gleysols		
Eutric and Dystric	Tropaquepts	Sols hydromorphes humifères à gley
Humic	Humaquepts	Sols humiques à gley
Plinthis	Plinthaquepts	Sols hydromorphes à accumulation de fer en carapace or cuirasse
Andosols	Andepts	Andosols
Planosols	Paleudalfs and Paleustalfs	Sols ferrugineux tropicaux lessivés (pro parte)
Cambisols		
Dystric	Dystropepts	Sols ferralitiques fortement et moyennement désaturés, rajounis (pro parte)
Eutric	Eutropepts	Sols ferrugineux tropicaux (non lessivés)
		Sols ferralitiques faiblement désaturés, rajeunis
Humic	Humitropepts	Sols ferralitiques fortement et moyennement désaturés, humifères, rajeunis
Luvisols	Alfisols	Sols ferrugineux tropicaux lessivés
Aerisols	Ultisols	Sols ferralitiques fortement désaturés
Ferralsols	Oxisols	Sols ferralitiques
Lithosols	Lithic subgroups	Lithosols et sols lithiques

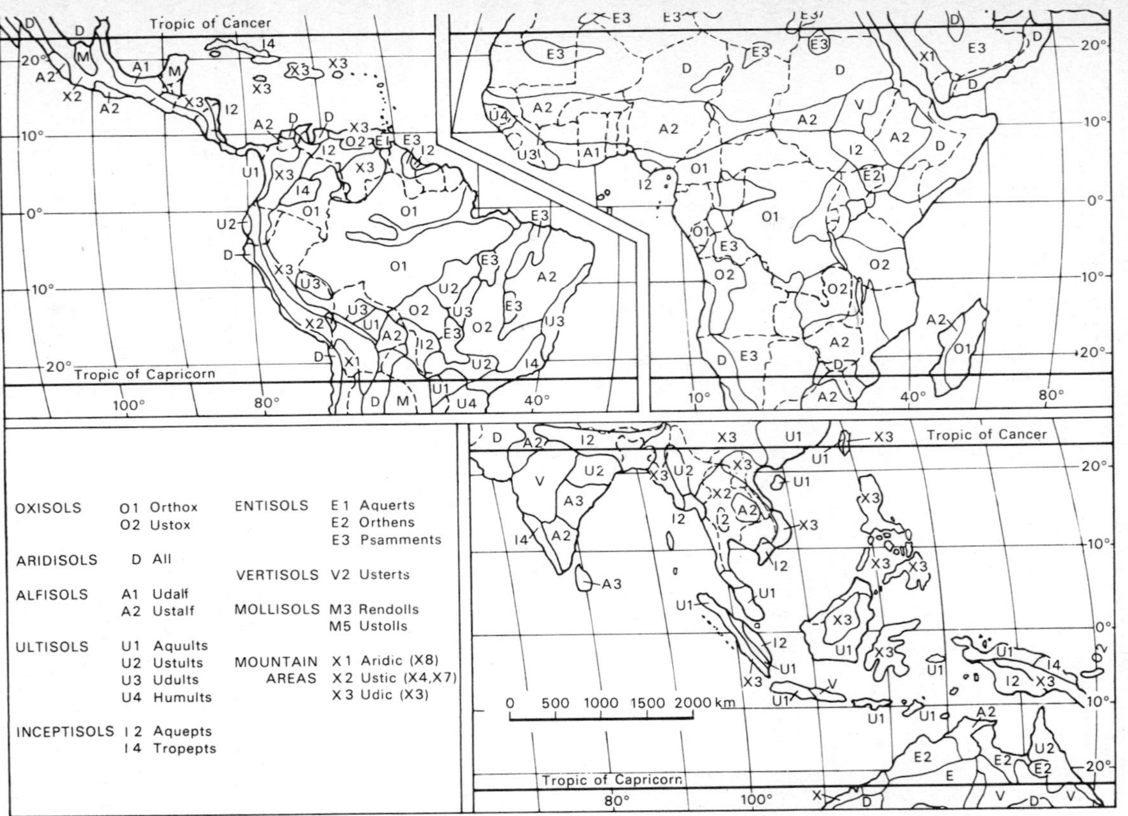

Fig. 2.4. Soils of the tropics (after Aubert and Tavernier (1972) from Sanchez (1976)).

Table 2.7. Approximate extent of major soil suborders in the tropics (million ha).
(See Sanchez (1976); calculated from maps in Aubert and Tavernier
(1972))

Order	Suborder	Africa	America	Asia	Total area	Percent
Oxisols	Orthox	370	380	0	750	15.0
	Ustox	180	170	0	350	7.5
		550	550	0	1100	22.5
Aridisols	All	840	50	10	900	18.4
Alfisols	Ustalfs	525	135	100	760	15.4
	Udalfs	25	15	0	40	0.8
		550	150	100	800	16.2
Ultisols	Aquults	0	40	0	40	1.0
	Ustults	15	35	50	100	2.2
	Udults	85	125	200	410	8.2
		100	200	250	550	11.2
Inceptisols	Aquepts	70	145	70	285	6.0
	Tropepts	0	75	40	115	2.3
		70	225	110	400	8.3
Entisols	Psamments	300	90	0	390	8.0
	Aquents	0	10	0	10	0.2
		300	100	0	400	8.2
Vertisols	Usterts	40	0	60	100	2.0
Mollisols	All	0	50	0	50	1.0
'Mountain areas'		0	350	250	600	12.2
Total		2450	1670	780	4900	100.0

including important groups such as the andepts. More accurate data is
becoming available as units of the FAO/UNESCO soil map of the world
are published.

The most important orders of the tropical soils are now briefly
discussed (adapted from Aubert and Tavernier (1972)).

Oxisols

In the humid tropics, oxisols generally occur on very old tablelands,
where weathering products have been protected against erosion during

long periods. Four suborders and their approximate equivalents in other classifications systems follow.

Aquox-oxisols of wet places. Aquox are dominantly grey, or mottled dark red and grey, and are seasonally saturated with water. They include some soils called in French *Sols à accumulation de fer en carapace ou cuirasse* and soils formerly called 'ground water laterite soils' in the United States.

Humox-oxisols of relatively cool moist regions with a large accumulation of organic carbon. Humox include soils called *Sols ferrallitiques fortement désaturés humifères* in French and formerly called 'black red latosols' and 'humic latosols' in the United States.

Orthox-oxisols of warm humid regions with short (less than 90 days) dry seasons or no dry season. Orthox include soils called in French *Sols ferrallitiques*. They were formerly called 'red latosols', 'earthy red latosols', or 'red–yellow latosols' in the United States.

Ustox- oxisols with well marked wet and dry seasons. Ustox include soils called *Sols ferrallitiques faiblement désaturés* in French and some soils formerly called 'low humic latosols' in the United States.

Oxisols are the dominant soils of continental shields in the permanently moist and alternately wet and dry climates. Orthox soils are largely found in the south of West Africa, in the central part of the African equatorial belt, in the Asian humid tropics (Sri Lanka, Thailand, Vietnam) and in the low-altitude areas of the humid tropics of South America. Ustox soils are extensive in the Brazilian plateau and East Central Africa. Humox soils occur in Africa in the eastern Congo basin, Central Madagascar, South Asia, tropical Andes, and in many islands of Oceania. Aquox soils, with more or less indurated plinthite, are found over large areas in some regions, for instance the central, southern, and eastern part of Senegal; the central and southern part of Upper Volta; northern part of Guinea; the Ivory Coast; Ghana; the northern part of the Central African Republic; the southern part of the Sudan; and the plateaux of Katanga. Near the equator, the large continental basins may have been filled also with materials that have acquired oxic properties. The Congo and the Amazon basins are examples.

Oxisols depend mostly on the quality and the amount of the organic matter for retention of cations. Without fertilizers, they can support extensive agriculture only under shifting cultivation or with tree crops that protect the soil, or extensive grazing with poor-quality grass. When soft plinthite occurs near the soil surface, clearing of the natural vegetation may accelerate the irreversible induration of the free sesquioxides and alter the soil into a hard rock.

Alfisols and ultisols

Two distinct but not necessarily exclusive processes are recognized in the more mature alfisols and ultisols. The first is the formation of an argillic horizon, which is the result of the accumulation of clay in a subsoil horizon. Deep weathering and clay destruction may be associated with the formation of some argillic horizons. The second process concerns the weathering of minerals into kaolinitic clays and free sesquioxides, which reaches maximal dimensions in what is called the oxic horizon. Both argillic and oxic horizons are diagnostic horizons in tropical soils.

The parent materials from which alfisols and ultisols have developed by definition have not reached the oxic weathering stages. As a rule, they still contain some weatherable minerals, and the clays are not exclusively composed of kaolinite and free sesquioxides. The argillic horizons seem to reach maximum development in climates that have a dry season.

Ultisols

Ultisols characteristically form under forest vegetation in climates with slight to pronounced seasonal moisture deficits and surpluses. In the humid tropics, the soils have a relatively low cation exchange capacity. The bases not held in the plant tissue are depleted.

Suborders of ultisols are defined in terms of their moisture regimes and organic matter contents. Four suborders and their approximate equivalents in other classification systems follow.

Aquults

Ultisols of wet places. A few aquults have black surface horizons on grey subsoils. They include some soils called in French, *Sols a amphigley* and *Sols à gley lessivés*, and some soils formerly called 'humic-gley' and 'low-humic gley' in the United States.

Humults

Ultisols with large accumulations of organic matter formed under relatively high, well-distributed rainfall in subtropical and tropical regions. Humults include some soils formerly called 'reddish-brown lateritic' and some 'humic latosols' in the United States. The approximate French equivalents are *Sols ferrallitiques fortement désaturés humifères*, and *Fertisols humifères*.

Udults

Ultisols of humid subtropical to tropical regions. Udults have moderate to very small amounts of organic matter, reddish or yellowish B horizons and no periods or only short periods when some part of the soil is dry (less than 90 cumulative days in most years). They include some soils formerly called 'red–yellow podzolic' and 'reddish-brown lateritic' in the United States. Some soils called in French *Sols jaunes et rouges blanchis* belong to this suborder.

Ustults

Ultisols of warm places with long periods (more than 90 cumulative days in most years) when the soil is dry. Ustults are brownish to reddish throughout. They include some soils formerly called 'red yellow podzolic' in the United States. *Sols ferrallitiques fortement désaturés lessivés (pro parte)* is the approximate French equivalent.

Ultisols with plinthite or thick B horizons are associated with mid-Pleistocene or older surfaces. The other great groups are largely restricted to late- or post-Pleistocene surfaces.

Alfisols

The order of alfisols becomes important at the transitional zone to arid climates. They characteristically form under forest or savannah vegetation in climates with a slight to pronounced seasonal moisture deficit. At some season evapotranspiration exceeds precipitation and stored soil moisture is depleted or exhausted.

Aqualfs

Alfisols of wet places. Aqualfs are dominantly grey throughout and are seasonally saturated with water. They include soils called in French, *Sols lessivés hydromorphes, Sols ferrugineux Tropicaux lessivés*, or *remaniés à pseudogley*. Soils formerly called 'low humic gley' and 'planosols' in the United States are included in this suborder.

Udalfs

Alfisols of humid temperate or warm regions. Udalfs are brownish throughout and have no periods or only short periods (less than 90 cumulative days in most years) when part or all of the soil is dry. They were formerly called 'gray-brown podzolic' soils in the United States.

Ustalfs

Alfisols of warm places with long periods (more than 90 cumulative days in most years) when soil is dry. Ustalfs are brownish to reddish throughout. They include soils called in French *Sols ferrugineux tropic-*

aux lessivés ou appauvris. They were formerly called 'reddish-brown soils' in the United States.

Great groups that are important in the tropics include paleudalfs and paleustalfs with strongly developed B horizons and rhodustalfs that have dark reddish colours and are usually developed from basic rocks or hard limestone.

Inceptisols

The term 'inceptisols' was coined for the class of soils that show some profile development but are relatively immature. They may show strong gleying as a result of reduction by water, release of free iron oxides as a result of weathering, or the formation of soil structure under conditions of wetting and drying. Within the inceptisols, one suborder of soils, andepts, deserves special attention. Andepts have been used intensively in the tropics, especially in areas affected by volcanic activity. They are derived from volcanic ash, which weathers rapidly and produces a peculiar kind of amorphous clay, allophane. In combination with organic matter, allophane gives the soil its most striking characteristics, namely high water-holding capacity and incomplete dispersibility. Allophane is known to trap organic matter, protecting it against rapid decomposition. Andepts, therefore, are usually darker than adjacent non-volcanic soils; on the other hand, they become rapidly depleted of bases, and once they have been leached and become acid, they often fix large quantities of phosphorus. At a low pH in the field, they have a low cation exchange capacity, which is strongly pH dependent. The richest members of the group, or those that have not been leached, are generally excellent soils.

Andepts are extensive in Central and South America along the Andes Cordillera. Indonesia and the isles of Oceania have large areas of volcanic ash soils. In Africa, andepts have been recognized in Cameroon, Kenya, Madagascar, Tanzania, Uganda, and Zaire. Many soils of Hawaii and Central America, known locally as 'latosols', are actually andepts.

The value of the other inceptisols in the tropics (tropepts) is primarily governed by the quality of the parent rocks. This is not surprising, since the properties of these incipient soils are closely related to the rocks that have generated them. For example, tropepts from basic rocks are generally productive soils, whereas parent material from acidic rocks as a rule develops into poor soils often mixed with stone-lines that may restrict root penetration at a shallow depth.

Entisols

Entisols are recent deposits that have not been altered or have been only slightly altered by soil-forming processes. On small-scale maps of the low humid tropics, they would hardly be represented, since rock outcrops are rare, recent sand dunes are not extensive, and young alluvial sediments (fluevents) occupy only a narrow strip bordering the waterways. Quartz sands (psamments) may be extensive in some areas, for instance in the south of the Congo basin. The value of entisols depends on many properties: shallowness of the stony soils, which limits the soil volume; sand dunes, which usually restrict water storage capacity; and alluvial sediments, which may suffer from periodic flooding. On the other hand, these soils often have not undergone the adverse action of tropical weathering and leaching and, when all other factors are favourable, they may be among the most productive soils of the world.

Soil survey and mapping

For a detailed discussion of soil survey and mapping procedures refer to the *Soil survey manual of the United States Department of Agriculture* (Soil Survey Staff, USDA 1951) and for a general discussion of the situation in the tropics to Aubert and Tavernier (1972).

Adequate land-use planning is based on knowledge of the properties of the soil. These properties are best related to soil survey that identify the soil features and show the geographic distribution of the kinds of soils on maps. Depending on the purposes to be achieved, soil surveys in the tropics have been carried out at varying levels of intensity. Broad-scale maps are mainly of use to governments for resource inventory, planning, etc., whereas more detailed maps can be of use to the research or extension worker or to an individual farmer.

An important point to realize about a soil map is that because of the usual variability of soils in the field the mapping unit used is usually some combination of taxonomic units (sub-order, series, etc.). This means that to interpret a soil map at a particular location in the field requires a good knowledge of the nature of the mapping unit used.

Most surveys are carried out for practical agricultural purposes. The survey report normally includes chapters dealing with the capability and the suitability of soils for different agricultural practices. The mapped soils may be placed in capability classes according to their potential to support certain types of land use. Subclasses may be framed on the basis of the limiting factors that are to be corrected in order to obtain maximum productivity. Thus soil survey and mapping may form part of a wider objective, i.e. a survey of land capability.

Land-capability surveys

This type of survey sets out to provide information on potential land use for maximizing production in the long term and usually from the biological point of view rather than the economic point of view.

The most common form of classification is based on the USDA Land Capability System concentrating on separating-out land which can be safely cultivated without erosion damage and that land which should not be cultivated. Limitations to land use are then listed according to water or nutrient problems.

The following is an example of the system as modified and used by the FAO.

Class I – very good land that can be cultivated safely with ordinary good farming methods. It is nearly level and easily worked. Some areas need clearing, water management, or fertilization. Usually there is little or no erosion.

Class II – good land that can be cultivated safely, with easily applied conservation practices. These include such measures as contouring, protective cover crops, and simple water-management operations. Customary practices are rotation and fertilization. Moderate erosion is common.

Class III – moderately good land that can be cultivated safely, with such intensive practices as terracing and strip cropping. Water management is often required on flat areas. Common requirements are crop rotation, cover crops, and fertilization. Usually Class III land is subject to moderate to severe erosion.

Class IV – fairly good land that is best suited to pasture and hay, but can be cultivated occasionally – usually not for more than one year in six. In some areas, especially those of low rainfall, selected land may be cultivated more than one year if adequately protected. When the land is ploughed, careful erosion–prevention practices must be used.

Class V – land suited for grazing or forestry, with slight or no limitatins; needs only good management. It is nearly level and not subject to any appreciable erosion.

Class VI – land suited for grazing or forestry, with minor limitations; needs careful management and protective measures.

Class VII – land suited for grazing or forestry, with major limitations; needs extreme care to prevent erosion or other damage.

This system is fairly broad and can be used for a rapid assessment of a region by someone experienced in assessing the erosion hazards of soils and the control of soil erosion. Often planners are not concerned initially with emphasis on the soil-erosion hazard but rather overall development possibilities.

In West Malaysia, for example, land-capability surveys are rather more complicated exercises requiring detailed maps of land alienation; aborigine, game, and veterinary reserves; mineral development potential; of soil, forestry, and water resources; and a detailed compilation of all relevant data. However, the classification itself has only five classes.

Class I — land possessing a high potential for mineral development and therefore best suited to mining.

Class II — land possessing a high potential for agricultural development with a wide range of crops and therefore best suited to diversification agriculture.

Class III — land possessing a moderate potential for agricultural development with a restricted range of crops and therefore best suited to agricultural development with crops having a wide range of soil tolerance.

Class IV — land possessing a potential for productive forest development and therefore best suited to commercial timber exploitation.

Class V — land possessing little or no mineral, agricultural, or forest development potential but suitable for development as protective reserves for conservation, water catchment, game, recreation, or similar purpose, or possibly suitable in the future for productive forest plantations with introduced species.

In order to produce a land-capability map which is more detailed, it is necessary to have a subtantial amount of information and research data as well as a detailed soil map.

An example of land classification of this type has been used in Ghana (Obeng 1968).

(a) *Land considered suitable for both mechanized and hand cultivation and for other uses*

(i) Class I land — this consists of very good soils with very minor or no physical limitations to mechanized cultivation. The soils are deep to very deep, well-drained, and medium to moderately heavy textured, on level to very gently sloping topography. They are moderately permeable and have a moderate to moderately high water-holding capacity, medium to low inherent fertility, but a good capacity to utilize added fertilizers. They are subject to no more than very slight erosion and are not subject to damaging inundation.

Class I is suitable for intensive agriculture involving any crop which the climate of the area allows and can sustain moderate to high crop production with management practices such as mulching, manuring, addition of commercial fertilizers, the inclusion of leguminous crops in the rotation, and contour ploughing.

(ii) Class II land — this land is made up of good soils with a few physical limitations which can be easily corrected. The soils are suited to cultivated crops, pastures, and for forestry purposes.

The possible limitations which make Class II not as good as Class I may be generally subdivided on the basis of (1) erosion — denoted by IIe, (2) wetness — denoted by IIw, and (3) other soil properties — denoted by IIs. The first and third subclasses commonly relate to upland soils and the second to lowland soils. The limitations may include one or more of the following effects within each of the undermentioned subclasses:

(a) Subclass IIe — 1. moderate erosion hazard; 2. gentle slopes.
(b) Subclass IIw — 1. imperfect to poor drainage; 2. heavy textured soils; 3. occasional inundations.
(c) Subclass IIs — 1. moderately deep soils; 2. moderately well-drained soils; 3. moderately permeable soils; 4. low to moderate water-holding capacity; 5; low inherent fertility; 6. very fair capacity to utilize added fertilizers.

The limitations of Class II restrict the selection of crops, time and ease of cultivation, and the amount of water and frequency of irrigation. Moderate productivity can best be maintained on Class II land by (1) mulching, manuring, addition of commercial fertilizers, liming and the inclusion of legumes in the rotation, (2) terracing, (3) strip cropping, (4) contour ploughing, and (5) water-control structures.

(iii) Class III land — the soils in Class III are moderately good but have more limitations to mechanized cultivation than Class II land. When used for cultivated crops, conservation practices are usually more difficult to apply and maintain. They are suited to arable and pasture crops, and for forestry purposes.

The limitations of soils in Class III that restrict the degree of cultivation, choice of crops, and the time of planting and harvesting may include one or a combination of the following effects in the under-mentioned subclasses:

(a) Subclass IIIe – 1. moderate to high erosion hazard; 2. sloping or undulating topography.

(b) Subclass IIIw – 1. poor to very poor internal drainage; 2. excessive wetness; 3. heavy textures; 4. moderate hazard of inundation.

(c) Subclass IIIs – 1. moderate depth to bedrock, claypan, indurated later, concretions or ironpan; 2. low moisture-holding capacity; 3. light or heavy textures; 4. wide range of permeability; 5. low inherent fertility; 6. limited capacity to utilize added fertlizers.

Moderate productivity of Class III land can best be maintained by following management practices recommended for Class II land with additional elaborate water control measures such as grassed waterways, close strip cropping and broad contour terraces or bunds.

(iv) Class IV land – this land consists of fairly good soils best suited for perennial vegetation. They can with great care be mechanically culti-vated for field crops. Hand cultivation and/or bullock farming can be practised. The possible limitations that may singly or in combination render Class IV not as good as Class III, may generally be described under subclass IVe, IVw, and IVs as follows:

(a) Subclass IVe – 1. Moderate to high erosion hazard; 2. sloping to hilly topography.

(b) Subclass IVw – 1. either poorly to very poorly or excessively drained soils; 2. heavy textured soils; 3. moderate hazard of inundation.

(c) Subclass IVs – 1. moderately shallow depth to bedrock, ironpan, concretionary layer or claypan; 2. light or heavy textured and/or gravelly or stony soils; 3. wide range of subsoil permeability; 4. low water hol-ing capacity; 5. low inherent fertility; 6. low capacity to utilize added fertilizers.

The productivity of Class IV can be raised by following more intensely the management practices recommended for Class II land but rotations to be followed should include long periods of forage or tree-crop production. Class IV is very extensive in Ghana and is mainly cultivated for food crops and tree cash crops.

(b) *Land considered unsuitable for mechanized cultivation but well
suited for hand cultivation and for other uses*

(i) Class V land — soils in Class V are unsuitable for mechanized cultivation because of severe limitations. They are best suited to limited clearing, grazing, and land cultivation for the production of perennial crops. These soils commonly occupy rocky areas or hilly to steep topography and are subject to moderate or severe erosion. Deep, heavy textured, and poorly drained soils on nearly level topography which are very difficult to drain will also fall within this class. Upon draining, however, the heavy textured soils can be raised to Class III or IV levels and thus be suitable for the production of rice, irrigated pastures, etc. Mechanized cultivation on Class V land is not feasible because of one or a combination of the following limitations described under subclasses, Ve, Vw, and Vs, below.

(a) Subclass Ve — 1. moderate to severe erosion hazard; 2. either steep or rocky slopes; 3. excessive drainage.

(b) Subclass Vw — 1. wetness; 2. heavy textures; 3. annual hazard of inundation.

(c) Subclass Vs — 1. extremes in textures; very light or very heavy; 2. wide range or permeability; 3. shallow depth and stoniness; 4. low or very low moisture-holding capacity; 5. low inherent fertility; 6. fair or poor capacity to utilize added fertilizers.

The productivity of Class V for perennial crops can best be raised by mulching, manuring, application of commercial fertilizers, and the inclusion of leguminous crops in the rotation. The establishment of stringent erosion control measures is essential even when such soils are hand-cultivated. In this respect, permanent sod, grassed waterways, broad-based terraces, and cover crops are recommended.

Class V is very extensive in the interior savannah and forest zones of Ghana. In the interior savannah zone it is mainly represented by moderately shallow phases of soils which are suited to livestock grazing. Most of Class V land in the forest zone is suited to tree crops such as cocoa, coffee, oil palm, and rubber.

(ii) Class VI land — soils within this class differ from those in Class V mainly because they are slightly shallower and occur on steeper slopes. They are marginal for hand cultivation of arable crops but are considered to be suitable for pasture grazing and tree crops such as cocoa, oil palm, coffee, and citrus.

In addition to subclasses VIs, VIe, and VIw, combinations of two or more subclasses can also occur under this class, e.g. VIse, VIsw, and VIwe. Management practices are the same as recommended for Class V.

(c) *Land very limited in use*

This land is considered unsuitable for cultivation and pasture grazing; it must be devoted to forestry purposes and may be the site of rural building development.

(i) Classes VII and VIII — soils within these classes are subject to very severe erosion hazard and are too shallow, steep, droughty, or too wet to be suitable for any type of cultivation. Soils in Class VII are underlain by loose gravel, soft weathered rock, or incipient claypan. They occur both in the forest zones and savannah zones of Ghana. Soils in Class VIII are very shallow to bedrock, solid ironpan, or impermeable claypan. They may be permanently waterlogged.

A more generalized version of the above system has been used with success in evaluating large areas earmarked for agricultural development in many parts of Ghana. This type of system is mainly of value in areas where no agricultural development has taken place. If the needs are more specific, e.g. if interest is just in land suitable for pasture development, it may be more efficient to design a land use classification system and survey for that specific need.

The economic gain from pasture production is usually less than cropping. This is an important factor in the types of recommendations for land use that are made, as in the systems described. This frequently means that land used for pasture production has serious physical and chemical limitations (see following sections); emphasizing the need for the understanding of these limitations and ways that they may be corrected.

2.2 SOIL FERTILITY AND ITS MAINTENANCE

Basic fertility factors

The function of the soil in relation to plant growth is to provide a roothold, requiring adequate soil depth and absence of physical impedance factors, and to supply water, nutrients, and oxygen to the roots. In addition heat must be transferred to allow adequate temperature for growth and there needs be an absence of toxic materials which may affect root growth.

Soil fertility can be defined in a general way as the capacity of the soil to produce high quality crops and pastures and to support good quality animals. More specifically a soil's fertility can be viewed as its capacity to produce the 'crop' desired in its particular environment. In this broad view of fertility any soil factor affecting growth and hence yield of the desired species is a legitimate aspect of fertility. The potential productivity of a plant community at a particular site is determined primarily by climate but the actual productivity is often determined by soil fertility.

The crop or species desired is important since there is no absolute fertility standard. At any particular site the conditions required for optimal growth of one species may be quite unsuitable for another and are frequently different in detail. Thus in the pasture context we have the different needs of legumes v. non-legumes and in general cropping the special requirements of rice, tobacco, etc. The environment is clearly critical since, as environmental characteristics change, so will plant requirements and so fertility. For example requirements under irrigation or high rainfall may be quite different to the situations under drier conditions. Thus as the water supply increases so does yield potential and hence requirement for nutrients. In such situations mobile nutrients are readily leached and in the case of high rainfall soils nutrient supply may be low, widening the gap between supply and requirement for optimum yield.

A further characteristic of a fertile soil is that its fertility is stable over time. Of particular importance here is resistance to erosion. Other problems which develop with time in less fertile soils are nutrient shortages and also toxicities, e.g. salinity. Useful general reference texts on soil fertility are Black (1968) and Russell (1973), and more recently Sanchez (1976) for soils in the tropics.

Physical factors
Factors such as soil depth, texture, structure, and also the slope at a particular site are basic soil features affecting many aspects of fertility. They are usually difficult to manipulate and are therefore extremely important in terms of the initial assessment of the capability of land for crops, pastures, etc. Plants can be affected by the physical characteristics of soils in a number of ways, (see Taylor and Ashcroft (1972) for a detailed treatment).

Soil depth

Adequate depth is essential to give roothold for the plant, to provide an adequate supply of essential nutrients, and to provide a store of water. Shallow soils often occur in upper, steeply sloping positions so that erosion is a further hazard. Some parent materials such as ash are readily penetrated by roots so that the expected problems are less serious.

Impedance effects

A variety of effects may occur in the soil surface, with particular influence on seedling germination. Stones or rock fragments in or on the topsoil may permanently restrict the soil volume available for growth. Clods can cause surface germination problems in seedbed preparation. Crusts are hard layers that may form on the soil surface as a result of the beating action of raindrops and the subsequent drying of the compacted layer preventing seedling emergence, particularly in plants with small seeds. Sandy loams to clay loams low in organic matter often show the most crusting effects. Compaction layers at depth in the profile may prevent root penetration and cause problems similar to those of shallow soils.

Aeration

Plant roots require an adequate supply of oxygen to maintain growth. If the oxygen concentration becomes too low — possibly below 10 per cent — or the concentration of carbon dioxide is too high, then plant growth may be affected. The soil air is usually richer in carbon dioxide and lower in oxygen than atmospheric air. Composition is governed by the activity of biological processes producing carbon dioxide and the rate of renewal of the soil air. Rate of renewal is governed by a number of factors of which the volume and the size of the pores are most important. In most soils, the volume of the pores, i.e. the 'air capacity', is of the order of 50 per cent — average pore size however is much smaller in some soils than others, e.g. sands tend to have more large pores than clays and therefore have fewer aeration problems.

In most normal soils, lack of aeration does not become a problem until they become wet. Water reduces the air space and exchange of air becomes slow, allowing oxygen to decrease and carbon dioxide to increase. Thus poor growth under conditions of waterlogging is due to poor aeration. Some plants are much less sensitive to waterlogged conditions than to others. Some of course grow quite well under waterlogged conditions.

Temperature

This is one of the primary factors controlling the growth of plants and their distribution. The quantity of radiant energy reaching the soil surface and its effect on soil temperature depends on environmental factors, such as latitude and whether vegetation is present or absent, and soil factors such as bulk density and moisture content which affect heat capacity and heat conductance. In addition to the direct effect of temperature on plant roots or germinating seedlings, temperature affects many important physical and chemical processes such as water movement, diffusion, ion exchange, phosphorus fixation, etc. It also significantly affects the activity of other biological components of the soil which may give rise to beneficial or detrimental effects on plants.

Moisture penetration and conservation

Moisture penetration (infiltration) is important in retention of rainfall and control of erosion. The lower the infiltration rate, the higher the run off and the greater the possibility of erosion by water. Infiltration is usually most rapid with sandy soils because of their predominance of large pores. Clay soils, particularly when they become wet, may have a very low infiltration, e.g. clays of the montmorillonite type found in vertisols. High clay of the kaolinite type, e.g. in oxisols gives good surface structure and good infiltration.

The ability of a soil to conserve water and to supply it later to a growing plant is a most important agricultural characteristic. When water enters the soil under the influence of gravity it might be expected that it would flow right through as though a sieve. If this was the case then plant growth over any significant time would not be possible. Some water does flow through if enough water enters but much is held in the soil. It is held both by attraction to particle surfaces and by capillary forces in the interstices (pores) formed between the particles within the soil. Soils high in colloidal organic matter and clay hold more water since they have a high surface-area per unit of soil material and often a higher percentage of small pores which tend to hold more water.

Soil water can be divided into three categories depending on the energy status of the water. First there is gravitational water of such high energy that it goes right through the profile; secondly there is water held by the soil but with high enough energy to be used by plants (available water); and thirdly water held strongly by the soil and as a consequence of such low energy that it cannot be used by plants (unavailable water). The traditional names for the soil moisture points for separating these divisions are field capacity (the moisture content at which gravitational flow ceases) and wilting point (the moisture content at which plants cannot maintain turgor).

The categories of soil water and the critical moisture points are shown in Fig. 2.5. Soil water content is usually expressed as a percentage by weight (% weight) (g water per 100g oven-dry soil) or a percentage by volume (% volume) (volume of water per unit soil volume) and if the water percentage is known for field capacity and wilting point the quantity of available water in centimetres per metre of soil profile can be calculated. It is usual for % water contents at wilting point and field capacity to increase as the quantity of clay and organic colloid in the soil increases (see Table 2.8). The available water (either as a percentage or as centimetres per metre of profile) also tends to increase with the quantity of clay but in some tropical soils, especially oxisols, the available water content is low despite a high clay content (see Table 2.8). On the other hand the energy with which water is held at the various moisture points is the same for all soils and is described as the water potential; various methods of expression for water potential at the moisture points are shown in Fig. 2.6. The relationship between water potential and % water by weight is shown for a sand and loam in Fig. 2.6.

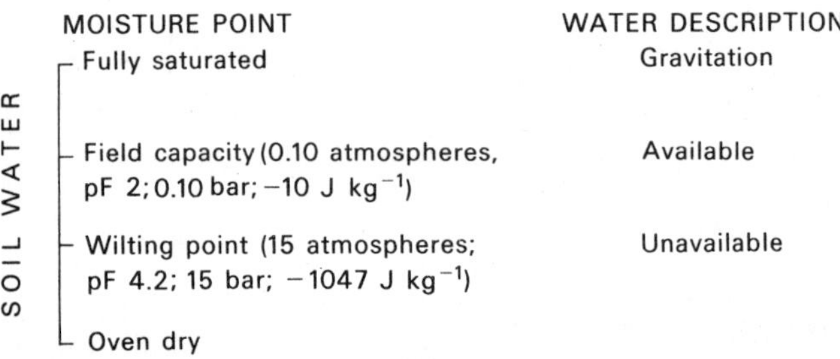

Fig. 2.5. Categories of soil moisture and critical moisture points.

Structure

The ability of the soil to produce and maintain aggregates of a suitable size (0.5 to a few millimetres), especially in the soil surface, is an important soil characteristic affecting particularly the physical properties of the soil. Such aggregates will give a sufficient proportion of large

Table 2.8. Typical moisture percentages at field capacity, wilting point, and for available water in soils of different texture

Soil texture	% water by weight at		Available water (%)
	Field capacity	Wilting point	
Sand	6	3	3
Sandy loam	13	4	9
Clay loam	27	14	13
Clay (Vertisol)	34	17	17
Clay (Oxisol)	44	36	8

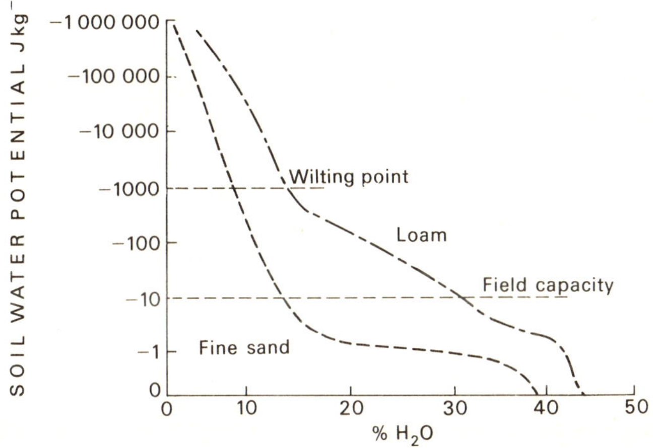

Fig. 2.6. Relationship between soil water potential and % water by weight for a sand and a loam.

pores to allow free entry of water into the soil and subsequent drainage through the profile; they will also allow rapid gaseous exchange between the soil and the atmosphere ensuring good aeration. Organic matter is usually closely related to structural development and soils high in organic matter, such as those under permanent pasture or forest vegetation usually have good structure. However organic matter decline in soils, for example following cultivation, often leads to rapid deterioration in soil structure, (refer to Sanchez (1976) for examples of structural deterioration after cropping in tropical regions).

Chemical factors

Adequate nutrient supply

Some 15 elements (excluding carbon, hydrogen, and oxygen) have been shown to be essential to the growth of plants, i.e. nitrogen, phosphorus, potassium, sulphur, calcium, magnesium, iron, manganese, zinc, copper, molybdenum, chlorine, boron, cobalt, sodium, and an additional two for animals (selenium and iodine) with the role of further elements still under investigation. These are usually divided into the macronutrients, required in relatively large quantities by plants (nitrogen, phosphorus, potassium, sulphur, calcium, magnesium) and the remaining micronutrients or trace elements required in lesser amounts, (refer to Epstein (1972) for a general treatment of mineral nutrition of plants).

The total supply of nutrients in soils is extremely variable. Important factors affecting this are the nature of the parent material and the intensity and duration of weathering and soil formation. Igneous rocks are usually better supplied than sedimentary and basic igneous usually higher in calcium, magnesium, phosphorus, iron, manganese, copper, and zinc, whereas acidic igneous are poorer in these but richer in potassium. Boron is often higher in sedimentary than in igneous rocks. Young soils may closely resemble the elemental composition of the parent material whereas old soils are usually leached so that significant differences develop between soil and parent material. Nitrogen supply in soils is not related to the very low value in rocks but seems correlated with general soil fertility, the supply building up gradually by processes such as atmospheric fixation.

Total supply of an essential nutrient may or may not be a good guide to the quantity present in the soil in a form available to the plant. It may occur mostly as a primary rock mineral and weathering and release to the plant may be slow, e.g. potassium in felspar. Alternatively, it may occur in a secondary mineral form of low solubility and hence availability, e.g. iron and aluminium compounds with phosphorus in oxisols or in humic compounds that are slowly mineralized. As a result much effort has been put into trying to ascertain the plant-available fraction of the essential elements.

Concepts of capacity, intensity, and rate parameters of availability have developed. Capacity referring to the pool of nutrient accessible to the plant over a time period; intensity to the differences in strength of retention or ease of withdrawal, i.e. the activity or less accurately the concentration in the solution; rate factors affecting the speed of replenishment of nutrients taken up by the plant, e.g. rate of movement of nutrient out of aggregates that roots do not penetrate. More recently the approach of measuring buffer capacity is being increasingly applied, i.e. the extent of change in the availability as nutrient is withdrawn. A wide range of soil test procedures have been developed which try to assess the available fraction of the essential elements (refer to Walsh and Beaton 1973).

The macronutrients

Nitrogen

Growth of plants is probably limited more often by a deficiency of nitrogen than of any other nutrient. Nitrogen is fundamental to plant biochemistry, the main effect of shortage being interference with protein synthesis and hence growth. Mature leaves are affected first, showing a general yellowing or chlorosis due to inhibition of chlorophyll synthesis. For a more detailed discussion of deficiency symptoms refer to the specialized texts such as Chapman (1965).

Surface soils (0–15 cm) usually contain 0.10 to 0.40 per cent nitrogen declining with depth in the profile. Most of the nitrogen (90–5 per cent in surface soils) occurs in organic forms so that the level of total nitrogen closely parallels that of organic matter. Organic matter and so nitrogen levels, tend to be higher under grassland or forest and to decline rapidly with cultivation. High levels are also favoured by low temperature, high rainfall, and high clay content. The remaining mineral nitrogen consists of so called 'fixed' ammonium (NH_4^+) held between clay sheets (4–8 per cent) and as ammonium (NH_4^+) nitrite (NO_2^-) or nitrate (NO_3^-) in the soil solution or adsorbed onto the soil colloids (1–2 per cent). Nitrite levels are usually very low.

Plant uptake of nitrogen occurs as NH_4^+ or NO_3^- with the additional special case of fixation of gaseous nitrogen by the legume. Fixed NH_4, unless it occurs as a result of recent application of fertilizer, is not available to plants. Nitrogen in organic form is also not available to plants, requiring the process of net mineralization to convert it into available mineral forms. This process is the fundamental one in terms of the availability of soil organic nitrogen.

Mineralization of organic nitrogen is carried out by the soil micro-organisms, net mineralization occurring if the nitrogen content of the organic matter is greater than that needed by the micro-organisms for their own growth. If the nitrogen content is less than that required by the micro-organisms uptake of mineral nitrogen will exceed release and net immobilization will occur, leading to a shortage of mineral nitrogen for plants.

$$\text{Organic N} \quad \underset{\text{immobilization } (I)}{\overset{\text{mineralization } (M)}{\rightleftarrows}} \quad \text{Mineral N } (NH_4^+)$$

Whether M or I predominate is determined by the balance of energy supply (C) and the nitrogen supply in the organic matter. This balance may be indicated from the carbon:nitrogen ratio, with M predominating below a ratio of about 15 and I predominating above about 22. In the carbon:nitrogen range 15–22 whether M or I predominates will depend on the environment and the substrate. Under permanent vegetation plant residuals with a high carbon:nitrogen ratio are constantly entering the soil so that I tends to predominate, mineral nitrogen levels are low and plants may be short of mineral nitrogen. Where soils are bare fallowed M tends to predominate leading to accumulation of mineral nitrogen in the soil. If nitrogenous fertilizer is added to the soil it leads to a rapid increase in mineral nitrogen. Some of this will be taken up by the plant if they are present, but some will be immobilized into organic nitrogen making it unavailable for plant growth.

The availability of NH_4^+ in the soil may be affected by further processes. Thus in normal aerobic soils NH_4^+ tends to be oxidized to NO_3^- by specific soil micro-organisms, a process called nitrification; i.e.

$$NH_4^+ \longrightarrow NO_2^- \longrightarrow NO_3^-$$
$$\text{Nitrification}$$

The conversion of NO_2^- to NO_3^- is usually very rapid so that NO_2^- levels are usually very low. Once in the NO_3^- form nitrogen is very mobile in most soils since it is not held by the negatively charged colloids, moving freely with the soil water. Such movement may improve availability if NO_3^- is moved from dry soil surface into a moist subsoil containing active roots. However if movement occurs to the deep subsoil availability may be reduced or net loss may occur if the NO_3^- is removed from the rooting zone. A further possibility is that under saturated or water-logged conditions NO_3^- may be converted to gaseous N_2 and N_2O by the soil micro-organisms (denitrification) resulting in net loss of nitrogen.

Gaseous loss of nitrogen as NH_3 may occur where levels of NH_4^+ build up under alkaline conditions allowing the reaction $NH_4^+ OH^- \rightleftharpoons NH_3 + H_2O$ to move to the right. Such loss may become substantial where a fertilizer which induces alkaline soil conditions, e.g. urea, is applied to the soil surface. Under grazing, loss as NH_3 may be substantial from urine patches under relatively dry conditions. Once urine or fertilizer is leached into the soil following rain, losses tend to be reduced. Recent evidence also indicates that NH_3 may be lost in significant amounts from growing plants. Losses may also occur from the removal of animal products (2–20 kg N ha^{-1} a^{-1}) or plant products (30–150 kg N ha^{-1} a^{-1}) from fire when vegetation is burnt and from erosion.

Net gains of nitrogen occur mainly as a result of fertilizer application or from symbiotic fixation by the legume. Gains by crop legumes are usually considered to be small but pasture legumes under favourable conditions may fix 3–400 kg N ha^{-1} a^{-1} and figures in the range of 50–200 are relatively common. With time the rate of gain tends to decrease leading to a new equilibrium level for total nitrogen and soil organic matter. Non-symbiotic fixation is usually considered to be of lesser importance except under particular conditions, e.g. by algae under lowland rice culture. Small gains may occur from rainfall ($<$10 kg ha^{-1} a^{-1}); larger gains may occur where animals are fed with produce brought on to a farm.

The various mechanisms discussed result in a cycling of nitrogen through the soil, plant, animal, and atmosphere as indicated in Fig. 2.7.

Nitrogen shortage in plants can frequently be detected by the characteristic symptoms (see above). Soil tests may be of value, particularly under cropping, where the common measurements are either the actual mineral nitrogen (or nitrate nitrogen) present or the potential release of mineral nitrogen by M/I processes. The latter is usually measured by an incubation test or less commonly by a chemical test such as total nitrogen. Under pastures nitrogen requirements are often met at least partially by the legume; nitrogen shortage is normal and fertilizer application is dependant mainly on whether it is economical in relation to the response expected in the particular environment.

Phosphorus

As a limiting factor in plant growth phosphorus is probably second in importance only to nitrogen. Phosphorus is a constituent of a number of biologically important organic compounds such as nucleoproteins, sugar phosphates, adenosine triphosphate, etc. Deficiency symptoms include chlorosis and death of older leaves (phosphorus is phloem mobile), greatly reduced plant size, and purple coloration due to anthocyanin production. Toxicity is possible, e.g. in legumes where it

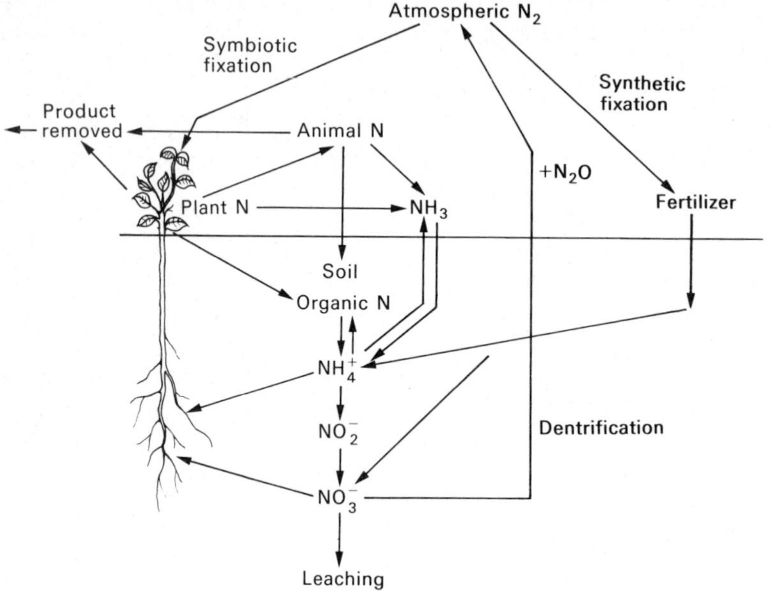

Fig. 2.7. Major components of the nitrogen cycle.

may cause leaf shedding and marginal leaf scorch. Plant contents are much lower than for nitrogen, often being in the range of 0.15–0.30 per cent.

The total soil content is relatively small, usually much lower than for nitrogen (a mean of 0.06 per cent has been shown for a range of United States soils and 0.03 per cent for Australian soils). Soil levels are usually related to the level in the soil parent material. Thus levels in soils derived from basic igneous rocks are usually relatively high, whereas they are low if derived from acidic igneous rocks such as granite. Soil phosphorus may be divided into the two categories of inorganic and organic, the inorganic usually predominating. The ratio of organic to inorganic is much higher in the soil surface than in the subsoil. Inorganic soil phosphorus occurs mostly as secondary combinations with iron and aluminium in acid soils and calcium and magnesium in alkaline soils. Very small amounts occur in the soil solution (often 0.1–0.2 mg l⁻¹ or less) mainly as $H_2PO_4^-$ and HPO_4^{2-}. The relative proportion of these ionic forms varies with pH, $H_2PO_4^-$ predominating in the pH range of most soils. Levels of organic phosphorus are indicated by the values for soil organic matter. The phosphorus content of organic matter is much lower than that of nitrogen and N/P ratios are variable.

Plant uptake is probably mainly as $H_2PO_4^-$ and to a lesser extent HPO_4^{2-}. The low values of these ions in the soil solution compared with plant requirements means that continuous replenishment is required. This replenishment comes mainly from inorganic sources since although mineralization of organic phosphorus occurs it is not significant in the short term. Experiments with tracers of radioactive phosphorus indicate that the inorganic phosphate ions in the soil solution are exchanging rapidly and continuously with a pool of solid phase inorganic phosphate. This fraction of the inorganic phosphorus therefore is relatively available to the plant and is sometimes described as the 'exchangeable' phosphorus. It is however only a small proportion of the total inorganic phosphorus.

A feature of the chemistry of soil phosphorus is the rapid reaction that occurs between soil and water-soluble phosphorus such as $Ca(H_2PO_4)_2$, as added when fertilizing with superphosphate, producing relatively insoluble phosphorus compounds of low availability. This reaction is commonly known as 'fixation' or 'reversion' of phosphate. Subsequent availability depends on the nature of the fixation reaction. In highly weathered soils such as the oxisols, strong combination with aluminium and iron occurs and subsequent availability is low. Over a longer time period continuous plant uptake of phosphorus applied as fertilizer can result in substantial accumulation of organic phosphorus of relatively low availability. A further consequence of the fixation reaction is that applied phosphorus fertilizer usually remains close to the point of addition and net loss by leaching is very small.

Net loss may also occur due to removal of plant or animal products but such losses are usually small. Loss may also occur due to erosion. Net gains in the short term will occur solely due to fertilizer addition with slow long-term gains from weathering.

A wide variety of soil tests have been proposed as measures of phosphorus availability. Many of these employ extractants such as acids, e.g. sulphuric, acetic, lactic, etc., which dissolves some of the phosphorus or reagents containing anions which are presumed to replace the phosphate, e.g. sodium hydroxide, ammonium fluoride, sodium bicarbonate, etc. Depending on the concentration and time of extraction they may be considered as reflecting the intensity or the capacity of the available soil phosphorus. The success of a particular method often can be related to the chemical form in which the inorganic phosphorus is held. Recently phosphorus sorption has shown promise as an availability index. Usually it is necessary to establish separate critical levels for each group of soils and each plant species. Plant analysis is a useful index of phosphorus deficiency and critical plant values have been published for many grasses and legumes, e.g. 0.24 per cent for siratro. However application of such values needs to be done with care since they are known to vary with such factors as season and plant density, even within one species.

Potassium

Deficiency tends to be less common than for nitrogen and phosphorus owing to larger soil reserves. Potassium is required as a co-factor in over 40 enzyme systems, deficiency leading to disruption of both carbohydrate and nitrogen metabolism. It is phloem mobile so that symptoms appear first on older leaves, the tips and margins becoming firstly chlorotic and then necrotic. Plant content is relatively high (1–3 per cent).

Soil contents show a very wide range, United States data indicating a range of 0.3–2 per cent, and averaging 0.83 per cent. Most of the potassium ($>$ 99 per cent) is classed as non-exchangeable, i.e. it is not held in exchangeable positions on the soil colloids consisting of felspars and rock micas found in the sand and silt fractions and clay micas in the clay fraction. The rock micas weather relatively easily but the felspar and clay micas weather slowly. The exchangeable (including water soluble) potassium is a minor component of the total potassium ($<$ 1 per cent) but a significant component of the exchangeable cations in most soils (often in the range 0.05–1.00 milliequivalents per cent).

Plant uptake of potassium occurs as K^+. Values of K^+ in the soil solution are extremely variable (0.2–400 mg l^{-1}) but usually small compared with plant needs over any period so that continuous renewal is required to maintain growth. Renewal occurs mainly from the exchangeable K^+ and is very rapid. This exchange of K^+ is only partial and affected by such factors as the nature of the soil colloid and the complementary cations present so that these factors affect its availability.

Continuous cropping experiments indicate that exchangeable K^+ is not the only source of potassium for renewal, i.e. that non-exchangeable sources make a contribution. It appears that if the solution K^+ value becomes sufficiently low from plant uptake soils that contain significant amounts of 2:1 type clays, particularly the clay micas, will release

K^+ from between the clay sheets to the soil solution making it available for plant uptake. When potassium fertilizer is added to soil some of the K^+ may take part in the reverse process of potassium 'fixation' which reduces the availability of the applied fertilizer. Decomposition of plant residues may also cause significant release. Over a period of years significant release to the soil solution may occur from the weathering of felspars and rock micas. These processes are summarized in Fig. 2.8.

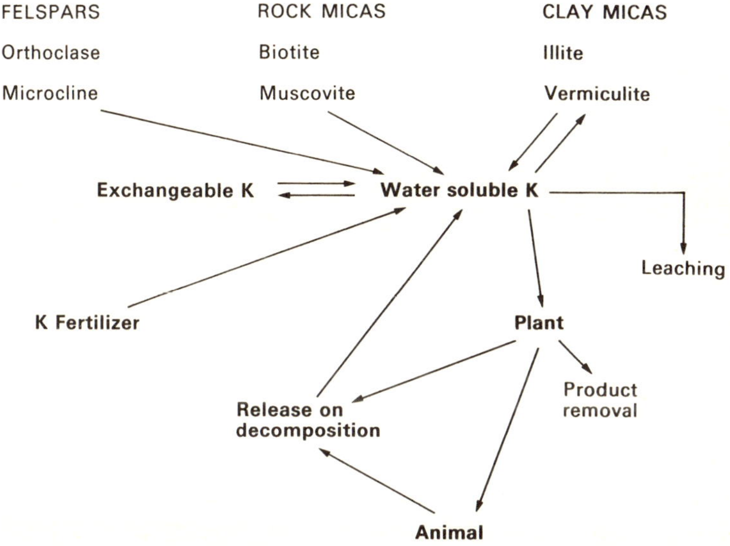

Fig. 2.8. Relationships between forms of potassium in the soil.

Short-term net gains of potassium will occur almost exclusively from fertilizer addition. Some gain to a farm can occur from the import of feedstuffs and the animal can cause significant redistribution within a farm. Net losses of potassium may occur from removal of large amounts of plant or animal products high in potassium (e.g. in root crops or fodder) and to a lesser extent as a rule from leaching of K^+ out of the profile or from erosion.

The most common and most successful soil test for potassium is that of exchangeable potassium with critical values often about 0.12 milliequivalents per cent. Less commonly the intensity of the exchangeable potassium (e.g. critical ion activity ratios) or empirical extractants such as HNO_3 have been used which attempt to assess the availability of the non-exchangeable potassium.

Sulphur

Sulphur is required by plants for the synthesis of essential components in protein synthesis so that deficiency affects nitrogen metabolism. Deficiency symptoms are pale green to yellow leaves with a more uniform chlorosis than with nitrogen deficiency. The sulphur content of plants is relatively high but variable with legumes usually being in the range of 0.2–0.7 per cent.

Soil contents are somewhat comparable to phosphorus and rather variable, values in the range of 50–500 p.p.m. being common for the surface of acidic soils. In such soils most of the sulphur occurs in organic combinations ($>$ 90 per cent) and a close relationship has been shown between total carbon, nitrogen, and sulphur, e.g. Williams (1974) reports 140:10:13 for Australian soils. The remaining inorganic sulphur occurs mostly as adsorbed and soluble sulphate (SO_4^{2-}), the values usually being quite low (often 5–20 p.p.m.). In leached acidic soils adsorbed sulphate may accumulate in the subsoil with values in the range of 50–500 p.p.m. Similar high values of sulphate in the form of gypsum may occur in the subsoil of arid soils.

Plant needs for sulphur are normally met by uptake of SO_4^{2-} from the soil solution but direct intake of gaseous SO_2 occurs and may become significant in industrial regions owing to the discharge of SO_2. Replenishment of solution SO_4^{2-} may come from a rapid exchange of adsorbed sulphate, if it is present, or from the slower dissolution of fertilizer such as gypsum if present. In the longer term replenishment may occur from the net mineralization of organic sulphur (soil humic material, plant or animal residues), in a process similar to nitrogen mineralization. Organic release is most likely to be important where land use changes quickly from permanent vegetation (forest, grass-land) to cropping. As with nitrogen if permanent vegetation is present net immobilization will predominate tending to keep solution SO_4^{2-} values low, even after application of sulphur fertilizer, and lead to substantial build-up of relatively unavailable organic sulphur. Root exploration is often an important mechanism in improving sulphur availability since the supply of sulphate is often better in the subsoil than in surface soils.

Net gains of soil sulphur occur from mineral weathering, the atmo-sphere or from addition of fertilizer. Significant short-term gain requires the addition of fertilizer, unless atmospheric SO_2 is high. Net losses may occur from erosion, leaching, and animal and plant product removal. Of these leaching is probably the most serious since in many soils SO_4^{2-} is not held by the soil colloids.

Soil tests for sulphur cover a wide range of extractants which recover some or all of the solution, adsorbed and organic sulphur. The most common is probably phosphate-extractable sulphate, with critical values commonly in the range of 8–12 p.p.m. For plant analysis total sulphur is most commonly used (also N/S ratio) critical levels for the whole plant ranging from 0.12 to 0.25 per cent for a range of pasture species and field crops.

Calcium

Calcium is required in plants for ion transport and various cellular activities. It is phloem-immobile so that young growth is affected and the growing-point may be lost. Root tips may die, leading to short, abnormally branched root systems. Plant contents are relatively high, particularly in legumes (often 1–2 per cent) but deficiency is not widespread because soil reserves are usually good.

Soil contents are quite variable ranging from less than 1 per cent in most humid soils to 25 per cent, or more in the subsoils of arid soils where accumulations of secondary $CaCO_3$ occur. Exchangeable Ca^{2+} usually dominates the exchange fraction of humid soils with values most often in the range of 1–25 milliequivalents per cent, much higher than exchangeable K^+, the ratio of exchangeable Ca^{2+}/exchangeable K^+ usually being in the range of 5–10.

Plant uptake of calcium occurs as Ca^{2+} from the soil solution. Replenishment following uptake is mostly due to rapid release of exchangeable Ca^{2+} from the soil colloids. The extent of this release is variable depending on such factors as the complementary cations and the nature of the soil colloid. Slow replenishment may occur from the decomposition of plant and animal residues or from the weathering of calcium-rich primary minerals. Short-term net gains will occur mostly from fertilizer addition, e.g. gypsum, superphosphate, or liming materials. Net losses due to the removal of plant and animal products, leaching, and erosion, are variable, but usually not great.

Calcium deficiency is not common, but if a soil test is indicated, exchangeable Ca^{2+} is the most appropriate. Soil pH offers a rough guide. Within a particular soil type as pH decreases so does the supply of exchangeable calcium.

Magnesium

Magnesium is required in plants as a co-factor of several enzyme systems and is a constituent of chlorophyll. Low magnesium concentrations in herbage can lead to hypomagnesaemia 'grass tetany', a serious and often fatal condition in livestock. Deficiency symptoms are interveinal chlorosis on older plant parts leading to necrosis. Plant contents are relatively high although generally lower than calcium, with legumes higher (often about 0.3 per cent) than non-legumes. Deficiency is uncommon owing to the usually good soil reserves.

Soil contents of magnesium are similar but rather lower than calcium. Mostly magnesium occurs in non-exchangeable form in a wide variety of primary and secondary minerals. Exchangeable Mg^{2+} is quite high, usually being the next most abundant exchangeable cation after Ca^{2+} and it may exceed Ca^{2+} in subsoils.

Plant uptake occurs as Mg^{2+} from the soil solution, with rapid replenishment coming from exchangeable Mg^{2+}, the extent being affected by similar factors are discussed for Ca^{2+}. Slow replenishment and net gain and loss mechanisms generally are similar to those for calcium. Exchangeable Mg^{2+} is the most appropriate soil test.

The micronutrients

Micronutrients are important in plant biochemistry as constituents of enzymes and hormone systems which control processes such as respiration, energy exchange, and synthesis of chlorophyll, carbohydrates, and proteins. Deficiencies can often be identified by characteristic plant symptoms (see below). Plant contents are low to very low, and soil values are also very low except for iron which is very high and

Table 2.9. Micronutrient concentrations in plant tissues and in soils (Mortvedt, Giordano, and Lindsay 1972)

| Element | Plant tissues | | | Soils (p.p.m.) |
	Deficient (p.p.m.)	Normal (p.p.m.)	Toxic (p.p.m.)	
Iron	<50	50–250	–	10 000–100 000
Manganese	<20	20–500	>500	20–3000
Copper	<4	5–20	>20	10–80
Zinc	<20	25–150	>400	10–300
Boron	<15	20–100	>200	7–80
Molybdenum	<0.1	0.5–100	–	0.2–10

manganese which is occasionally high (refer Table 2.9). Micronutrient transitions in soils are summarized in Fig. 2.9. The availability of micronutrients depends on their abundance in forms which can be assimilated by the plant. Mechanisms by which the concentration in solution can be maintained include chelation with suitable ligands (e.g. zinc, copper, and manganese), reduction to more soluble forms (e.g. manganese) and desorption from clay minerals, iron and aluminium oxides, and organic matter. Availability is often strongly influenced by soil pH (see Fig. 2.10).

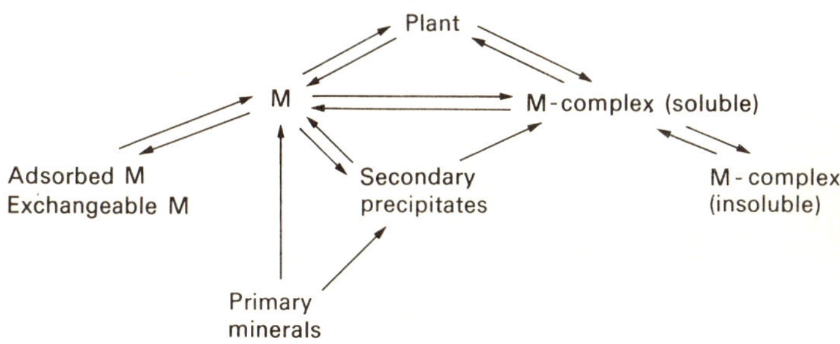

Fig. 2.9.

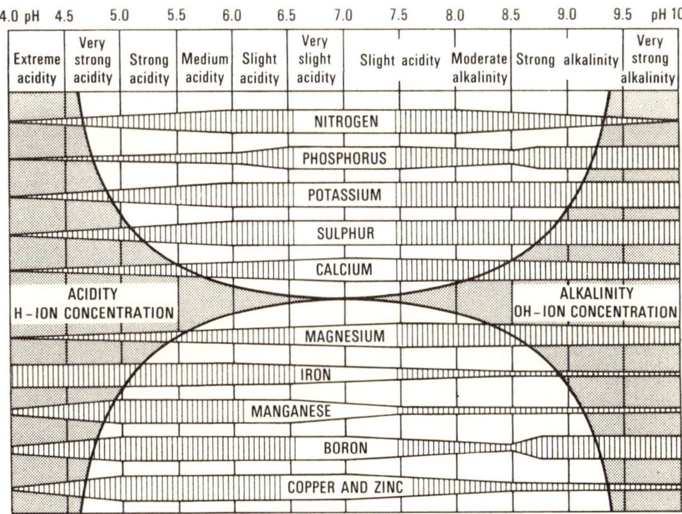

Fig. 2.10.

Iron

Deficiency symptoms are a yellowing of leaves with chlorosis of interveinal areas of the younger leaves, veins often remaining green.

Iron in the soil occurs mostly as hydrated iron oxides, adsorbed on clay surfaces or complexed with organic matter. Total supply is rarely a problem but in the soil the availability is dependant on the solubility of the iron oxides, and this becomes very low as pH increases so that deficiency may occur in strongly alkaline soils (sometimes due to over-liming). Correction by applying soluble iron may be difficult because of its rapid conversion in the soil to insoluble forms. Because of this iron chelates with organic molecules (e.g. Fe-EDDHA) or foliar application may be desirable.

Soil tests using exchangeable iron and extraction with an iron chelate such as DTPA (critical range 2.5–4.5 p.p.m.) have been proposed but have not gained wide acceptance.

Manganese

Deficiency symptoms in broad-leafed plants are interveinal chlorosis of younger leaves similar to that found in iron deficiency except that the green zone adjacent to the veins is usually wider and more diffuse. In cereals and grasses greyish or brownish spots and streaks may occur in the middle or basal part of younger leaves. Toxicity symptoms are chlorosis of younger leaves in tropical legumes and in cereals and grasses red, brown, or black spotting of the older leaves.

Manganese occurs in soils mainly as the higher oxides (oxidation state +3 or +4) which are relatively insoluble. Under waterlogged or strongly acid conditions reduction to Mn^{2+} occurs which raises the solution concentration, the excess being adsorbed. Thus manganese resembles iron in that deficiency may occur in alkaline soils. It is also found in soils high in organic matter. Toxicity may occur in waterlogged soils or strongly acidic soils (see section on acidity) particularly if the soil is high in manganese. Correction of deficiency by application of soluble manganese may be difficult because of its oxidation to unavailable oxides. Foliar application is effective.

Soil testing for manganese is based on the concept of determining 'active manganese', i.e. exchangeable Mn^{2+} plus the higher oxides that can be readily reduced. A variety of reagents such as quinol, phosphate, etc. have been used. Manganese excess can be predicted from a combination of a soil analysis and pH. Plant analysis is useful in both diagnosis of deficiency and excess.

Copper

Deficiency symptoms include twisting, curling, or rolling of leaves and abnormal curvature of petioles and stems, often with particular symptoms characteristic for the species (e.g. Andrew (1963) for tropical legumes).

Copper in soils is mainly adsorbed on clay minerals and hydrous oxides or associated with organic matter. Availability decreases with increasing soil pH so that unavailability problems may occur on alkaline soils. Availability is also low in soils high in organic matter. Deficiency may occur on acid sandy soils of low total copper content and in these soils copper fertilizer must be used with care since excess is toxic.

Soil tests involving extraction with reagents such as ammonium acetate, dilute acids, and chelating agents have had varying success.

Zinc

Deficiency symptoms are shortening of internodes, 'little leaf' and interveinal chlorosis, resembling those produced by manganese and iron deficiency.

Zinc in soils occurs as secondary hydroxides or carbonates, chelated with organic matter or adsorbed on clay minerals and hydrous oxides. Availability falls markedly as soil pH increases with low concentrations of Zn^{2+} in solution in calcareous soils. Deficiency in such soils is fairly common. Deficiency may also occur in soils with a low total supply (e.g. acid sandy soils) and in soils receiving high applications of phosphate. Foliar application of fertilizer is more efficient than soil application on calcareous soils.

Soil tests using acids and complexing or chelating agents have had some success. Recently DTPA has shown promise and has the advantage that it can also be used for iron, manganese, and copper. Critical levels are: low – 0-0.5 p.p.m., marginal – 0.5-1.0 p.p.m., and adequate > 1.0 p.p.m.

Molybdenum

Deficiency symptoms vary greatly often being characteristic for the species. In legumes the most usual symptoms are those of nitrogen deficiency. Although plant requirements are very small, molybdenum is needed both for utilization of nitrate nitrogen and for biological nitrogen fixation.

Molybdenum occurs in soils in insoluble minerals such as calcium molybdate, in organic matter complexes and as the adsorbed anion MoO_4^- on layer silicates and hydrous oxides. The solubility and hence the availability of molybdenum is low in acid soils, liming of such soils improving availability. Organic soils may also be deficient, also weathered soils containing high iron oxides which reduce molybdenum availability through adsorption.

Soil testing is complicated by the minute plant requirement and the consequent importance of seed reserves in supplying crop needs, and the difficulty of determining low levels of molybdenum. Hot water, ammonium oxalate, and exchange resins have had success as extractants. Microbiological assay with *Aspergillus niger* gives comparable results to ammonium oxalate. Correction of deficiency may be made by applying fertilizer to the soil, the plant, or to the seed.

Boron

Deficiency symptoms are similar to those of calcium deficiency affecting root and shoot apices because of its immobility in the plant. Toxicity symptoms are chlorosis and necrosis of older leaves. Plants vary widely in their tolerance of both deficiency and toxicity.

Boron occurs in the soil as the primary mineral tourmaline or adsorbed on hydrous oxides, clay minerals, and organic matter. In the soil solution it may occur as undissociated H_3BO_3 or as the anions $B(OH)_4^-$ or $(B_4O_7)_4^{2-}$. These forms are susceptible to leaching so that deficiency occurs in acid and sandy soils low in organic matter. It also occurs in alkaline soils containing free calcium carbonate. Care must be exercised in applying boron fertilizer to the soil since the ratio of toxic to normal levels is smaller than for other nutrients. Toxic levels of boron may also occur in natural soils high in boron (e.g. in soils derived from marine sediments high in boron or in arid region soils where leaching is minimal).

Hot-water extraction is the most successful soil test for boron with levels of <1.0 p.p.m. indicating probable response and >5.0 p.p.m. possible toxicity. Correction of deficiency is by soil or foliar application.

Cobalt

Essential for animal health and shown to be required for symbiotic nitrogen fixation in legumes. Occasional field responses have been shown by legumes grown on low-cobalt soils.

Chlorine

Shown to be essential for some plant species but no reports of deficiency under field conditions.

Sodium

Essential for animals, rarely for plants. A partial substitute for potassium in a number of plants. Plant uptake extremely variable so that animal health may be affected by the pasture species present.

Freedom from adverse chemical conditions

Acidity

A comprehensive discussion of the nature, measurement, and fertility aspects is given by Black (1968). Acidity develops in natural soils mainly as a result of the leaching of bases and their replacement by hydrogen and aluminium leading to an unsaturated exchange complex. The greater the leaching action the higher the unsaturation and the greater the acidity (values of about 4 are the lower limits). In acid-sulphate soils, acidity in the range of 1–2 may arise from the oxidation of FeS_2 following drainage. Fertilizers can also reduce pH substantially. Thus continued use of ammonium sulphate as fertilizer can reduce the pH by a unit or more.

1. Toxic substances

An early theory was that soil acids were harmful and were neutralized by lime. This is now discounted in soils with a pH above 4 since nutrient culture experiments in this pH range show growth to the satisfactory level as long as the nutrient supply is good and manganese and aluminium are not present in excessive quantities.

Aluminium and manganese concentrations in the soil solution increase sharply as the pH drops below about 5.0. Solution culture experiments confirm that plant growth is reduced at the levels found in soils so that liming may ameliorate these toxicities. Aluminium toxicity is characterized by inhibition of root growth, with aluminium accumulating on or in the roots. Phosphate applications have been shown to reduce toxicity in some soils. In contrast to aluminium, manganese is taken up readily by the plant, producing characteristic symptoms (see above) and plant analysis can be diagnostic. Sensitivity to both manganese and aluminium toxicity is quite variable between species.

2. Nutrient availability

A number of nutrients may be affected by low pH. Response may simply be due to a low supply of calcium, since the exchangeable bases including calcium decrease as pH decreases, but such a response is relatively uncommon (see above notes). The increase in pH from liming acid soils increases phosphorus availability so that in phosphorus deficient soils this may be part of the reason for the response. Similarly the supply of micronutrients may be improved, particularly molybdenum (see above notes).

3. Effects on the soil micro-organisms

Acidity has a pronounced effect on many micro-organisms affecting aspects of fertility. Thus liming of acid soils increases organic matter decomposition and hence may release nitrogen and possibly sulphur and phosphorus. Nitrification may increase leading to more of the mineral nitrogen occurring as nitrate. Nitrogen fixation, both symbiotic and non-symbiotic is affected by pH.

Plant disease incidence may be affected by soil pH. Thus the broader adaptability of fungi to acid conditions makes fungal disease more prevalent under acid conditions and less so after liming.

Soil testing for acidity is very useful with rapid laboratory and field methods available (see practical notes).

Salinity and excess sodium

These effects are grouped since they frequently occur together. A detailed discussion of the fertility implications is given in Black (1968) and Kovda, van den Berg, and Hagan (1973). Definitions and methods of measurement are given in Richards (1969). A classification of soils affected by salinity and excess sodium (salt-affected) is given in Table 2.10. Other groupings have been proposed more recently. Thus Northcote and Skene (1972) have grouped Australian soils into six classes depending on salinity, alkalinity, and sodicity. They give evidence for sodic effects below and ESP (exchangeable sodium as a percentage of the cation exchange capacity) of 15 and so have a sodic grouping, i.e.

Table 2.10. Classification of salt-affected soils according to their chemical properties (Richards 1969)

Soil group	Specific conductivity of saturation extract at $25\,^{\circ}C$ ($mS\ cm^{-1}$)	Saturation of cation exchange capacity with sodium (%)
Saline – non-sodic soils	>4	<15
Saline – sodic soils	>4	>15
Non-saline – sodic soils	<4	>15
Non-saline – non-sodic (normal) soils	<4	<15

soils with ESPs between 6 and 14, and a strongly sodic group, i.e. with ESPs $\geqslant 15$.

Salt-affected soils occur naturally in situations of low leaching, the salt arising usually from slow atmospheric input or as part of the parent material. Man-induced salting is relatively common in semi-arid areas owng to land-use changes such as clearing trees and replacing them with pasture, resulting in more groundwater flow and salting developing in lower slope positions. Also irrigation often results in salinity problems due to the use of salty water or to rising water-tables where the subsoils are salty.

In classifying waters for irrigation the major hazards recognized are salinity and sodium. The development of salinity or sodicity from the use of salty water will depend on the degree of concentration of the water in the particular environment, as well as the initial levels.

Causes of plant damage

1. Salinity

Plants grown on saline soils show reduced growth, often without distinctive foliage symptoms. A non-specific osmotic effect of salts on growth can readily be shown in solution culture or soil. This appears to be due to the need for plants to take up salts so that water uptake can be maintained and the reduction in growth during this adjustment period. Specific ions in the salts may affect plants. Thus boron levels may be toxic; sodium and chloride have specific toxic effects on sensitive species, and bicarbonate may cause damage to plants. Tolerance to salinity is quite variable between plant species.

2. Sodicity

This may occur with salinity, e.g. in the saline-sodics; the additional effects of sodium as such are better seen in the non-saline-sodics (see Table 2.10). Effects on plants in these soils may be physical or nutritional. The physical effects may occur as a result of the sodium ion dispersing soil structural units, the dispersed material filling pores and reducing water conductivity or forming surface crusts on drying which may affect seedling germination. Nutritional effects may be due to the effects of alkalinity, since sodic soils are usually alkaline to strongly alkaline. This may reduce micronutrient availability (see earlier section) or have a deleterious effect on the soil micro-organisms which carry out nitrogen fixation, etc. High soil solution values for sodium may also affect the availability of nutrient cations such as potassium, calcium, or magnesium owing to competition for plant uptake sites. If sodicity is high enough, direct toxic effects may occur (see earlier section).

Soil testing for salinity and sodicity is effective with critical levels such as those shown in Table 2.10. However such values only give a general guide since the plant species, soil, the climate and the soil management techniques will affect critical levels in particular situations.

Other chemical excesses

A number of less common chemicals may affect plants and/or animals. Thus excess arsenic, lead, and the gases H_2S, SO_2, and NH_3 are injurious to vegetation and excesses of copper, molybdenum, fluoride, and arsenic may affect animals.

Maintenance of fertility

Of basic importance is the ability of a soil to maintain its fertility over a time period, as well as its initial fertility. The most significant aspect is probably the susceptibility of a soil to erosion since it affects so many fundamental fertility aspects. Thus water erosion for example affects soil depth, structure, water penetration and retention, nutrient status, etc. In many tropical regions water erosion is a major problem because of the quantity and nature of the rainfall. This is particularly the case where cultivation is attempted, on land with relatively steep slopes. Under permanent pasture susceptibility to erosion is much reduced but even here significant erosion can occur particularly during the establishment phase when the soil is bare.

Maintenance of an adequate nutrient supply may be difficult. Leaching of relatively mobile nutrients such as nitrate and sulphate can be a problem in permeable soils where the rainfall is high. The problem is most acute during the establishment phase of a crop or pasture. Contiued removal of plant tops may induce deficiency of nutrients such as potassium and sulphur, where the initial soil supply is marginal. Fixation of nutrients by their reaction with the soil may require annual applications, e.g. phosphorus in some strongly fixing tropical soils.

Maintenance problems may develop in soils susceptible to excess chemical effects. Thus salinity may develop over time in soils in lower slope positions requiring reclamation procedures such as drainage. Acidity problems can develop due to the use of fertilizers such as ammonium sulphate so that amelioration with lime may be required.

Because of its importance in long-term production soil erosion will be considered in more detail.

Soil erosion

We usually refer to soil erosion as that influenced by man. Natural erosion occurs which would have occurred if man had been present or not. New volcanic ash deposits undergo natural erosion. Refer to Hudson (1971) for a detailed treatment of erosion.

Processes of erosion

Two processes are involved in normal erosion. A soil particle is first detached from a soil aggregate and then transported. Thus we have the processes of – (i) detachment and (ii) transportation. Generally detachment occurs which produces a particle small enough for transportation.

Detachment requires a force which could be either the raindrop or water flow. The raindrop possesses energy depending on its drop size. A rainfall has energy which is the sum of all the energy possessed by the raindrops and therefore depends on the rate of rainfall or intensity as well as the drop size. The *erosion index* has been used to express this and is the product of

Total kinetic energy of the rainfall × Maximum intensity for a 30 minute period.

As the erosion index is correlated with soil loss it is important to measure intensity. It is generally expressed in millimetres per hour. The prediction of intensity can be made using rainfall records gathered over a long period so that maps can be constructed showing the intensities likely to be expected, say, once in one year or once in ten years for a particular duration (ten minutes or one hour).

It is of interest to note the following statement by Dawes in his *Soils of east-central Java*.

Of great importance for agriculture and for the erosion problem is the highly intensive character of the rainfall. Diurnal rainfall of 50–100 mm is very common. The average for the wettest month ranges from 232–1,034 mm throughout the area. Generally, 20% of the rain in the wettest month falls in one single day and with decreasing monthly rainfall this percentage increases. In the driest month of the year the average rainfall in one single day even ranges from 30–70%. The lower the monthly rainfall figure, the higher is the percentage on the wettest day. All this proves that a considerable part of the rain falls in heavy showers, which is especially true for the driest regions and the dry seasons. Without exaggeration it may be stated that the whole area has an extremely erosive type of rainfall. Under such unfavourable conditions of rain distribution irrigability is of the utmost importance.

Water flow due to run-off from the soil surface also possesses energy which can detach particles from aggregates. If the water is clear turbulence and kinetic energy causes detachment but if the water is turbid the abrasion by suspended particles can cause detachment. Once the particle is detached, there is no erosion unless transportation takes place. The agents causing transportation are the raindrop, surface flow, and gravitation.

The raindrop causes particle transportation by 'bouncing' the particle downhill; surface flow possesses kinetic energy and carries the particle in suspension a distance which will depend on the velocity of surface flow and the size of the particle. The velocity of surface flow will, in turn, largely depend on the surface slope and roughness.

In order to prevent erosion then there has to be a resistance to the forces of detachment and transportation.

Factors affecting resistance to detachment are mainly those soil factors associated with stable structure, e.g.

1. The higher the clay content the stronger the aggregate.
2. Iron oxide coatings produce stronger aggregates.
3. Organic matter produces bonds between particles and hence stronger aggregates.
4. Excess sodium ions on the exchange complex will produce weak aggregates.

The resistance to transportation depends on the size and weight of the aggregate. If aggregates are large and dense then the possibility of transportation is less.

Man can influence this process by reducing the forces of detachment and transportation and increasing the resistance to those forces in order to control erosion or increasing the erosive forces and reducing the resistance to those forces. In other words man is responsible for the rate of erosion. There is not much information available for the tropics in general but the following shows that 36 per cent of the East-Central Java region has what is classed as severe erosion.

Erosion class	Extent (ha)	Percentage of total area
Severe	570 000	36.0
Moderate	170 000	10.5
Slight	71 000	4.5
None	785 000	49.0

In describing the severe erosion class Dawes stated

The severely eroded soils comprise the lateritic soils of the usually steep Southern Mountains, which have been wholly deforested in the last century. Shifting cultivation when still sufficient virgin forest land was available, without soil-conservation measures on the one hand, and the greater demand for soils as a result of the rapidly increasing population since 1900 on the other hand have resulted in a wellnigh total destruction of these areas, notwithstanding the slight erodibility of these soils. Measures to preserve the soil, such as terracing, were only taken when the deterioration was under way, and moreover these measures are very one-sided and do not cover the whole problem of soil management. At the best they only result in some slowing down of the continuous soil deterioration. Large parts of these areas are too steep for agriculture. Moreover, the fact that the water conditions in these steep and dissected terrains made the establishment of irrigated 'sawahs' impossible was unfortunate, as the 'sawah' may be considered the ideal form of soil conservation.

It is obvious from these statements that vegetation plays a major part in modifying the erosion process. It forms a barrier to the raindrop force causing detachment and transportation. It provides resistance to surface flow and thus reduces velocity and the abrasive load. It provides organic matter which causes a higher soil resistance, and a root system which attaches soil particles. It is well known for example that a long period under pasture growing vigorously with a high soil fertility, produces surface soils with considerable resistance to erosion. However the soil must be susceptible to erosion while the ground cover is low (i.e. during the preparation and emergence stages).

Stocking management will obviously effect the ground cover and ground cover will, in turn, effect run-off and erosion. This is shown in Fig. 2.11 which is typical of the information gained over many experiments. Stocking rate should be matched with pasture production so that safe ground covers can be maintained especially during periods of high rainfall intensity.

The greater the erosion 'hazard' the greater the need for careful management and maintenance of ground cover. This would be the situation on steep slopes with poorly structured soils.

If soils remain exposed under conditions of high erosion hazard, it may be necessary to introduce mechanical means of modifying the hazard. In this case the transporting force of surface flow can be reduced by constructing barriers to its path in the form of banks or terraces. Normally this is costly and is only justified when a high-return crop is used.

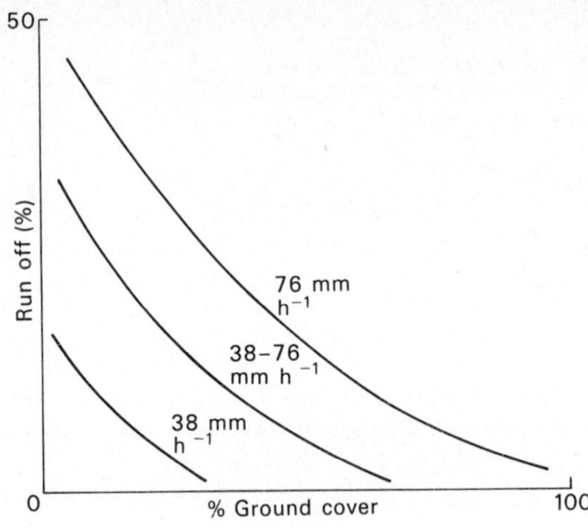

Fig. 2.11.

Assessment of erosion hazard

This can be determined if all those factors which affect the erosion process are recorded and taken into account. For example, the relationship could be put in this form — $E = f(RGSV)$, where R = rainfall characteristics (intensity frequencies can be mapped); G = ground features (this refers mainly to the steepness of slope which can be mapped); S = soil erosiveness (this can be assessed by taking into account all those factors associated with aggregate stability); and V = vegetation. The most important feature is the ground cover.

Visual observations will also reveal those areas of active gullying or those streams which are carrying high sediment loads. This information is often mapped for use in erosion control planning. Emphasis must be placed, however, on mapping active sites or foci.

2.3 ASSESSMENT AND CORRECTION OF SOIL FERTILITY IN RELATION TO TROPICAL PASTURES

Most techniques of fertility assessment deal with nutrient deficiencies as these can usually be corrected by the application of fertilizers or by the selection of tolerant species or varieties. Excess chemical effects are often readily diagnosed but not always easily corrected. Adverse physical factors are even more difficult to alter.

1. *Assessment techniques*

The most widely used methods for assessing nutrient deficiencies will be discussed. No matter what scale of programme is being undertaken it is essential that it be properly planned, taking soil differences into consideration and providing a basis for extrapolation of the results to a wider area.

The first step is to recognize different types of soil in the area of interest. The ideal is to have a soil survey report and map as a guide. If this is not available soils can still be recognized and classified even if their boundaries are not mapped in the first instance.

Valuable information can be obtained from experience with other crops in the area or from known animal health problems. These will give an indication of nutritional problems which may arise for pasture plants.

Glasshouse experiments

Pot experiments in the glasshouse are a widely used initial step for obtaining information on soil nutrient deficiencies and excesses. Many soils can be screened at one time under conditions of greater environmental control and freedom from pests and diseases so that only soil factors limit plant growth. There are also advantages of speed and cost of labour and materials. A standardized approach is necessary and is discussed in the practical classes. Andrew and Fergus (1976) have also described techniques.

The information from pot experiments is very useful but it must be remembered that the results cannot be applied directly to the field. Incidences of deficiencies require confirming in the field and rates in pots cannot be extrapolated to field rates. However the design of field experiments is simplified following pot experiments and the combination of detailed experiments at a few field sites with the screening of a larger number of sites in pot experiments gives a good basis for formulation of regional soil fertility statements.

Field experiments

Field experiments are the standard procedure as they more nearly approach commercial practice. They have the advantage of dealing with the soil *in situ* under the prevailing seasonal conditions. However they suffer from disadvantages of soil and plant population variation, climatic limitations, pests, and diseases and are expensive and labour-intensive if conducted properly.

Field experiments are essential for deciding rates of nutrients required and for calibrating soil and plant analysis for diagnostic use. Properly calibrated soil and plant analyses provide a basis for extrapolating the site-specific information from field experiments to other locations in the region.

Laboratory methods

Soil analyses

The chemical analysis of soils has attracted the attention of soil chemists for over a century yet there is no universal acceptance of methods of analysis or interpretive indices. The first problem is the availability concept whereby only a small proportion of the total nutrient supply is available to the plant; the second problem is that soils differ in their abilities to supply available nutrients and in their reactions with applied fertilizers; while the third problem is that plants differ in their abilities to absorb nutrients from the soil.

The advantage of soil analyses is that they are rapid and inexpensive and that they provide information on soil fertility before a crop is planted. Many simple soil measurements provide information that allows the behaviour of the soil as a medium for plant growth to be anticipated, e.g. soil depth, slope, moisture characteristics, texture, pH, conductivity, organic matter content. Other measurements when properly interpreted enable nutrient recommendations to be made, e.g. extractable nutrients and exchangeable cations.

There is no substitute for conducting a large number of field experiments and relating the response of plants to the amounts of nutrients measured by soil analyses. The 'best' method of analysis is the one which correlated best with plant response. It is misleading to take soil test interpretations developed in one region and to apply them to another region without first checking them against field experiment responses.

Results obtained from a programme where field responses are correlated with soil chemical analyses are given in Fig. 2.12. These are from the work of Rayment, Bruce, and Robbins (1977) and show the relationship between relative yield of Siratro (relative yield is yield in the absence of phosphorus as a percentage of maximum yield) and the bicarbonate extractable phosphorus soil test of Colwell (1963). Two different methods of deciding the soil test critical level are shown.

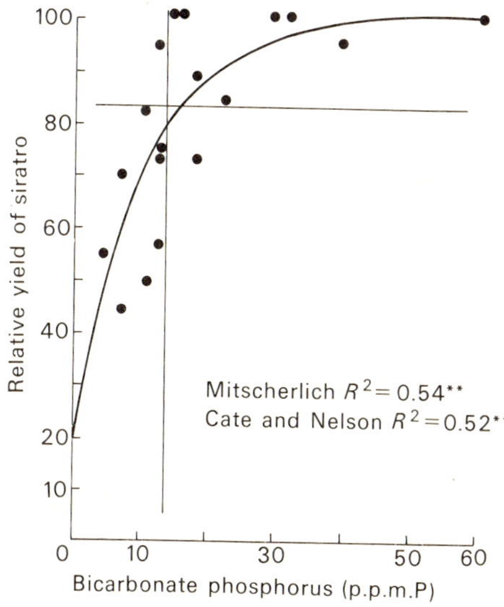

Fig. 2.12. The relationship between Siratro relative yields and soil bicarbonate extractable phosphorus, showing both Cate and Nelson separation and Mitscherlich response curve (Rayment, Bruce and Robbins, 1977).

Firstly a continuous mathematical function has been fitted to the data — in this case it was $Y = 100.43 - 79.73$ exp. $(-0.095\ X)$ where Y is relative yield and X is the soil test value. In this approach the next step is to read off the soil test value corresponding to a relative yield regarded as 'optimum'. The second method illustrated is that of Cate and Nelson (1971) whereby the points are divided into four quadrants, with most of the points falling into two quadrants — a low soil test, low relative yield group and a high soil test, high relative yield group. The vertical dividing line between the groups cuts the x-axis at the critical soil test value.

An important concept which users of soil tests should understand is that in most cases the analysis is an empirical one and the amount of nutrient extracted does not represent a definite or discrete fraction of the soil. In other words, the results of the analyses are only indices whose successful use and interpretation depends on correlation with plant response criteria.

Plant analysis

The diagnostic use of plant analysis is appealing because one is dealing

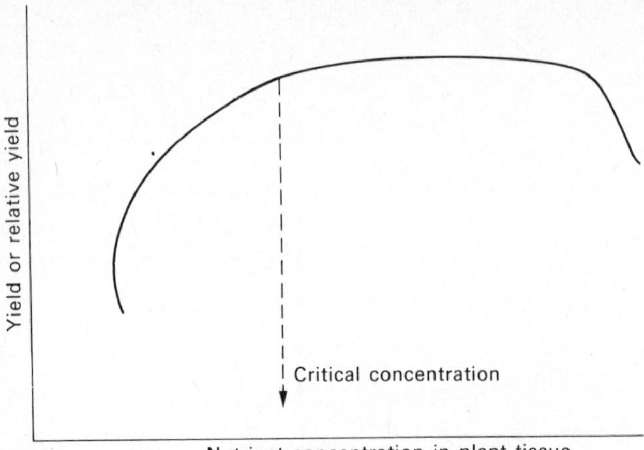

Fig. 2.13. Diagrammatic relationship between dry matter yield and nutrient concentration in plant tissue.

directly with the plant of interest. The approach depends on the 'critical nutrient concentration' concept. This relies on the concentration of a nutrient reflecting the sufficiency of that nutrient for plant growth. This is shown diagrammatically in Fig. 2.13, where yield is plotted against internal nutrient concentration in a particular plant tissue. Above the critical concentration yield does not increase and luxury uptake occurs; below the critical concentration both yield and concentration are reduced.

Unfortunately there are many factors which influence critical nutrient concentrations in plants, e.g. part of plant sampled, age of plant, physiological stage of growth, season of year, previous rainfall. As for soil analyses a programme of calibration experiments is required and strictly standardized procedures must be followed. So far this method has been most useful for tree crops and other perennial or longer term crops but has been applied to some annual crops. Table 2.11 gives critical nutrient concentrations for some tropical legumes.

Applications
The techniques outlined above have been used in soil fertility research for tropical pastures in Australia by a number of workers and the integrated approach has developed over a number of years. Some recent publications reporting the use of these techniques are mentioned below.

Table 2.11. Critical concentrations of phosphorus, potassium, and sulphur in
some tropical pasture legumes (Andrew and Robins 1969a, b;
Andrew 1977)

Species	Critical concentration		
	% P	% S	% K
Centrosema pubescens	0.16	–	0.75
Lotononis bainesii	0.17	0.15	0.9
Stylosanthes guyanensis	0.17	–	0.7
Stylosanthes humilis	0.17	0.14	0.6
Macroptilium lathyroides	0.20	0.15	0.75
Macroptilium atropurpureum	0.24	0.17	0.75
Desmodium intortum	0.22	0.17	0.72
Desmodium uncinatum	0.23	0.17	0.8
Glycine wightii	0.23	0.17	0.8

Glasshouse techniques have been used to screen soils for nutrient deficiencies by Isbell, Jones, and Gillman (1976), Webb (1975), Teitzel and Bruce (1972), Kerridge, Andrew, and Murtha (1972), and Crack (1971).

A combination of glasshouse and field nutrient screening experiments have been reported by Kerridge and Everett (1975), Teitzel and Bruce (1971), and Jones (1973).

Field experiments to determine rates of fertilizer required have been conducted by Teitzel and Bruce (1971), Blunt and Humphreys (1970), Tietzel (1969), Jamieson (1969), Jones (1968), and 't Mannetje (1967).

Soil phosphorus analyses have been correlated with the response of newly planted pastures to phosphorus by Bruce and Bruce (1972) and White and Haydock (1968).

Critical nutrient concentrations in tropical pasture species have been published for phosphorus by Andrew and Robins (1969a, 1971), Jones (1968), Bruce (1974), Smith (1975); for potassium by Andrew and Robins (1969b); for sulphur by Jones and Robinson (1970) and Andrew *et al.* (1974); and for copper by Andrew and Thorne (1962).

2. *Correction techniques*

Fertilizers

Once they have been identified, factors limiting plant growth need to be corrected. There are many fertilizer compounds available for correcting nutrient deficiencies in soils. Some of the more common fertilizers and their chemical compositions are given in Table 2.12.

Most countries have laws whereby guaranteed concentrations must be shown on a label attached to each bag of fertilizer. In Australia concentrations are expressed as percentages of elements present, e.g. % P, % K, but some countries follow a convention of expressing concentrations as oxides, e.g. % P_2O_5, % K_2O.

Phosphatic fertilizers

Fertilizers containing phosphorus are the most important for pasture legumes. This is because legumes require relatively large amounts of phosphorus for growth and nitrogen fixation and because they are usually grown on soils low in phosphorus.

Single superphosphate. This is the most commonly used source of phosphorus. It contains approximately 10 per cent P, 10 percent S, and 20 per cent Ca and is manufactured by the action of sulphuric acid on rock phosphate. Most of the phosphorus is water soluble and readily available to plants.

Double or triple superphosphate. This is manufactured by separating the phosphoric acid produced by treatment of rock phosphate with sulphuric acid and reacting it with another lot of rock phosphate. The product contains two to three times the amount of water-soluble phosphorus as single superphosphate but much less sulphur, e.g. 19 per cent P, 1.6 per cent S, 16 per cent Ca.

Rock phosphate. The amount of phosphorus in rock phosphates varies with the source of the material but it is always practically insoluble in water. When finely ground and used at high rates on acid soils it slowly becomes available. It has some value for long-term crops and pastures.

Potassium fertilizers

There are two commonly used fertilizers containing potassium:

Muriate of potash. This is potassium chloride. It is crystalline, soluble, easy to handle, and contains 50 per cent potassium.

Sulphate of potash: This compound is more expensive than muriate of potash but is preferred for certain crops when chloride additions need to be minimized. It contains 42 per cent potassium and is also crystalline and soluble.

Nitrogen fertilizers

There are many fertilizer compounds containing nitrogen. They are not

Table 2.12. Some useful data on fertilizers

Element	Formula	Molecular weight	Solubility in water	Percentage element present	Other nutrients
Boron	B	10.8	Insol.	–	–
Borax	$Na_2B_4O_1.10H_2O$	381	Sol.	11.3	–
Sod. metaborate	$Na_2B_2O_4.4H_2O$	204	Sol.	10.6	–
Fertilizer borate	$Na_2B_4O_7.4H_2O$	–	Sol.	13–14	–
Borated superphosphate	–	–	Sol.	0.4–0.8	–
Colemanite	$Ca_2B_6O_{11}.5H_2O$	–	SH. sol.	10.1	–
Boron frit	–	–	Insol.	4–5	–
Calcium	Ca	40	Dec.	–	–
-carbonate	$CaCO_3$	100	V. sli. sol.	40	–
-hydroxide	$Ca(OH)_2$	74	Sil. sol.	54	Hydrated lime may contain 4.5 kg S t^{-1}
-oxide	CaO	56	Sli. sol.	71	–
-magnesium carbonate (dolomite)	$CaMg(CO_3)_2$	184	V. sli. sol.	22	10–13% Mg
-sulphate (gypsum)	$CaSO_4.2H_2O$	172	Sli. sol.	23	14–19% S
Carbon	C	12	Insol.	–	–
Chlorine	Cl	35.5	Sol.	–	–
Chromium	Cr	52	Insol.	–	–

(continued)

Table 2.12. (*cont'd*)

Element	Formula	Molecular weight	Solubility in water	Percentage element present	Other nutrients
Sod. dichromate	$Na_2Cr_2O_7.2H_2O$	298	V. sol.	35	–
Pot. dichromate	$K_2Cr_2O_7$	294	Sli. sol.	35	26.5% K
Cobalt	Co	59	Insol.	–	–
-sulphate	$CoSO_4.7H_2O$	281	Sol.	21	11% S
-ized super-phosphate }				1.3 / 0.6	Ca and P
Copper	Cu	63.6	–	–	–
-carbonate (basic)	{ approx. formula $CuCO_3 Cu(OH)_2$) }	221	Insol.	About 55	–
-sulphate (bluestone)	$CuSO_4.5H_2O$	250	Sol.	25	12.8% S
-oxychloride	{ Oxychloride base }	–	Insol.	50	–
-cuprous oxide	Cu_2O	143	Insol.	89	–
-ized superphosphate	–	–	–	0.6	Ca. P
Hydrogen	H	1	Sol.	–	–
Iron	Fe	56	Insol.	–	–
-sulphate	$FeSO_4.7H_2O$	278	Sol.	20	11.5% S
-chelates	(for spray application)		–	6–8	–

Magnesium	Mg	24.3	Insol.	—	—
-carbonate					
(amorphous)	$MgCO_3$	84.3	V. sli. sol.	29	—
-sulphate					
(Epsom salts)	$MgSO_4.7H_2O$	246.5	V. sol.	10	13% S
Magnesite	$MgCO_3$	—	—	20–9	Occurs in natural deposits
Talc. Magnesite (approx.)	$\{H_2SiO_3\,MgSiO_3\\\;\;\;\;+MgCO_3\}$	— —	good samples poor samples	20 2–10	— —
Dolomite	$MgCO_3\,CaCO_3$	184	V. sli. sol.	10–13	22% Ca
Kieserite	$MgSO_4.H_2O$	138	Sol.	17	23% S
Serpentine approx.	$3MgO2SiO_2.2H_2O$	—	—	18–21	—
Dunite	—	—	—	20–4	—
(Serpentine or Dunite if used alone would need to be very finely ground, as these materials are not readily soluble.)					
Serpentine superphosphate	—	—	—	5	Ca. P
Manganese	Mn	55	Dec.	—	—
-acetate	$Mn(C_2H_2O_2)_2.4H_2O$	245	Sol.	22.5	—
-sulphate	$MnSO_4.4H_2O$	223	Sol.	25	14% S
-manganite	$Mn_2O_3.H_2O$	166	—	62	—
Molybdenum	Mo	96	Insol.	—	—
trioxide	MoO_3	144	Insol.	66.7	—
sod. molybdate	$Na_2MoO_4.2H_2O$	242	Sol.	40	—
-ized superphosphate	$\{(0.04\%\,Na_2MoO_4\,*)\\(0.06\%\,Na_2MoO_4\,*)\}$		— —	0.016 0.024	Ca. P

*Standard levels

Table 2.12. (*cont'd*)

Element	Formula	Molecular weight	Solubility in water	Percentage element present	Other nutrients
Nickel	Ni	59	Insol.	—	—
-carbonate	$NiCO_3$	119	V. sli. sol.	50	—
-sulphate	$NiSO_4.6H_2O$	263	Sol.	22	12% S
Oxygen	O	16	Sli. sol.	—	—
Phosphorus	P	31	Insol.	—	—
-pentoxide	P_2O_5	142	V. sol.	44	—
Superphosphate	—	—	Sol.	$9 \left\{\begin{array}{l}(20\text{–}22\% \\ \text{as } P_2O_5).\end{array}\right\}$	10–12% S 20% Ca
Potassium	K	39	Dec.	—	—
-chloride (muriate of potash)	KCL	74.6	Sol.	52 (60K_2O)	—
-sulphate	K_2SO_4	174	Sol.	45 (48K_2O)	18% S
-nitrate	KNO_3	101	Sol.	39	14%N
Nitrogen	N	14	Sol.	—	—
Ammonium nitrate	NH_4NO_3	80	V. sol.	33	—
Ammonium sulphate	$(NH_4)_2SO_4$	132	Sol.	20	24% S
Monoammonium phosphate	$NH_4H_2PO_4$	—	Sol.	11	21% P
Diammonium phosphate	$(NH_4)_2HPO_4$	—	Sol.	21	23% P

Nitrolime	35% precipitated chalk	–	Sol.	20–1	13% Ca
Nitrochalk	{ 44% nitrate, 48% precipitated chalk	–		15.3	20% Ca
Potassium nitrate	KNO_3	101	Sol.	14	39% K
Sodium nitrate	$NaNO_3$	85	Sol.	16	–
Urea	NH_2CONH_2	60	V. sol.	46	–
Ureaform	formalin + urea	–	–	20–40	–
Dried blood	–	–	Insol.	10–14	–
Blood and bone	–	–	Insol.	4–8	5% P
Sodium	Na	23	–	–	–
Sulphur	S	32	Insol.	–	–
Superphosphate	–	–	Sol.	10–12 S	20% Ca 9% P
Gypsum	$CaSO_4.2H_2O$	172	Sli. sol.	14–19	23% Ca

(See also sulphates of ammonia, copper, magnesium, potassium.)

Tungsten	W	184	Insol.	–	–
Sod. tungstate	$Na_2WO_4.2H_2O$	330	Sol.	56	–
Scheelite	$CaWO_4$	288	Insol.	50–65	8% Mo
Vanadium (see also basic slag)	V	51	Insol.	–	–
Ammonium metavanadate	NH_4VO_3	117	Sol.	43.5	12% N
Sod. orthovanadate	$Na_3VO_4.16H_2O$	472	V. sol.	10.8	–
Zinc	Zn	65.4	Insol.	–	–
-carbonate	$ZnCO_3$	125.4	V. sli. sol.	52	–
-sulphate	$ZnSO_4.7H_2O$	287.5	V. sol.	23	11% S

Zinc powder can also be used for pastures and crops.

usually applied to legume-based pastures but in some circumstances it may be economic to apply nitrogen to grass pastures.

The most widely used nitrogen fertilizers are ammonium sulphate (20 per cent N), urea (46 per cent N), and ammonium nitrate (34 per cent N). Two ammonium phosphate compounds are also available. They provide soluble nitrogen and phosphorus. Mono ammonium phosphate (MAP) contains 11 per cent N and 22 per cent P, while diammonium phosphate (DAP) contains 19 per cent and 20 per cent.

Many factors influence the choice of which fertilizer to use but relative costs of available materials are usually the first consideration. Most fertilizers whether inorganic or organic supply more than one nutrient, e.g. superphosphate contains phosphorus, sulphur, and calcium. Sometimes nutrients other than those guaranteed are present as impurities. Fertilizer mixtures are also sold, the commonest ones are those which supply nitrogen, phosphorus, and potassium, but fertilizers with added trace elements are also common. Mixtures often have the advantage of ease of handling but may have the disadvantage of higher cost. It is wasteful to use a mixture if only one nutrient requires correction. In experimental work it is difficult to work out rates of nutrients required if mixtures are used and it is difficult to compare the effectiveness or relative economics of mixtures containing different amounts of various elements.

There are limits to the types of different materials that can be mixed and there has been much research in fertilizer technology to develop suitable mixtures. *If mixed and used quickly*, almost all commercial fertilizers, with the exception of nitrates with elemental sulphur, can be mixed with safety and with little or no loss of plant nutrients. However, with storage some mixtures may become hard or may form a damp product which is difficult to spread evenly.

Most problems associated with the compatability of fertilizers in mixtures can be linked to the inclusion of nitrogenous compounds in the mix. The acidifying effect of nitrogen fertilizer is well known. Phosphorus and potassium fertilizers have little effect on soil pH unless they are used in mixtures containing nitrogen. The acidifying effect of ammonium ions occurs mainly when they are nitrified. Nitrification is represented by the equation $NH_4^+ + 2O_2 \rightarrow 2H^+ + NO_3^- + H_2O$. The H^+ ions released reduce the pH of the soil solution and can also displace basic cations from the exchange complex which are removed with the NO_3^- ion by leaching. In addition the NH_4^+ ion itself can displace basic cations from the exchange complex

Liming

The practice of liming acid soils has been an accepted part of intensive agriculture in temperate climates for centuries. It is not a widespread practice in tropical areas but agricultural research is identifying situations where lime is beneficial.

The main reasons for liming soils are to adjust soil acidity to a range preferred by specific crops; to provide nutrients directly or to increase availability of other nutrients; or to reduce the solubility of toxic aluminium or manganese. The amount of lime required will depend on the reason for liming and on the composition of the soil being limed.

The preferred pH ranges for some crops is given in Table 2.13. There is a fair tolerance by crops on either side of the range provided that nutrients are in adequate supply. The effect of soil pH on nutrient availability is given in Fig. 2.10. It can be seen that maximum availability of nutrients is in the pH range 6.0 to 7.0.

Table 2.13. Some pH preferences of various plants (Reynolds 1971)

Plant	pH range	Plant	pH range
FOOD CROPS			
Asparagus	6.0–7.0	Melon (water)	5.0–5.5
Avocado	6.0–8.0	Onion	6.0–7.0
Banana	6.0–7.0	Passionfruit	6.0–8.0
Bean	6.0–7.5	Pea	6.0–8.0
Cabbage	6.0–7.0	Pecan	6.0–7.0
Carrot	5.5–6.5	Pepper	6.0–6.5
Celery	6.0–6.5	Pineapple	5.0–6.0
Citrus		Potato (sweet)	5.0–7.0
lemon	5.5–7.0	Pumpkin	5.5–6.5
orange	5.0–7.0	Radish	6.0–7.0
Cocoa	6.0–7.5	Rice	6.0–7.0
Coconut	6.0–8.0	Sugar cane	6.0–8.0
Coffee	5.0–6.0	Taro	5.5–6.5
Corn	6.0–7.0	Tea	5.0–5.5
Eggplant	6.0–7.0	Tomato	6.0–7.0
Groundnut	6.0–6.5	Walnut	6.0–8.0
Lettuce	6.0–7.0	Watercress	6.0–8.0
Melon (cantaloupe)	6.0–6.5		
OTHER PLANTS			
Acacia	6.0–8.0	Pine	4.5–6.0
Alfalfa	6.0–7.0	Pointsettia	6.0–8.0
Kudza	5.5–6.5	Rose	6.0–8.0
Orchid	4.0–6.0	Sorghum	5.5–7.0
Pangola grass	6.0–7.5	Tulip tree	6.0–7.0
Panicum	5.5–6.5		

The amount of lime required to provide the required pH change in a soil can be estimated from laboratory tests or from general 'rules of thumb' based on soil texture.

The laboratory tests are based on either measuring 'exchangeable hydrogen' at a specified pH (6.0, 7.0, 8.2 have been used) or by constructing a buffer curve by incubating soil with graded amounts of $Ca(OH)_2$ and water and then measuring the pH of the suspension and plotting a curve. More recently it has been suggested that in tropical soils exchangeable Al can be used as a guide to lime requirements (Kamprath 1970).

The commonest liming material is limestone (calcium carbonate). Naturally occurring limestone is finely ground and sold as pulverized limestone or agricultural limestone. Dolomite is a naturally occurring calcium–magnesium carbonate and its use is similar to that of limestone.

3. *Control of soil fertility in tropical pastures*

To maintain productive grass–legume pasture mixtures four important criteria have to be met. These are:

(i)   choice of suitable species,
(ii) use of appropriate *Rhizobium* inoculum,
(iii) an adequate supply of plant nutrients,
(iv) careful grazing management.

Criterion (iii) will be considered in this section. The provision of an adequate supply of plant nutrients requires the application of principles of soil chemistry and plant nutrition as well as fertilizer technology.

In fertilizing grass–legume pasture mixes, the philosophy followed is that if nutrient supply is adequate for the legume component, it is usually adequate for the grass component. This is because legumes are more sensitive to nutrient deficiency and because the grass depends upon the legume for its nitrogen supply. Retaining a high proportion of vigorously growing legume is necessary. Once legume content falls, the grass becomes nitrogen deficient, its growth rate declines, and weed species are able to invade the pasture. The end result is a weedy pasture which is unable to maintain animal production because of its low yields of poor quality feed. It is convenient to distinguish between the fertilizer requirements for establishment and the amounts that must be applied to the established pasture each year to maintain production (maintenance fertilizer requirements).

Establishment fertilizers

The methods discussed in Section 2.3.1 provide the answers to the questions of which nutrients are deficient, which fertilizers are required to correct these deficiencies, and what rates of fertilizers are required to supply adequate quantities for optimum yields on the various soil types. Regional fertilizer recommendations for soils take into account soil and climatic differences. Properly calibrated soil tests are useful at this stage for predicting fertilizer responsive situations before planting and allow a little more precision in fertilizer recommendations.

In most areas where tropical pastures are grown phosphorus deficiency has been found to be the major limitation to pasture production. Sulphur and molybdenum are widely deficient in tropical Australia so molybdenized superphosphate is the commonest fertilizer used for pasture establishment. In some areas copper, zinc, and potassium are also required. These differences are illustrated by the following fertilizer recommendations which are currently made for pasture establishment in some areas of Queensland (all recommendations per hectare).

Wet tropics, basaltic soils:
    250 kg molybdenized superphosphate (0.04% Mo)

Wet tropics, granitic soils; eucalypt forest vegetation:
    500 kg superphosphate
    100 kg muriate of potash
      8 kg copper sulphate
      8 kg zinc sulphate

Coastal lowlands, south-east Queensland
    500 kg molybdenized superphosphate (0.02% Mo)
    100 kg muriate of potash
    500 kg lime
      8 kg copper sulphate
      8 kg zinc sulphate

Maintenance fertilizers

The questions to be answered in developing a maintenance fertilizer strategy are which fertilizers are required, how much is required and how often are they required. In a perennial pasture we are dealing with established plants whose initial demands for nutrients have been met and whose root systems exploit a large volume of soil. Consequently it is likely that nutrient requirements will be different from those at planting time where large quantities of readily available nutrients are required near the seedling to provide the nutrients initially immobilized in both the above and below ground plant tissues.

Nutrients are needed to replace those lost through chemical reactions with soil constituents, through removal in animal products, through leaching loss in drainage water. The most important of these pathways is the reaction with soil. This field of residual value of fertilizers is not fully understood. The availability of most nutrients declines exponentially with time but the chemistry of the soil determines the rate of decline. As most pasture soils are initially low in fertility another 'reaction' that occurs in soil is the increase in soil organic matter that occurs with time under pasture. Nutrients, particularly nitrogen, phosphorus, potassium, and sulphur, are immobilized in the organic matter.

Andrew and Bruce (1977) have described three forms of experiments for studying fertilizer maintenance requirements. Firstly we have the full scale grazing experiment, with rates of nutrient addition as treatments, which includes the whole soil–plant–animal complex. These must necessarily be long-term experiments and are very expensive to conduct. Some examples are given by Andrew and Bruce (1977, Table 1). It is essential then that as much information as possible be obtained from them so that the results can have as wide an application as possible. Animal performance data are of local use but may be extended to larger areas where soils, soil fertility, and environmental conditions are similar. If animal performance can be related to soil, plant, and animal indices (particularly if indices derived under controlled conditions can be confirmed or modified) we have a basis for much wider application of the results.

The cost of such experimentation will be prohibitive and need to be complemented with a second form of experimentation. This is to apply fertilizer treatments to areas of grazed pastures and to assess sufficiency of the treatments in terms of dry matter production. Animals need to be kept off treated areas either by rotational grazing or by enclosing treated areas with cages or fences. This approach allows the soil–plant relationships to be studied. Rayment *et al.* (1977) have used this approach with pastures based on Siratro (*Macroptilium atropurpureum* cv. Siratro). They showed that Siratro responded to maintenance applications of phosphorus if soil bicarbonate-extractable phosphorus was less than 10 p.p.m. phosphorus.

The third form of experimentation is to take soil from the pasture for pot experiments in the glasshouse. This technique has limitations for maintenance fertilizer studies but can be used to identify nutrient deficiencies which develop in the soil after a few years of pasture and which may not have been present when the initial fertility screening experiments were done.

Once fertilizer recommendations have been formulated and critical soil and plant indices decided the ideal application of the techniques is in monitoring the fertilizer programme through soil and plant analyses (soil and crop logging). This is shown diagrammatically in Fig. 2.14. Fertility strategy A has led to luxury levels of soil or plant nutrient and fertilizer inputs should be reduced after year 3 but strategy B is providing inadequate fertilizer inputs.

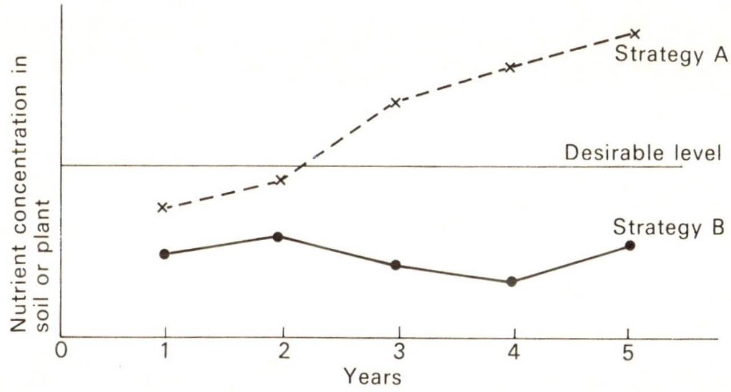

Fig. 2.14. Diagrammatic representation of monitoring two fertilizer programmes.

The results in Fig. 2.15 are from a long-term small plot fertilizer experiment with different establishment and maintenance rates of phosphorus (Lowe and Bruce, unpublished data).

The maintenance rate of 24 kg P ha^{-1} is just adequate to maintain the effect of an initial 96 kg P ha^{-1} but is more than adequate at maintaining the effect of initial 24 kg P ha^{-1}.

A research programme should initially cover the full biological potential and measure maximum plant and animal production. Maximum plant production at high fertilizer rates may not necessarily coincide with maximum animal production and neither may give the maximum economic return per unit area. The system of grazing management and pasture utilization needs to be considered in deciding whether

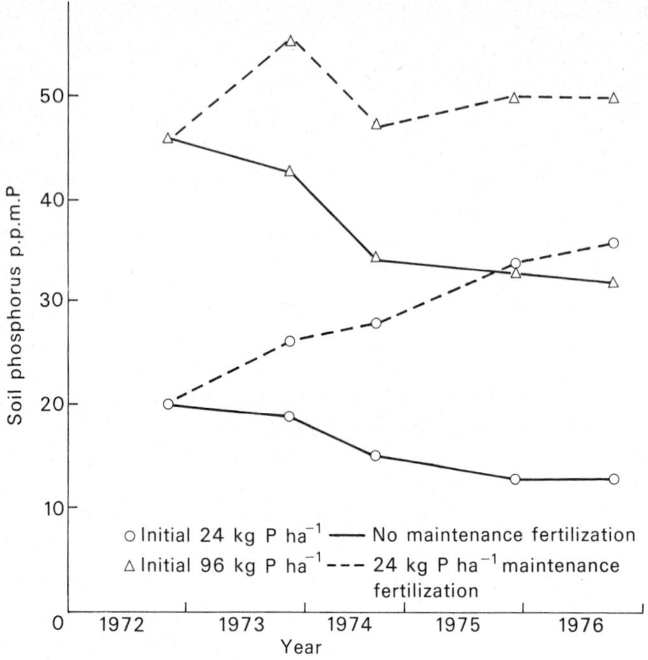

Fig. 2.15. The effect of initial and annual maintenance applications of phosphorus fertilizer on soil acid extractable phosphorus over a 5 year period. (Bruce and Lowe, unpublished data).

maximum pasture yield is the aim. Unless all feed produced is eaten efficient use is not being made of fertilizer. It may be that say 70–80 per cent maximum yield is the most economic return on fertilizer or that additional areas of pasture compensate for lower stocking rates. However if maximum plant and animal production have been measured in experiments the data can be used to decide which inputs would be the most suitable under different prevailing economic conditions.

Another alternative is that species more tolerant of lower fertility can be grown with consequent savings on fertilizer costs. There are differences between tropical legumes in their tolerance of low fertility or high acidity. For example, Andrew and Thorne (1962) have shown the following tolerance of copper deficiency — *Desmodium uncinatum* > *Macroptilium lathyroides* = *Centrosema pubescens* > *Stylosanthes guianensis*; while for excess manganese the tolerance is (Andrew and

Hegarty 1969) – *Centrosema pubescens* = *S. humilis* = *L. bainesii* >
M. lathyroides = *D. uncinatum* > *Glycine javanica* = *M. atropurpureum*.

Another way of utilizing the differences between species in response
to soil fertility status is to regard some legumes as pioneer species which
will grow best during the first few years of a pasture but as soil fertility
improves will be gradually replaced by legumes which grow best at
higher fertility. This has been observed in Queensland where stylo is
gradually replaced by centro and where molasses grass is replaced by
guinea grass.

REFERENCES

Andrew, C. S. (1963). Copper deficiency of some tropical and temperate
pasture legumes. *Aust. J. Agric. Res.* **14**, 654.
— (1977). The effect of sulphur on the growth, sulphur and nitrogen
concentration, and critical sulphur concentration of some tropical
and temperate pasture legumes. *Aust. J. agric. Res.* **28**, 807.
— and Bruce, R. C. (1977). Evaluation of fertilizer requirements of
tropical legume based pastures. *Trop. Grassl.* **9**, 133–9.
— and Fergus, I. F. (1976). Plant nutrition and soil fertility. In *Tropical pasture research – principles and methods* (eds N. H. Shaw and
W. W. Bryan) p. 101. Common. Agric. Bur. Bull. No. 51.
— and Hegarty, M. P. (1969). Comparative responses to manganese
excess of eight tropical and four temperate pasture legume species.
Aust. J. agric. Res. **20**, 687–96.
— and Robins, M. F. (1969*a*). The effect of phosphorus on the growth
and chemical composition of some tropical pasture legumes. I.
Growth and critical percentages of phosphorus. *Aust. J. agric. Res.*
20, 665–74.
— — (1969*b*). The effect of potassium on the growth and chemical
composition of some tropical and temperate pasture legumes. II.
Growth and critical percentages of potassium. *Aust. J. agric. Res.*
20, 999–1007.
— — (1971). The effect of phosphorus on the growth, chemical
composition, and critical phosphorus percentages of some tropical
pasture grasses. *Aust. J. agric. Res.* **22**, 693–706.
— and Thorne, P. M. (1962). Comparative responses to copper of
some tropical and temperate pasture legumes. *Aust. J. agric. Res.*
13, 821–35.
— and Vanden Berg, P. J. (1973). The influence of aluminium on
phosphate sorption by whole plants and excised roots of some
pasture legumes. *Aust. J. agric. Res.* **24**, 341.
—, Crack, B. J., and Rayment, G. E. (1974). Evaluation of plant and
soil sulphur tests in Queensland. In *Handbook on sulphur in Australian
agriculture* (ed. K. D. McLachlan) p. 55. CSIRO, Melbourne.

Aubert, G. (1968). Classification des sols utilisée par les pédologues francais. *FAO World Resources Rept.*, Vol. 32, pp. 78–94.

— and Tavernier, R. (1972). Soil survey. In *Soils of the humid tropics*, pp. 17–44. US National Academy of Sciences, Washington.

Black, C. A. (1968). *Soil–plant relationships*, 2nd edn. John Wiley, New York.

Blunt, C. G. and Humphreys, L. R. (1970). Phosphate response of mixed swards at Mt. Cotton, south-eastern Queensland. *Aust. J. exp. Agric. Anim. Husb.* **10**, 431–41.

Brams, E. (1971). Continuous cultivation of West African soils: organic matter diminution and effects of applied lime and phosphorus. *Pl. Soil* **35**, 401–14.

Bruce, R. C. (1974). Growth response, critical percentage of phosphorus, and seasonal variation of phosphorus percentage in *Stylosanthes guyanensis* cv. Schofield topdressed with superphosphate. *Trop. Grassl.* **8**, 137–44.

Bruce, R. C. and Bruce, I. J. (1972). The correlation of soil phosphorus analyses with response of tropical pastures to superphosphate on some north Queensland soils. *Aust. J. exp. Agric. Anim. Husb.* **12**, 188–94.

Buol, S. W., Hole, F. D., and McCracken, R. J. (1973). *Soil genesis and classification*. Iowa State University Press, Ames.

Buringh, P. (1970). *Introduction to study of soils in tropical and sub-tropical regions*, 2nd edn. Centre for Agricultural Publishing and Documentation, Wageningen.

Cate, R. B. Jr. and Nelson, L. A. (1971). A simple statistical procedure for partitioning soil test correlation data into two classes. *Soil Sci. Soc. Am. Proc.* **35**, 658–60.

Chapman, H. D. (1965). *Diagnostic criteria for plants and soils*. Homer D. Chapman, 830 South University Drive, Riverside, California. 92507.

Colwell, J. D. (1963). The estimation of the phosphorus fertilizer requirement of wheat in southern New South Wales by soil analysis. *Aust. J. exp. Agric. Anim. Husb.* **3**, 190–7.

Crack, B. J. (1971). Studies on some neutral red duplex soils (Dr. 2.12) in north-eastern Queensland. 2. Glasshouse assessment of plant nutrient status. *Aust. J. exp. Agric. Anim. Husb.* **11**, 336–42.

D'Hoore, J. L. (1968). The classification of tropical soils. In *The soil resources of tropical Africa* (ed. R. P. Moss) pp. 7–28. Cambridge University Press, London.

Dudal, R. (1968). Definitions of soil units for the soil map of the world. FAO World Soil Resources Rept. 33.

— (1970). *Key to soil units for the soil maps of the world*. FAO, Rome.

Epstein, E. (1972). *Mineral nutrition of plants: principles and perspectives*. John Wiley, New York.

Greenland, D. J. and Nye, P. H. (1959). Increases in carbon and nitrogen contents of tropical soils under natural fallows. *J. Soil Sci.* **9**, 284–99.

Hudson, N. (1971). *Soil conservation*. Batsford, London.

Isbell, R. F. Jones, R. K., and Gillman, G. P. (1976). Plant nutrition studies on some yellow and red earth soils in northern Cape York Peninsula 1. Soils and their nutrient status. *Aust. J. exp. Agric. Anim. Husb.* **16**, 532–41.

Jackson, M. L. and Sherman, G. D. (1953). Chemical weathering of minerals in soils. *Adv. Agron.* **5**, 219–318.

Jamieson, G. I. (1969). Effect of superphosphate application rate on pasture establishment on Queensland's wet tropical coast. *Qd J. agric. Sci.* **26**, 529–36.

Jenny, H. (1941). *Factors of soil formation*. McGraw-Hill, New York.

Jones, R. K. (1968). Initial and residual effects of superphosphate on a Townsville lucerne pasture in north-east Queensland. *Aust. J. exp. Agric. Anim. Husb.* **8**, 521–7.

—— (1973). Studies on some deep sandy soils on Cape York Peninsula, North Queensland. 2. Plant nutrient status. *Aust. J. exp. Agric. Anim. Husb.* **13**, 89–97.

——, and Crack, B. J. (1970). Studies on some solodic soils in north-eastern Queensland. 2. Glasshouse assessment of plant nutrient status. *Aust. J. exp. Agric. Anim. Husb.* **10**, 342–9.

—— and Robinson, P. J. (1970). The sulphur nutrition of Townsville lucerne (*Stylosanthes humilis*). Proc. XIth Int. Grassld Congr., Surfers Paradise, p. 377.

Kerridge, P. C. and Everett, M. L. (1975). Nutrient studies on some soils from Eungella and East Funnel Crk., Mackay Hinterland, Qld. *Trop. Grassl.* **9**, 219.

—— Andrew, C. S. and Murtha, G. G. (1972). Plant nutrient status of soils of the Atherton Tableland, North Queensland. *Aust. J. exp. Agric. Anim. Husb.* **12**, 618–27.

Kovda, V. A., van den Berg, C., and Hagan, R. M. (1973). *Irrigation, drainage and salinity*. FAO/UNESCO–Hutchinson, London.

't Mannetje, L. (1967). Pasture improvement in the Eskdale district of south eastern Queensland. *Trop. Grassl.* **1**, 9–19.

Mendez-Lay, J. (1973). Effects of lime on phosphorus fixation and plant growth in various soils of Panama. M.Sc. thesis, Soil Science Department, North Carolina State University, Raleigh.

Mohr, E. C. J., Van Baren, F. A., and Van Schuylenborgh, J. (1972). *Tropical soils: a comprehensive study of their genesis*, 3rd edn. Mouton, The Hague.

Mortvedt, J. J., Giordano, P. M., and Lindsay, W. L. (1972). *Micronutrients in agriculture*, pp. 9 and 332. SSSA, Madison, Wisconsin.

Northcote, H. H. and Skene, J. K. M. (1972). Australian Soils with Saline and Sodic Properties. Soil Pub. No. 27. CSIRO, Australia.

Obeng, H. B. (1968). Proceedings of the ISSS Conference, Adelaide, p. 215.

Rayment, G. E., Bruce, R. C., and Robbins, G. B. (1977). Response of established siratro (*Macroptilium atropurpureum* cv. Siratro) pastures in south east Queensland to phosphorus fertilizer. *Trop. Grassl.* **11**, 67–77.

Reynolds, S. G. (1971). A manual of introductory soil science and simple soil analysis methods. Information Document No. 24, South Pacific Commission, Noumea, New Caledonia.

Richards, L. A. (1969). *Diagnosis and improvement of saline and alkaline soils.* Agriculture Handbook No. 60. USDA.

Russell, E. W. (1973). *Soil conditions and plant growth.* 10th edn. Longman, London.

Sanchez, P. A. (1976). *Properties and management of soils in the tropics.* John Wiley, New York.

Smith, F. W. (1975). Tissue testing for assessing the phosphorus status of green panic, buffel grass and setaria. *Aust. J. exp. Agric. Anim. Husb.* **15**, 383.

Soil Survey Staff, United States Department of Agriculture (1951). *Soil survey manual,* USDA Handbook No. 18. US Government Printing Office, Washington, DC.

— (1975). *Soil taxonomy: a basic system of soil classification for making and interpreting soil surveys.* USDA Handbook No. 436. Government Printing Office, Washington, DC.

Sys, C. A., Van Wambeke, R., Frankart, F. L. (1961). *La cartographie des sols au Congo, ses principes et ses methodes.* INEAC, Sér. Tech. 66, Institut National pour l'Étude Agronomique du Congo, Brussels.

Taylor, S. A. and Ashcroft, (1972). *Physical edaphology: the physics of irrigated and non-irrigated soils.* W. H. Freeman, San Francisco.

Teitzel, J. K. (1969). Responses to phosphorus, copper and potassium on a granite loam of the wet tropical coast of Queensland. *Trop. Grassl.* **3**, 43–8.

— and Bruce, R. C. (1971). Fertility studies of pasture soils in the wet tropical coast of Queensland. 2. Granitic soils. *Aust. J. exp. Agric. Anim. Husb.* **11**, 77–84.

— — (1972). Fertility studies of pasture soils in the wet tropical coast of Queensland. 3. Basaltic soils. *Aust. J. exp. Agric. Anim. Husb.* **12**, 49–54.

Thorp, J. and Smith, G. D. (1949). Higher categories of soil classification: order, suborder and great soil groups. *Soil Sci.* **67**, 117–26.

Truog, E. (1946). Soil reaction on availability of plant nutrients. *Soil Sci. Soc. Am. Proc.* **11**, 305–8.

Walsh, L. M. and Beaton, J. D. (1973). *Soil testing and plant analysis,* revised edn. SSSA, Madison, Wisconsin.

Webb, A. A. (1975). Plant nutrient status of soils of the Maryvale Land System, North Queensland. *Qd J. agric. Sci.* **32**, 19–26.

White, R. E., and Haydock, K. P. (1968). Phosphate availability and phosphate needs of soils under Siratro pastures as assessed by soil chemical tests. *Aust. J. exp. Agric. Anim. Husb.* **8**, 561–8.

Williams, C. H. (1974). The chemical nature of sulphur in some New South Wales soils. In *Handbook on sulphur in Australian agriculture* (ed. K. D. McLachlan) pp. 16–23. CSIRO, Melbourne.

3

The tropical pasture species

3.1 PLANT INTRODUCTION AND EVALUATION

Pasture improvement and development programmes may be initiated through a number of different approaches. The first approach is based on improved management and utilization of existing natural pasture resources. The second approach is to replace existing natural vegetation with introduced pasture species, while a third method is a combination approach, where an introduced species may be oversown into existing native pasture.

The improved management and utilization of natural pastures finds particular application in the extensive rangeland grazing areas, usually in the drier regions. The basic philosophy is that the existing species in the region are well adapted to the environment and research and management seeks to increase or at least maintain the most productive species for animal production. In the extensive rangelands, where productivity per hectare is usually low, pasture improvement methods must be relatively low cost. Forage resources may be improved by grazing animal management practices such as control of stocking rate and hence grazing pressure, rotational grazing systems aimed at allowing valuable species to set seed and regenerate (Anderson 1967), improved distribution of animals by fencing and placement of water points (Lange 1969) and shade (McIlvain and Shoop 1971). A major increase in ground forage resources is achieved by clearing or thinning standing trees (Walker, Moore, and Robertson 1972; Beal 1973). Other vegetation-management practices include the use of mechanical, chemical, and burning practices to control the ingress of woody weeds.

In regions better favoured than the extensive rangelands in terms of moisture or soil fertility or topography or infrastructure, intensive pasture improvement has been based on replacement of natural vegetation with higher yielding species. Introduction of improved species is usually associated with a marked change in the ecology of the area through clearing, cultivation, addition of fertilizer, improved soil–water relations, and changes in grazing management. The introduced or sown species are better able to utilize the improved growing conditions. The oversowing of an introduced species, usually a legume, into existing natural pastures also requires a change in management such as application of fertilizer, minimal cultivation or burning, or changes in grazing management to favour the oversown species. The development of a pasture improvement project based on sown species usually requires the introduction, evaluation, and field testing in the region of a range of potentially adapted pasture species.

Hartley and Williams (1956) point out that few if any of the important cultivated pasture grasses or legumes are constituents of the flora of the major grassland regions of the world. None of the dominant species of the great temperate grasslands, the steppes, prairies, or pampas, are of importance in cultivated pastures. Most cultivated pastures have been developed in areas of higher rainfall on soils of higher or improved fertility, than those which support natural grasslands. In both the temperate and tropical regions the typical improved pasture grasses are allied to and probably derived from woodland and forest margins. Temperate examples include *Poa* spp. *Festuca* spp, *Lolium* spp, and *Dactylis* spp, while tropical examples in the grasses are *Pennisetum* spp, some *Panicum* spp, and *Setaria* spp. The same appears true of the tropical legumes such as *Glycine* spp, *Macroptilium*, and *Desmodium* spp. Of the 10 000 grass species known to occur, a high proportion of which contribute to natural grazing, only about 40 species are widely used in improved pastures. Of these 25 species are wholly temperate, while about 12 are tropical. A similar picture is also true of the legumes. Of the 11 000 species listed (6000 are tropical), only about 15 species are widely used for pasture purposes in the tropics (Hartley and Williams 1956).

The marked difference between selected cultivated species and natural species is particularly evident in Southern Australia. The native grasses (*Danthonia, Poa, Themeda*), native legumes and shrubs (*Atriplex*) were susceptible to high grazing pressure, and responded poorly to improved fertility. Introduced pasture legumes were more responsive to superphosphate than the existing native legumes (Begg 1963). Introduced grasses such as *Lolium* spp and *Phalaris* sp, and legumes *Trifolium* and *Medicago* spp gave much higher animal production under the improved fertility.

Characters required in pasture plants

In developing a plant introduction and evaluation programme the first stage should be to define the characteristics required in pasture plants for the particular region. This requires an analysis of the climatic and edaphic limitations, the limitations of the existing species, and current land use, and a definition of the characters sought in introduced pasture species. These can be defined in general terms as:

(1) high yield of good quality forage;
(2) persistence;
(3) ability to associate with other species; and
(4) ease of propagation.

(1) *Yield*

Total dry-matter yield is an important characteristic, determining to a large extent the animal-carrying capacity of a pasture. Many selection programmes aim to select those species which give the maximum dry-matter yield within the limitations of the environment. However, production of large amounts of dry matter of low nutritive value may be of little use in animal production, particularly where stocking rate is limited by an annual period of stress. Also selection for high yield in grasses may limit the ability of associated legumes to compete and remain in the sward. Thus selection of grasses on the basis of yield alone, without taking account of associative ability and pasture quality factors is not recommended. In legumes, however, selection for high yield is important, since nitrogen fixation and nitrogen yield is closely related to total dry matter yield. Some of the characters which will contribute to yielding ability include — rapid establishment and seedling vigour, tolerance of stress conditions, ability to take up and utilize soil nutrients and competitive ability with sown and weed species.

(2) *Persistence*

A pasture, in contrast to an annual crop, generally implies a degree of permanence and an ability for longer term survival under fluctuating seasonal conditions and the rigours of regular defoliation by grazing animals. The characteristics associated with persistence include — tolerance of grazing; the ability to survive water stress or waterlogging; the ability to withstand low or high temperatures, even perhaps burning; resistance to disease and insects; and the ability to regenerate from stolons, rhizomes, or seed. The latter is particularly important in annual species.

(3) *Associative ability*

Although monospecific pasture swards are widely used, particularly in grass-plus-nitrogen-fertilizer pasture systems, mixed pastures based on grass plus legume species have wide application. Mixed pasture swards have advantages as different species have different seasons of growth and different chemical and nutrient composition, while the legume component also contributes a higher crude protein status in the feed, and through nitrogen fixation provides a nitrogen input into the pasture system.

Competition within pasture swards is a complex of many factors and influenced by competition for light, water (Donald 1963), and nutrients (Hall 1971), and modified by the growth form of the species, season of growth, palatability and animal grazing selection relative to other species in the sward, and the rate of regrowth following defoliation (Jones 1974a).

(4) *Ease of propagation*

The ability readily to establish a pasture sward from seed or vegetative cuttings is an important consideration in large-scale pasture sowings. Since seed is easier to transport and sow, the availability of seed is often an important limitation in pasture development. Adequate seed production is also important in pasture persistence particularly in annual species. Seed production is also important in perennial species. Firstly, high seed production may ensure regeneration following some natural catastrophe, e.g. drought, excessive frost, fire, or overgrazing. Secondly, seed production is most important from the point of view of producing seed for establishing new pastures. Many of the tropical legumes and grasses are recent additions to agriculture. They have not had long periods of selection for seed yield as have the common crop plants. A number of promising species such as *Arachis* sp and *Psoralea* sp cannot be adequately evaluated because of problems of seed production. The poor seed production of many new species has a secondary effect of causing high seed prices, which in turn limits their use, or causes plantings to be made at low seeding rates. However, even if seed production is poor in new introductions they should still be evaluated, as methods of harvesting may be improved, or seed production improved by selection or breeding.

Approaches to plant introduction

Introduction of pasture species for evaluation in a region can be undertaken either by correspondence with other research institutions or agencies (such as the Food and Agriculture Organization of the United

Nations), or by plant collection in areas likely to yield plants adapted to the region. For most regional introduction programmes in new pasture-development areas, evaluation of existing species and cultivars held by major institutions in regions with similar environmental conditions will be most appropriate. Plant exploration and collecting programmes are a long-term process, as well as costly, and will largely be the province of major plant introduction centres.

In both approaches it is necessary to assess the regions likely to yield species adapted to the area for introduction. This is based on either climatic comparisons or on the phytogeographical distribution of species.

Climatic comparisons aim to define areas of similar *homoclimate* defined as regions with similar agricultural potential with regard to amount of rainfall, season of precipitation, and temperature (Hartley 1960). Broad-scale comparisons can be made on the basis of climatic classifications, such as Köppen's, as outlined in Chapter 1. However, most of these classifications are on a large scale and cover areas of considerable heterogencity. More recent approaches have used computer techniques to classify climates from which introductions have been made (Williams, Burt, and Strickland 1976), to compare climates in different regions for degrees of homology (Russell and Moore 1970, 1976) or to compare climates in different areas in which particular species have been successful (Reid 1973).

The use of the homoclimate concept has particular application in seeking introductions for arid and semi-arid regions where plants must grow with minimum inputs from management. It has been particularly successful with introduction of grasses, where different subfamilies and tribes have characteristic relationships with climate. Ebersohn (1969) has used this approach in seeking introductions for semi-arid Queensland.

A major centre of distribution of the tropical grasses is East Africa from where species in the genera *Brachiaria, Cenchrus, Chloris, Cynodon, Panicum, Setaria*, and *Urochloa* have been introduced to many countries. Southern Africa has yielded important species of *Digitaria* and *Urochloa*. *Cenchrus* species are also distributed through Arabia, Pakistan, and northern India. The other major area for introduction of grass species is in subtropical South America where the genera *Axonopus, Bromus, Paspalum*, and *Sorghum* are important.

In the tropical legumes collection from areas of high floristic potential or centres of distribution where there is a high density of different species within a genus, has been more successful. Hartley (1960) has shown that for *Macroptilium*, *Desmodium*, and *Leucaena*, centres of high species density exist in Central America, Bolivia, Southern Brazil, and Paraguay. Also parts of central and southern America are rich in species of *Centrosema* and *Stylosanthes*. East Africa is an important area for *Glycine* and *Macrotyloma* species and Southern Africa of *Lotononis*. Other species in the genera *Pueraria, Vigna, Calopogonium*, and *Indigofera* have centres of distribution in South-East Asia.

Organization of plant introduction

A plant introduction group has a number of functions, the basic ones being to:

Acquire new plant accessions and organize their introduction into the country.
Organize the documentation of all introductions.
Provide initial screening, description, and evaluation.
Build up seed stocks or planting materials, and to maintain seed stocks for future testing or breeding programmes.

In larger organizations the plant introduction group will be involved in taxonomic studies, phytogeographical studies of species distribution, and plant-collecting expeditions.

The introduction and evaluation programme a newly introduced plant passes through will depend on whether it is a 'new' accession about which little is known, or a released cultivar described and used in another country. In both cases the first step should be a documentation and quarantine stage. All details about the introduction should be recorded, it would be given an introduction number, and the seed inspected and treated for possible insect or disease contamination. A released cultivar would then go directly into a regional testing prog-ramme, after a small sample of seed has been retained for future reference.

In the case of new accessions only a small amount of seed would be available. The aims of the first stage of evaluation are to provide a quarantine check of introductions, build up seed stocks or vegetative planting material, and to describe the morphology, phenology, and growth characteristics of the accession.

Firstly the plants are raised in a quarantine glasshouse and after checking are sown in rows in the field. The introduction area should be isolated and provided with irrigation. Even if varieties are intended for arid areas it is important at this stage to provide water and adequate soil fertility to ensure that they can be grown. The new introductions are usually compared with standard varieties.

Records are kept throughout the growing season and include:

Growth habit – erect, prostrate, bunch.
Leafiness.
Vigour – ground cover, size, bulk.
Time of flowering – length of flowering period.
Seed set – amount harvested.
Effect of low or high temperature.
Regeneration.
Incidence of pests and diseases.
Nodulation in the case of legumes.

The primary evaluation stage may last from one to three years. Accessions obviously unsuited to the proposed environment, and those showing undesirable characteristics may be eliminated at this stage. Legume accessions with obvious nodulation problems may be identified and further *Rhizobium* strain testing undertaken. Promising accessions would then go to second-stage evaluation at the introduction centre or at regional sites in the development regions.

The second stage of evaluation aims to examine the agronomic characteristics under field sward conditions. Competition and stress effects of water, nutrients, and pests and diseases will be greater than under nursery conditions. However, nutrient deficiencies should be eliminated as far as possible by correct basal and maintenance fertilizer applications.

Productivity is usually assessed at this stage by regular quadrat cuts from replicated sward trials. It is important to measure the distribution of yield as well as total annual yield. Plots may be subjected to defoliation after each yield sampling by either mowing or intermittent grazing. It is important to assess vigour of the sward at this stage through estimates of botanical composition to check on the ingress of weed species. Persistence of sown species can be estimated by plant counts at establishment and subsequent survival.

This stage of evaluation should demonstrate which species are the most productive and persistent in the particular area. Also some assessment of potential feeding value can be undertaken through *in vitro* digestibility studies, and intake studies with penned sheep. Some examples of second stage evaluation are given below.

(a) *Species evaluation in the Northern Wallum* (Evans 1967)
Thirty-nine tropical grass species and varieties and 40 legume species

and varieties were evaluated in pure stands in plots 5 × 5 m with two replications. All deficient nutrients were applied. Species were assessed as productive, non-productive, or non-persistent. Productive species were ranked according to dry-matter yield. Five grass species and five legume species were finally selected for further evaluation.

(b) *Assessment of African Clover introductions* ('t Mannetje 1964)
From initial nursery observations of 27 introductions, 12 introductions were selected for second stage sward trials in 4 × 10 m plots, with two replications. The *Trifolium* spp were inoculated and sown with *P. plicatulum*. After a one year establishment period plots were subjected to three heavy grazings. Persistence, dry-matter yield, and nitrogen content of legumes and associated grass was measured. Two perennial species (*T. semipilosum* and *T. burchellianum*) were selected as most promising for further trials.

(c) *Evaluation of Paspalum introductions* (Shaw, Ehlich, Haydock, and Waite 1965).
Fifty introductions were screened initially in nursery plots from which 18 introductions were selected for further testing in plots 2 × 2 m, with four replications. Plots received adequate nitrogen, phosphorus, and irrigation. Total and seasonal dry matter yields, persistence, and frost tolerance were measured. From this study four introductions were selected for further testing.

These examples demonstrate how a large number of introductions can be screened and the most promising species selected for final assessment. At the final stage the important objective is to assess the species potential in terms of animal production, as well as to examine its compatibility with associated grasses or legumes and to define special nutrient requirements. Since grazing trials are expensive to operate only the most promising species can be evaluated, and trials must be carefully designed.

When a new introduction, species, or variety has been tested under a range of environmental conditions and has demonstrated its value as a pasture plant it may then be released for commercial use. In Australia all new cultivars are released through the Herbage Plant Registration Authority, after being recorded in the Herbage Plant Register. The principal purposes of the Register are to record and preserve in authoritative descriptions, the origins and identification of herbage plant cultivars of economic importance or promise to Australian agriculture, and to obviate duplication and confusion in naming new cultivars. It aims to provide informative descriptions of promising new cultivars for all those concerned with their development and commercial utilization (Barnard 1972). A new cultivar must be different from, and possess some character of merit in comparison with previously registered cultivars; it must also have a reasonable degree of uniformity or genetic stability. The release of new cultivars is usually made on the advice and recommendations of State Herbage Plant Committees to the Registrar.

3.2 AGRONOMIC CHARACTERISTICS OF THE TROPICAL PASTURE SPECIES

The search for better tropical pasture species is progressing at an increasing rate, both through plant introduction programmes and through plant breeding and selection. Even so, there are only about 12 to 15 genera of tropical grasses which are widely used in sown pastures. However, most genera have more than one important sown species. The number of widely used legume genera is similar, and again some genera are represented by more than one important species.

Different genera and species demonstrate different broad adaptations to regional climatic and soil conditions. Within species, pasture agronomists and plant breeders have selected plant types which are better adapted to particular conditions in the general environment and these have been released as named cultivars. Most commonly cultivars differ in their time to flowering, but cultivars have been selected on the basis of adaptations to particular soil or climatic conditions, or on plant characters such as higher leaf production, higher dry-matter digestibility, better seed yields, and so on.

In this chapter the main aim is to describe the important features of the sown grasses and legumes at the genus and species level. The Australian cultivars within a species are listed, but detailed differences between cultivars are not discussed. These are best available in the authoritative descriptions in the Australian Herbage Plant Register (1967, 1969-) or in a more general way in Bogdan (1977), Humphreys (1974), or O'Reilly (1975). In Section 3.3 the important adaptations of the species to climate, soils, and stress conditions are analysed in some detail.

The tropical grasses

1. *Brachiaria* spp
This genus is a native of central and eastern Africa and is best adapted

to the humid tropics. Perennial and annual species occur with either tufted or creeping habits. Bogdan (1977) states that confusion exists in the literature regarding naming of *Brachiaria* spp. Other *Brachiaria* spp. have been described as *B. brizantha* and data from earlier publications should be treated with caution. Three species are important in sown pastures:

(i) *Brachiaria decumbens* (signal grass) cv. Basilisk

Signal grass is a perennial stoloniferous species, best adapted to the humid tropics with rainfall above 1250 mm, but has produced well in subtropical latitudes with a summer growing season with rainfall between 1000 and 1250 mm per annum. In cutting trials cv. Basilisk responded markedly to nitrogen fertilization in dry matter and protein content. Its annual dry matter production ($33\ 000\ \text{kg ha}^{-1}$) significantly exceeded that of pangola, para, and guinea grasses (Barnard 1972; McKay 1974*b*). Animal production on signal grass pastures has been good. In North Queensland at a stocking rate of 4.6 beasts ha^{-1} and 196 kg N $\text{ha}^{-1}\ \text{a}^{-1}$ liveweight gains of 1030 kg $\text{ha}^{-1}\ \text{a}^{-1}$ have been recorded. At 3.5 beasts ha^{-1} and 196 and 0 kg N $\text{ha}^{-1}\ \text{a}^{-1}$ liveweight gains were 740 and 592 $\text{ha}^{-1}\ \text{a}^{-1}$ (Harding, personal communication). Centro-signal grass pastures have maintained 30 per cent legume and produced 565 kg ha^{-1} liveweight gain at 4.8 yearlings ha^{-1} in Bali (Nitis, Rika, Supardjata, Nurbudhi, and Humphreys 1976). In the Solomon Islands 20 per cent legume was maintained giving 546 kg ha^{-1} liveweight gain at 4.5 beasts ha^{-1} (Watson and Whiteman, unpublished data).

Because of its vigorous growth, difficulty in maintaining a legume component may be experienced. *Desmodium heterophyllum* combines well with signal (Gutteridge, Whiteman, and Watson 1976; Bogdan 1977).

Signal grass produces good seed yields, particularly in the first year. Germination is low unless the seed is stored for one year or acid scarified.

(ii) *Brachiaria mutica* (para grass)

Para grass is a perennial stoloniferous grass widely grown in the humid tropics and subtropics with rainfall in excess of 1250 mm per annum. It is highly tolerant of waterlogged and flooded conditions. It forms a more open sward than signal and if not flooded combines well with legumes, particularly *Centrosema pubescens*. It is well accepted by cattle and may even be selectively grazed.

Bogdan (1977) reports data from Florida recording liveweight gains in steers reaching 900 kg ha^{-1}. No details of stocking rate or fertilization are quoted.

Centro-para grass pastures have given 640 kg live-weight gain (LWG) ha^{-1} at 4.5 beasts ha^{-1} in the Solomon Islands (Watson and Whiteman, unpublished data).

Para grass is normally established from runners. Although seed production is low seed supplies are commercially available.

(iii) *Brachiaria ruziziensis* (ruzi grass) cv. Kennedy

Ruzi grass is a leafy semiprostrate rhizomatous species adapted to humid frost free conditions. It is less vigorous than para or signal and combines well with legumes. It appears more shade tolerant than signal or para and may be better adapted for use in pastures under coconut plantations (Watson, personal communication). In this regard it appears to be similar to *Brachiaria brizantha* which has been widely recommended for use in coconut plantation pastures in Sri Lanka (Santhirasegaram and Ferdnandez 1967). Ruzi produces high seed yields, but seed requires storage or acid scarification to reduce seed coat impermeability for germination.

2. *Cenchrus* spp

This genus is native to the dry areas of East Africa. It is widely valued for its drought tolerance and high nutritive value. The members of this genus are tufted, often rhizomatous perennials or annuals. The twenty five species are distributed in the warm predominately tropical and subtropical areas, mostly in arid or semi arid areas (Bogdan 1977). Two species are important pasture plants:

(i) *Cenchrus ciliaris* (buffel grass) cvs Biloela, Molopo, Boorara, Lawes,
   Nunbank, Tarewinnabar, Gayndah, American, West Australian

Buffel grasses are tufted, rhizomatous perennials, rhizomes if present are short, stout, and not numerous (Bogdan 1977).

This species shows considerable drought tolerance. This is believed to be due to the deep rooting habit (Humphreys 1974; Bogdan 1977).

A number of cultivars are available suited to a variety of climatic and edaphic conditions (Barnard 1972). Buffel is best suited to the drier inland regions of the tropics and subtropics with rainfall between 350 and 900 mm per annum. Bogdan (1977) reports that in Kenya with two rainy periods growth occurs with an annual rainfall greater than 300 mm per annum.

Because of the relatively easy establishment and persistence of this species it is used for reseeding denuded or arid pastoral lands in Australia, India, Pakistan, and East Africa (Bogdan 1977).

In mixtures with legumes Buffel grass has been shown to be a vigorous competitor. Possible combinations with *Stylosanthes humilis* and *Macroptilium atropurpureum* are reported. 't Mannetje (1972) reported that a mixed sward of buffel and siratro produced equivalent liveweight gain of a monoculture buffel stand plus 168 kg ha^{-1} of nitrogen.

The cultivars have been grouped into three types (Humphreys 1974; Whiteman, Humphreys, and Monteith 1974; O'Reilly 1975):

(a) Tall (up to 1.5 m) rhizomatous types cvs: Biloela, Molopo, Boorara, Lawes, Nunbank, Tarewinnabar.

(b) Medium height (up to 1.0 m) non-rhizomatous types cvs: Gayndah, American

(c) Short (up to 0.75 m) non-rhizomatous types cv. West Australian

(ii) *Cenchrus setigerus* (birdwood grass)

This tufted perennial without rhizomes has marked drought resistance due to rapid maturity and seed production (6–8 weeks from seedling). It is best adapted to the arid and semi-arid tropical climates with a long dry season. Sandy, light textured soils are preferred. The annual rainfall should exceed 200 mm per annum. This species has been used successfully in the north of Western Australia to improve scalded and denuded pastures (Fitzgerald, personal communication). Humphreys (1974) reports that birdwood grass will combine well with *Stylosanthes humilis*.

3. *Chloris* sp

This genus of tufted, rhizomatous or stoloniferous perennials or annuals contains one important pasture species:

(i) *Chloris gayana* (Rhodes grass) cvs Pioneer, Katambora, Samford, Callide

This species is well adapted to subtropical climates with rainfall ranging from 650 to 1150 mm per annum. It is not well suited to high rainfall conditions.

Rhodes grass is often a pioneer grass in fallows, abandoned cultivations and along roadsides. It is a tufted, stoloniferous perennial growing to 1.5 m.

Major attributes of this grass are its ease of establishment, rapid ground cover, tolerance of high soil salinity, and heavy grazing.

Rhodes grass has been reported to combine with the following legume species: *Centrosema pubescens, Clittoria ternatea, Desmodium uncinatum, Glycine wightii, Macroptilium atropurpureum, Medicago sativa*, *Stylosanthes guianensis*, *Stylosanthes humilis*, and *Trifolium repens* (Bogdan 1977). Management procedures are important in maintaining the legume component in a Rhodes grass pasture.

The four cultivars provide a range of maturity, adaptation to a variety of soil conditions, and frost tolerance (Barnard 1972).

4. *Cynodon* spp

Considerable debate exists regarding the taxonomy of this species. Bogdan (1977) states that until recently reports of *Cynodon dactylon* in the tropics may have been mistaken for *C. nlemfuensis* or *C. aethiopus*. *C. plectostachyus* has also been confused with other *Cynodon* species.

(i) *Cynodon dactylon* (Bermuda grass, green couch)

This stoloniferous and rhizomatous perennial is found in tropical, subtropical, and warm temperate regions throughout the world but rare and perhaps absent in tropical Africa although common in southern and northern Africa (Bogdan 1977). It is a variable species and several botanical varieties have been described (Bogdan 1977). The species as known in subtropical Australia is a fine leafed creeping perennial spreading by both stolons and rhizomes. Green couch shows little cool season growth and is damaged by frost. Selections have been made with improved dry matter digestibility. Several hybrid cultivars of Bermuda grass have been developed in the United States.

(ii) *Cynodon plectostachyus* (African star grass)

African star grass is adapted to subtropical areas with rainfall of between 500 and 900 mm per annum. It is a tufted perennial species with stoloniferous habit and can form dense swards. High cyanide levels have been recorded but it is still used under grazing.

5. *Digitaria* spp.

Digitaria is a large genus with over 300 species of annual or perennial grasses, mainly tropical and subtropical but also of warm temperate areas (Bogdan 1977). Section *Erianthae* is almost exclusively of African origin from which the most important cultivated species *D. decumbens* (pangola grass) originates. Other important species in this genus that are being fully evaluated at this time are *D. pentzii*, *D. milanjiana*, *D. setivalva*, *D. smutsii*, and *D. valida*. These species grow in relatively dry parts of South and tropical Africa with an annual rainfall ranging from 500–1000 mm per annum with well-pronounced one or two dry seasons (Bogdan 1977). Strickland (1971) has collected a large range of material and is evaluating them in southern Queensland.

The most important species at this time is *D. decumbens*.

(i) *Digitaria decumbens* (pangola grass)

This species has become one of the most widely used improved pasture species in the humid and subtropics. Although best adapted to wet tropical conditions (> 1000 mm per annum) it has shown tolerance of colder, drier inland environments (Russell and Tothill 1974).

Pangola is a vigorous strongly stoloniferous perennial grass. No viable seed is produced thus propagation is by stolon cuttings. It shows good response to improved fertility and responds well to nitrogen. Yields of up to 44 tonnes dry matter per hectare per annum have been recorded at high levels of nitrogen (Vincente-Chandler, Silva, and Figarella 1959).

Pangola is well accepted by animals, and maintains its nutritive value when adequately grazed. Pangola maintains a higher sugar (fructosan) content in stems than most other tropical pasture grasses (Hunter, Mcintyre, and McIlroy 1970). Another factor contributing to higher levels of animal production on pangola grass swards may be related to the leaf density and high bulk density (kg dry matter per cm sward height per hectare). The highest levels of animal production recorded on pangola was 2760 kg liveweight gain ha^{-1} a^{-1} on irrigated swards with 670 kg N ha^{-1} and 12.4 animals ha^{-1} (Evans and Pulsford, personal communication).

In legume-based pasture systems the highest levels of production have been obtained with a pangola–*Desmodium intortum* pasture in Hawaii giving 1304 kg liveweight gain ha^{-1} a^{-1} at a stocking rate of 5.88 animals ha^{-1} (Younge and Pluncknett 1966). Pangola may combine well with *Lotononis bainesii* (Bryan 1961), and with *Centrosema pubescens*, *Pueraria phaseoloides*, *Desmodium heterophyllum*, and *Calopogonium mucunoides* in the tropics.

Productivity of pangola grass is markedly reduced in many areas of the world by pangola stunt virus. The rust *Puccinia ohauensis* reduces yield and nutritive value (Davis and Norton, personal communication).

(ii) *Digitaria pentzii*

This species shows promise as a grass adapted to similar regions as pangola. It has the advantage of seed propagation. Several accessions are being evaluated in Southern Queensland in comparative trials with *D. decumbens*. Results are promising (Peake, Strickland, and Hacker 1975).

6. *Melinis* spp

There are about 15–20 species in tropical and South Africa of which one important species is:

(i) *Melinis minutiflora* (molasses grass)

Molasses grass requires a frost free environment with good drainage. It requires rainfall in excess of 1000 mm per annum. Good growth is obtained from acid or poor, leached soils. It is a tufted perennial that will establish quickly, even on poorly prepared seedbeds. It is thus well suited as a pioneer species. In mixtures it combines with *Glycine wightii*, *Desmodium* spp, and *Macroptilium atropurpureum*.

Molasses grass is often sown in a mixture of grasses to provide a quick cover. The more competitive grasses then take over to form a stable pasture. This species is susceptible to fire and can be eradicated by burning.

7. *Panicum* spp

This is a very large genus with up to 500 species which occur in warm countries, mainly in the tropics, throughout the world (Bogdan 1977). Two major species are used widely in tropical and subtropical environments. Within each of these species are a number of subspecies and cultivars giving a wide range of types to suit many environments.

(i) *Panicum antidotale* (blue panic)

This species is a vigorous tufted perennial, indigenous in subtropical northern India, Pakistan, Afghanistan, and Iran, and also in southern tropical India (Bogdan 1977). It is not as important as a pasture plant as the other species.

The plant is very deeply rooted giving good drought tolerance and is adapted to the drier margins at 400–700 mm rainfall (Humphreys 1974). It is a vigorous summer grower capable of shooting quickly in response to rainfall. It is burnt by frost but survives through to the next growing period.

(ii) *Panicum coloratum* var. *makarikariense* (Makarikari grass)
  cvs Bambatsi, Pollock, Burnett

Makarikari grass is best adapted to subtropical regions with a rainfall between 500–1000 mm per annum. It is a robust tussocky perennial varying in habit from prostrate and actively stoloniferous to ascending and weakly stoloniferous. It is adapted to self mulching soils and has good resistance to frost so that it is grown in inland subtropical regions. In mixtures good results have been obtained from combinations of Makarikari grass and *Medicago sativa* and *Macroptilium atropurpureum*. The cultivars give a range of adaptation in terms of climate and soil requirements (Barnard 1972).

(iiia) *Panicum maximum*
Cultivars of *Panicum maximum* are well established throughout tropical countries where they play important roles in beef and dairy production. Australian experience suggests that a sensible agronomic and environmental grouping is to divide the commercially available *P. maximum* varieties into:

(1) the 'guineas' cvs: Riversdale, Hamil, Coloniao, Embu, Makueni, Coarse Guinea.
(2) the 'panics' cvs: Petrie, Sabi, Gatton.

The panics are suitable for parts of the subtropics or elevated moist tropics whilst the guineas are most productive on high rainfall tropical lowlands (McCosker and Teitzel 1975). The guineas perform well under high temperatures and in regions with more than 1300 mm rainfall per annum, whereas the panics have better drought tolerance and perform well in drier environments.

The cultivars available provide a choice of types suited to different conditions (Barnard 1972). Cv. Makueni was selected for higher yield and growth in the cool dry season of the wet tropics of Australia.

(b) *Panicum maximum* var. *trichoglume* (green panic) cv. Petrie
Green panic is adapted to a wide range of climatic conditions, rainfall between 635 and 1800 mm per annum. It is not as well suited to the wet tropical coast as cv. Riversdale. It is slightly frost tolerant with moderate drought tolerance with a marked response to rainfall. An important characteristic of green panic is its ability to tolerate shaded conditions. Green panic combines well with legumes such as *Macroptilium atropurpureum* and *Desmodium intortum*.

(c) *Panicum maximum* var. *typica* (common guinea) cv. Riversdale
Cv. Riversdale was selected from a wide range of 'common guinea' types in Queensland. The parent material was selected from a Department of Primary Industries substation in 1973 where guinea grass had been used in experiments since 1941. Selection of a pure line has allowed release of cv. Riversdale and made uniform seed commercially available. The selection of a uniform type for the export trade was important for continued seed sales. This cultivar shows best growth on tropical rainforest hillsides free of frost.

8. *Paspalum* spp
These plants are predominately perennials, tufted, stoloniferous, or

rhizomatous. Paspalum is a large genus of up to 250 species distributed in the tropics, subtropics, and, to a lesser extent, in warm temperate areas throughout the world but mainly in America (Bogdan 1977).

(i) *Paspalum commersonii* (scrobic) cv. Paltridge
This is a loosely tufted short lived perennial species adapted to areas with rainfall greater than 900 mm per annum. It is susceptible to frost and not a good competitor. It requires a good level of soil fertility and produces highly palatable and digestible forage. Scrobic combines well with *Macroptilium atropurpureum*.

(ii) *Paspalum dilatatum* (common paspalum)
This species is best adapted to humid subtropical areas receiving in excess of 900 mm rainfall per annum. It is valued mainly for its vigour, persistence, high herbage yields, the ability to withstand high grazing pressure, and a relatively high crude protein content (Bogdan 1977). Paspalum is frost tolerant with a strong underground rootstock giving drought tolerance.

In mixtures paspalum combines well with *Trifolium repens* in subtropical regions and has been grown with *Desmodium intortum*, *Centrosema pubescens*, and *Macroptilium atropurpureum*.

Paspalum grows best on heavier textured alluvial soils or red loams. It has high fertility requirements, and if fertility is not maintained the pasture is invaded by inferior grasses and weeds (Humphreys 1974). Paspalum is affected by an ergot, and damage to stock can result from ingestion of ergot infected heads.

(iii) *Paspalum notatum* (bahia grass)
A rhizomatous, stoloniferous, tufted perennial adapted to the subtropics and tropics. Bogdan (1977) records three subspecies. Extensive research into this species has been carried out in Florida, with the development of cv. Pensacola.

Bahia grass forms a dense cover and is valued for its productivity, relative ease of establishment, and persistence (Bogdan 1977). Being an aggressive species stable mixtures are difficult to maintain. *Trifolium repens* and *Lotononis bainesii* have been tried as mixtures with bahia grass. Bahia grass is free from ergot which is an advantage over common paspalum.

(iv) *Paspalum plicatulum* (plicatulum) cv. Rodd's Bay, Bryan
This species is well adapted to the wetter coastal regions receiving in excess of 750 mm rainfall per annum. It is an erect tufted perennial apparently well adapted to low fertility soils, waterlogging, and can also

withstand extended moisture stress. An earlier released cultivar Hartley is at present unavailable commercially (Loch 1976). Rodd's Bay has good seed production which is strongly synchronized. In mixtures plicatulum has been successfully combined with *Macroptilium atropurpureum, Desmodium uncinatum*, and *Stylosanthes guianensis.*

Cattle production from legume–plicatulum mixtures has not been as high as three other grass–legume combinations (Bryan 1968). This was related to inferior dry-season performance when compared with signal, hamil, and pangola (Halim and Whiteman, unpublished).

9. *Pennisetum* spp

This genus contains annuals and perennials, tufted or creeping, often with branched stems. Two important pasture species are:

(i) *Pennisetum clandestinum* (kikuyu grass) cvs Whittet, Breakwell

Kikuyu is a low growing perennial species bearing numerous long rhizomes and stolons. It thrives on the basaltic tablelands throughout the tropics but is restricted to the substropics if grown at low elevation. Kikuyu is a native of highlands and mountains of tropical East Africa where it occurs at altitudes ranging from 1500 to almost 3000 m (Bogdan 1977). It requires a well distributed annual rainfall above 1000 mm for good levels of production. This species is a palatable and nutritious grass able to withstand moderate frosting.

Until recently propagation by cuttings was widely practised but now seeding forms are available (cv. Whittet, Breakwell). The production and harvesting of kikuyu seed require special techniques (Wilson, Dale, and Spurrs 1975). High fertility conditions are required for stable pastures. Response to nitrogen fertilizer is good. In mixtures kikuyu has been shown to combine well with *Trifolium repens* (Coleman 1975). Careful management is required to maintain legumes in this vigorously growing species.

Other legumes that can combine with kikuyu are *Desmodium* spp, *Glycine wightii, Macroptilium atropurpureum*, and *Trifolium semipilosum*. Persistence in the sward is a problem unless management procedures are good (Coleman 1975). In a four-year experiment in northern New South Wales liveweight gain varied from 551 to 1624 kg ha^{-1} depending on the nitrogen rates and stocking rate. The nitrogen rates were 0, 134, 336, and 672 kg N ha^{-1} a^{-1} (Mears 1975). High stocking rates and intensive utilization are necessary for full utilization of this species. Kikuyu pastures are widely used in dairy production in subtropical and elevated tropical areas of Australia.

(ii) *Pennisetum purpureum* (elephant grass, napier grass) cv. Capricorn. This species is a robust perennial forming large broad clumps, spreading by stem bases rooting from nodes or by short rhizomes. Cv. Capricorn is shorter (1.2–2.4 m) and smaller in all its parts than the more robust forms of this species. It gives little growth in dry or cold conditions. Elephant grass is well suited to the wet tropical coast (up to 2500 mm rainfall per annum). Elephant grass is usually grown in pure stands and cut and hand-fed but cultivation in mixtures with *Pueraria phaseoloides* had good success in Belize and the West Indies (Bogdan 1977).

Bogdan (1977) records several grazing experiments where good liveweight gains from elephant grass pastures have been obtained; for example in Puerto Rico on heavily fertilized elephant grass, liveweight gains of up to 1200 kg ha^{-1} were recorded.

10. *Setaria* spp

This genus of annuals and perennials are mostly tufted but sometimes rhizomatous. It is a large genus of 140 species, mostly tropical, but with some in temperate regions (Bogdan 1977). The most important species as a pasture plant is:

(i) *Setaria anceps* (Setaria) cvs Nandi, Kazungula, Narok
This densely tufted perennial has both very short rhizomes and stolons. It is well adapted to subtropical regions with rainfall exceeding 750 mm per annum.

Setaria is relatively frost tolerant. Grazing must be heavy to maintain vegetative growth and maintain palatability. Some forms of the species have high oxalate levels which may affect animal production. Cv. Nandi has relatively low oxalate levels, while cv. Kazungula is relatively high.

In mixtures the following legumes have been grown with setaria: *Desmodium intortum*, *Macroptilium atropurpureum*, *Glycine wightii*, *Trifolium repens*, *Macroptilium lathyroides*, *Trifolium semipilosum*, and *Desmodium uncinatum*. The slow establishment of this species allows good legume establishment in the initial years of the pasture. After several years *Setaria* becomes more aggressive and management becomes increasingly important in maintaining adequate legume content (Bogdan 1977). Cv. Kazungula is a more vigorous plant and thus it is more difficult to maintain legume in the pasture.

11. *Sorghum* spp

The genus contains tall annuals or perennials which includes both pasture grasses and grain sorghum. The important pasture species is:

(i) *Sorghum almum* (Sorghum almum) cv. Crooble
This species is an erect, robust, tussocky short-lived perennial with numerous tillers and thick short rhizomes. It produces well in the 460–760 mm annual rainfall regions. Cv. Crooble is a good pioneer species with drought and salt tolerance. In mixtures a number of legumes have been tried but the stability of the pasture is not good. Legumes tested include: *Medicago sativa, Glycine wightii, Vigna* spp, and *Stylosanthes humilis*. Some problems for grazing stock on young growth have been recorded with the high levels of cyanide.

12. *Urochloa* spp
This genus of African and tropical Asian origin includes approximately 25 species of which *U. mosambicensis* is the most important pasture species.

(i) *Urochloa mosambicensis* (sabi grass) cv. Nixon
This creeping stoloniferous perennial has a low growth habit. It is drought-resistant and produces well in the 400–800 mm annual rainfall region. The species has low frost tolerance and cannot withstand water-logging.

In mixtures it has been shown to combine well with *Stylosanthes humilis*. Whiteman and Gillard (1971) report grass yields in mixtures ranging from 830 to 6520 kg dry matter ha^{-1} and those for the legume from 25 to 3438 kg dry matter ha^{-1}. Cv. Nixon is best adapted to tropical monsoon conditions with a rainfall of 600–1200 mm and a dry season of 5–9 months (McKay 1974*a*). Grazing trials have shown it to have a high carrying capacity in the Northern Territory where under set grazing at a stocking rate of one beast to 0.3 to 0.6 ha, cv. Nixon plus *Stylosanthes humilis* have given up to 450 kg liveweight gain per ha per annum (Austin 1970).

The tropical legumes

1. *Calopogonium* spp
This small genus originates in the tropics and subtropical regions of Central and South America. The one important pasture species is *C. mucunoides*.

(i) *Calopogonium mucunoides* (calopo) (cowpea)*
Calopo is a short lived perennial species densely covered with hairs on

---

*The terms 'cowpea' or 'specific' will be used to describe whether the species has a specified rhizobium requirement or can be inoculated with the general cowpea type *Rhizobium* such as CB756.

the leaves and stems. It is an extremely vigorous species that will climb on any available support. This feature has been utilized to control woody weeds in newly cleared land. This species is best adapted to the humid tropics with an annual rainfall exceeding 1250 mm. Seed production is good and plants establish from seeds annually.

Initial evaluation of this species indicated poor palatability possibly related to hairiness, but more recent work (Parker, personal communication) has shown the importance of the species as a pioneer legume in the Northern Territory at Adelaide River near Darwin, and its ready acceptance by cattle during the dry season.

Mixtures with calopo have been reported but the aggressive nature of this species appears to limit its use as a stable pasture component. Bogdan (1977) reports mixtures with *Melinis minutiflora*, *Chloris gayana*, and *Digitaria decumbens*.

2. *Centrosema* spp

This genus of 40-50 species is native to the American tropics. All species are climbing or prostrate herbs or subshrubs (Bogdan 1977). The most important species is *C. pubescens*.

(i) *Centrosema pubescens* (centro) cv. Common, Belalto (specific)

This species is one of the most widely distributed of the tropical legumes native to Central and South America (Teitzel and Burt 1976). For many years cv. Common was the most productive commercially available type and cv. Belalto (from Costa Rica) was the result of a plant collecting and testing programme aimed specifically at finding legumes with improved cool season growth (Teitzel and Burt 1976).

Centro is very well adapted to wet tropical coastal situations and shows relatively little growth when temperatures were reduced from 32 °C day/24 °C night to 26 °C day/15 °C night ('t Mannetje and Pritchard 1974). Frost causes severe damage to plant tissues. Grof and Harding (1970a) have shown the improved cool season growth of cv. Belalto. Centro has become one of the most popular tropical fodder legumes highly valued for its vigorous and productive growth, good quality of herbage, and the ability to form mixtures with grasses (Bogdan 1977).

Centro is very slow to establish and the land preparation must be thorough to produce a vigorous stand. Seed harvesting by mechanical means is difficult and yields can be quite low. Centro combines well with tufted tropical grasses (Humphreys 1974) particularly guinea grass. Mixtures with elephant, pangola, and para grasses have also been successful. Centro is moderately palatable and can withstand heavy grazing (up to 4 beasts ha^{-1}) after establishment (Grof and Harding 1970*b*).

Animal production from guinea grass–centro pastures in northern Queensland has been as high as 928 kg ha^{-1} a^{-1} in the first year. This production became relatively stable at 560 kg ha^{-1} a^{-1} after three years. The reduction was apparently related to declining nitrogen status possibly due to deficiencies of phosphorus and molybdenum in the soil reducing fixation of atmospheric nitrogen (Teitzel and Burt 1976).

3. *Desmodium* spp

Desmodium is a large genus of approximately 300 species which range from perennial, rarely annual, herbs, subshrubs, or shrubs to plants of a rambling or prostrate habit. Most species are tropical or subtropical but temperate members are known. The majority of species are native to Mexico and Brazil (Bogdan 1977). Three species provide important pasture legumes:

(i) *Desmodium heterophyllum* (hetero) cv. Johnstone (specific)
Hetero is a creeping perennial, native to south Asia and the islands of the Indian Ocean. It is well suited to the humid wet tropical coast where rainfall exceeds 1500 mm per annum. Hetero is particularly resistant to heavy grazing and has formed stable mixtures with the stoloniferous tropical grasses *Digitaria decumbens* and *Brachiaria decumbens*.

This species does produce viable seed but seed harvesting is difficult as flowering occurs over a prolonged period and the seed is formed close to the ground. Vegetative propagation techniques are used successfully. Animal production on *Brachiaria decumbens*–hetero pastures has been over 750 kg liveweight ha^{-1} a^{-1} (McKay 1973).

(ii) *Desmodium intortum* (Greenleaf Desmodium) cv. Greenleaf
    (specific)
This species occurs widely throughout Central and South America, generally above 1500 m altitude in the tropics and down to sea level in the subtropics (Bogdan 1977). The species is summer growing and is best adapted to areas with an annual rainfall exceeding 900 mm. It is a vigorous stoloniferous legume, forming a dense mat of stolons. It is frost-susceptible and also susceptible to *Amnemus* root weevil attack (Humphreys 1974). Greenleaf has combined well with a number of grass species, for example: *Setaria anceps* cv. Nandi, *Panicum maximum*, *Melinis minutiflora*, *Panicum maximum* var. *trichoglume*, *Paspalum*

plicatulum, Digitaria decumbens, Brachiaria mutica, and *Pennisetum clandestinum*.

Animal production on pastures with Greenleaf as a component have been up to 256 kg ha^{-1} a^{-1} at a stocking rate of 2.12 head ha^{-1} (Jones 1974*b*). Bogdan (1977) suggests this stocking rate may be too high for maximum production and suggests 1 head ha^{-1} may be more appropriate. Definitive evidence is unavailable. Bryan (1969) has presented a detailed review of this species and *Desmodium uncinatum*.

(iii) *Desmodium uncinatum* cv. Silverleaf (specific)
This species is a native of South and Central America in the warm lowlands up to 1000 m altitude (Bogdan 1977). It is a large trailing perennial which roots at the nodes on sandy and other friable soils. It is a summer grower in areas receiving more than 900 mm annual rainfall. Although growth is restricted by low temperatures (Whiteman 1968) and plant tops may be killed by frost, survival after frosting is noted by Jones (1969). This species prefers well drained sandy soils or loams of low salinity but can withstand short-term waterlogging.

Silverleaf is valued for its vigour, persistence, high herbage and seed yields, and relatively easy establishment, but it is susceptible to *Amnemus* root weevil attack. In mixtures silverleaf has been combined with a number of summer growing grasses; from sward forming types such as *Paspalum notatum* to erect types such as *Setaria anceps* and *Paspalum plicatulum*. Other species tried in mixture with silverleaf are: *Hyparrhenia* (in Kenya), *Chloris gayana, Panicum maximum, P. coloratum, Digitaria decumbens, Paspalum commersonii, P. dilatatum*, and *Pennisetum clandestinum*.

Seed pods of the *Desmodium* species are covered with dense, hooked hairs which aid spread of this species (Humphreys 1974). Liveweight gains reported in Bryan (1969) range up to 340 kg ha^{-1} for silverleaf–grass mixtures.

4. *Glycine* spp
This genus contains twining or procumbent herbs, except for soyabean (*G. max*) which is an annual oilseed and protein grain crop. It is a small genus of about ten species found in the East Indies, tropical Asia, and tropical East Africa (Humphreys 1974). The important pasture species is *G. wightii*.

(i) *Glycine wightii* (glycine) cvs Tinaroo, Cooper, Clarence, Malawi
   (cowpea)
This species is a climbing, trailing, or procumbent perennial herb. It is

well adapted to cooler, elevated parts of the tropics or subtropics than many other tropical legumes. It can be killed by severe frosting.

G. wightii is valued for its reasonable drought resistance, good yields of herbage, an ability to mix well with grasses, and reasonable persistence and longevity (Bogdan 1977). However it requires well-drained fertile soils, and is not as tolerant of low pH and low nutrient status as many other tropical legumes.

Glycine is best suited to areas receiving above 750 mm rainfall annually but is not well suited to the true wet tropics. It seems to do best on the red soils of volcanic uplands.

In mixtures Glycine has been shown to combine well with *Panicum maximum* cv. Petrie, *Chloris gayana, Cenchrus ciliaris, Setaria anceps, Melinis minutiflora, Digitaria decumbens*, and *Pennisetum clandestinum*.

Slow initial growth in this species may be a problem when sown in mixtures as the other components may provide too much competition for a good establishment of Glycine. A range of cultivars are available for a variety of climatic and edaphic conditions (Barnard 1972).

5. *Lablab* spp
A monospecific genus, closely related to *Dolichos*.

(i) *Lablab purpureus* (lablab) cvs Rongai, Highworth (cowpea)
This species is mainly used as an annual forage or green manure crop in Australia and does not form part of stable pasture mixtures. It is a summer growing, rampant, and vigorously twining herbaceous annual or short-lived perennial. It is moderately drought-resistant and can grow with as little as 500 mm annual rainfall, but best growth occurs under rainfall exceeding 1000 mm annual rainfall.

Lablab has replaced cowpea in some areas as a cover crop. It is more cold tolerant, more disease resistant, less prone to insect damage, and has some perenniality which can provide some grazing in spring if carefully managed (Humphreys 1974).

6. *Leucaena* spp
This genus of 50 species occurs almost exclusively in tropical America (Bogdan 1977). The important pasture plant is *L. leucocephala*.

(i) *Leucaena leucocephala* (leucaena) cvs Peru, El Salvador (specific)
Leucaena is a deep rooted perennial shrub or tree, and is native to Central and South America. This species is best suited to well drained neutral to alkaline soils in areas receiving more than 750 mm rainfall annually.

The plant can be grazed directly and is quite resistant to heavy frequent defoliations. Its height affords some measure of frost avoidance. The leaves and small stems are particularly well accepted by stock.

Some accessions of this species have high levels of the alkaloid mimosine in their foliage which can cause detrimental effects in monogastric animals. A breeding programme has produced low mimosine cultivars. High mimosine cultivars can cause problems in the reproductive system of cows if leucaena is the sole source of feed. Hardseededness is common in this species, this can be overcome by hot-water treatment.

When leucaena constitutes a major portion of the diet Jones, Blunt, and Holmes (1976) recorded reduced levels of animal production and enlarged thyroid glands in cattle. Leucaena must be grazed cautiously. A two-paddock rotational grazing system is recommended ('t Mannetje, Jones, and Stobbs 1976).

7. *Lotononis* spp

This genus of approximately 100 species is predominately South African in origin. They are prostrate or suberect annual or perennial herbs. The most important pasture species is *L. bainesii*.

(i) *Lotononis bainesii* (lotononis) cv. Miles (specific)

This species is a glabrous perennial herb with creeping stems that root freely at the nodes. It is well suited to areas receiving more than 875 mm rainfall per annum in the subtropics and tropics. Lotononis is especially frost tolerant, being one of the most tolerant of the commercially available tropical legumes. A further advantage of lotononis is the ability to grow on acid soils of poor drainage. It performs well on better drained soils and responds to phosphorus and potassium.

In mixtures lotononis is one of the few species to persist well in *Digitaria decumbens* pastures. It has been grown with a range of other species including: *Chloris gayana, Paspalum dilatatum, Paspalum commersonii, Paspalum plicatulum, Paspalum notatum*. The management of pastures containing lotononis is not fully understood if lotononis is to be a stable component in the mixture. Lotononis appears to be extremely productive in some years but the yield can be very low in other years. Considerable research effort has been commenced to improve the management techniques of lotononis pastures (Pott, personal communication). Lotononis provides high quality feed for stock with unusually high digestibility.

8. *Macroptilium* spp

A genus of two species closely related to *Phaseolus*. The genus is native of central America. Both species are utilized in pasture production.

(i) *Macroptilium atropurpureum* (siratro) cv. Siratro (cowpea)

A perennial species with trailing or creeping stolons rooting at the nodes, siratro was bred by Dr E. M. Hutton of CSIRO, Brisbane in the early 1960s and has since been widely grown throughout the subtropics and tropics. Siratro is a summer growing species adapted to the warm subtropics and tropics. It performs best in areas receiving more than 750 mm rainfall anually.

Siratro foliage is sensitive to frost leading to leaf fall. Severe frosts will kill the stolons but whole stands are not normally killed (Shaw and Whiteman 1977). It is not adapted to the wet tropics where productivity and survival is greatly affected by *Rhizoctonia* leaf rot, and other various root rots.

Ease of establishment of productive pastures has helped siratro reach its important place in tropical pasture expansion. It is widely used in Brazil and Mexico as well as many other countries.

Siratro is adapted to a wide range of soils with respect to texture and pH. It does particularly well on many sandy soils and can succeed on quite shallow soils. Problems have occurred on poorly drained soils (Shaw and Whiteman 1977). A continuing breeding programme will provide improved cultivars of this species better adapted to a variety of environments and resistant to diseases of siratro (Hutton and Beall 1977). Seed-production technology over recent years for siratro has improved with the development of mechanical and suction harvesters allowing good seed recovery (Hopkinson 1977). Hardseededness occurs in siratro but can be overcome with scarification. In mixtures siratro is compatible with a wide range of grasses. Success has been recorded with the following: *Chloris gayana, Setaria anceps, Paspalum dilatatum, P. commersonii, P. plicatulum, Panicum maximum* var. *trichoglume*, and others.

Trials in Queensland, Uganda, and Fiji have shown that the inclusion of siratro into native or improved pastures greatly increases liveweight gain, improves breeder performance, and increases weaning rate (Walker 1977). Stocking rate is the main management factor affecting liveweight gain on fertilized siratro pastures, due largely to its effect on siratro yields. Walker (1977) reviews productivity of siratro pastures and reports highest weight gains of 163 to 202 kg head^{-1} a^{-1} over a range of sites where rainfall, soil, associated grasses, breed and sex of cattle varied.

An excellent series of papers on siratro production are contained in a special edition of *Tropical Grasslands* (volume **11**(1), 1977).

(ii) *Macroptilium lathyroides* (phasey bean) cv. Murray (cowpea)
This species is an annual, biennial, or short-lived perennial herb (Bogdan 1977). The plant originated in tropical South America and is adapted to areas with rainfall in the range 600–2000 mm per annum. An important characteristic of phasey bean is its ability to persist and produce under extremely waterlogged conditions.

Phasey bean has been replaced by the newly introduced legume species but is still widely distributed throughout Queensland. Commercial seed supplies are scarce. Seed of phasey bean can contain a high degree of hardseededness which can be readily broken by scarification or acid treatment. This species is often used for pot trials because of ease of establishment and rapid growth.

In mixtures phasey bean has combined with *Paspalum commersonii* to provide good animal production. Other species tried in mixtures have been *Paspalum dilatatum*, *Panicum maximum* var. *trichoglume*, *Chloris gayana*, *Setaria anceps*, and *Digitaria decumbens*. Other legumes are generally included in pasture mixes with phasey bean as a pioneer species. It has been widely sown with para grass (*Brachiaria mutica*) in areas subject to flooding and waterlogging.

9. *Macrotyloma* spp
This genus, closely related to *Dolichos*, contains about 25 species distributed mainly in Africa and Asia. Two species are used as pasture plants:

(i) *Macrotyloma axillare* (Archer axillaris) cv. Archer (cowpea)
A trailing and twining perennial herb indigenous in tropical Africa, Madagascar, Mauritius, and Sri Lanka. It is best adapted to areas where the rainfall exceeds 1000 mm per annum. It has a marked drought tolerance but cannot tolerate frosting or waterlogging. In spring whilst many other legumes cannot respond to slowly increasing temperatures and moisture conditions cv. Archer will produce good vegetative yields (Barnard 1972). In mixtures this species combines well with the tall growing tropical grasses. Palatability is considered to be low but once cattle become accustomed to the pasture it is readily consumed. The resistance of this species to pests and diseases that affect many other genera is considered important by Humphreys (1974).

(ii) *Macrotyluma uniflorum* (horsegram) cv. Leichhardt (cowpea)
This species is also a native of tropical Africa and Asia and is a twining

annual or perennial herb forming a dense canopy. It is a pulse legume of some importance in India (Rachie and Roberts 1973).

The cv. Leichhardt is a short-season, summer-growing plant best suited to subtropical and tropical areas from 600 to 1150 mm annual rainfall (Barnard 1972).

This species has a good degree of drought tolerance. It cannot withstand waterlogging. Horsegram will combine with the tall growing tropical species but detailed information as to productivity of mixed swards is lacking.

10. *Medicago* spp

A native of the Mediterranean area, the most important species is *M. sativa*.

(i) *Medicago sativa* (lucerne) cvs. Hunter River, Du Puits, African,
    Siro Peruvian, Paravivo, Demnat (specific)

This species is an extremely important perennial fodder plant used for conserved fodder in many regions of the world. Lucerne has been used as a component in a pasture mixture with some success. It is an erect perennial with an extremely deep root system and a low-set crown. These characters make lucerne extremely resistant to drought and to high stocking rates. In pure stands lucerne is generally rotationally strip-grazed for maximum utilization. This species is grown most widely in the warm temperate regions but is also grown in subtropical areas. It is more frost-tolerant than many of the tropical legumes and because of its deep-rooted habit can grow in areas with as little as 500 mm annual rainfall. The plant is intolerant of waterlogging and low pH and therefore requires well drained soils. It is responsive to fertilizer applications, particularly phosphorus, sulphur, and potash (Humphreys 1974).

In pasture mixtures lucerne combines well with *Panicum maximum* var. *makarikariense*, *P. maximum* var. *trichoglume* and *Sorghum almum*. *Cenchrus ciliaris* and *Chloris gayana* have been tried with lucerne but good management is required to maintain the lucerne component in these vigorous grasses. Rotational grazing of lucerne based pastures is essential to maintain the legume component. Set stocking, especially early in the growth stage, will eliminate lucerne in the pasture. A number of cultivars are available adapted to a range of environments. The potential of interspecific crosses of *Medicago* to produce a creeping form that will root at the nodes is being investigated.

11. *Peuraria* spp
This genus contains about twenty species of robust, vigorous climbers indigenous to tropical Asia and Pacific Islands.

(i) *Pueraria phaseoloides* (puero) (cowpea)
Puero is an extremely vigorous climbing species with long runners. It is densely covered with hairs. Puero is a summer growing species adapted to the humid tropics receiving annual rainfall between 1200 and 1500 mm. It is relatively drought-resistant. It is often used as a pioneering legume or as a cover crop in the wetter tropics. The plant is highly palatable and in this character is superior to *Calopogonium mucunoides*. The relatively high palatability may lead to selective grazing and it is thus not as persistent as calopo.

Several grass species have been tried in mixtures with puero including *Brachiaria mutica*, *Panimum maximum*, *Pennisetum purpureum*, *Melinis minutiflora*, *Hyparrhenia rufa*, and *Digitaria decumbens* (Bogdan 1977).

12. *Stylosanthes* spp
This genus is relatively small with about 25 species distributed throughout the tropics. The genus contributes about one third of all commercial varieties suitable for tropical and subtropical areas. It is the only genus to contribute species of widespread use in both the wet and dry tropics (Burt and Miller 1975). Burt and Miller (1975) have provided an excellent review titled 'Stylosanthes — *a source of pasture legumes*'.

(i) *Stylosanthes guianensis* (stylo) cvs. Schofield, Cook, Endeavour,
    Oxley (cowpea except cv. Oxley which has a specific requirement)
A herbaceous erect perennial species containing a wide range of markedly different forms coming from different climatic regions. Due to this difference in forms the cultivars will be considered separately.

(a) Cv. Schofield   This cultivar is well adapted to the wet coastal climates of the world. It requires frost-free conditions and is highly productive under a long growing-period. Schofield is well adapted as a pioneer legume due to easy establishment and high first-year yields. It is not well adapted to continuous heavy grazing.

(b) Cv. Cook   This cultivar is similar to cv. Schofield but flowers earlier and competes better with companion grasses and weeds. It makes relatively good cool season growth.

(c) Cv. Endeavour   This cultivar is intermediate between cv. Cook and cv. Schofield and shows better early summer growth than Schofield.

(d) Cv. Oxley (fine stemmed stylo)   This cultivar is adapted to the drier regions of the subcoastal subtropical regions. It can withstand fires, frosts, and drought but is intolerant of waterlogging (Burt and Miller 1975). It requires a specific strain of *Rhizobium* for effective nodulation. The adaptation to the stresses listed above may be related to the presence of an underground crown.

Grasses grown successfully with stylos include: *Digitaria decumbens, D. smutsii, Chloris gayana, Cenchrus ciliaris, Melinis minutiflora, Setaria anceps, Andropogon gayanus, Heteropogon contortus, Hyparrhenia rufa*, and *Panicum maximum*. Animal production from pastures containing stylo is greater than from pure grass stands (Bogdan 1977).

(ii) *Stylosanthes hamata* (Caribbean stylo) cv. Verano (cowpea)
This species is adapted to the dry tropical areas acting as a perennial or an annual–perennial mixture. It is a much branched herb occurring naturally in South and Central America. Cv. Verano resembles *Stylosanthes humilis* in its general appearance, ability to produce large amounts of seed, and phosphorus response. Verano differs markedly from 'typical' *S. hamata* and Burt and Miller (1975) suggest that it may be a hybrid between *S. hamata* and *S. humilis*.

This cultivar has only recently been released and as yet details of animal production and persistence are not available. Preliminary results show that this is a plant with considerable potential to replace *S. humilis* in many situations.

(iii) *Stylosanthes humilis* (Townsville stylo) cvs Gordon, Lawson,
    Patterson (cowpea)
This species is an erect or subprostrate annual occurring naturally in Central America, Venezuela, Brazil, and also the Ivory Coast of Africa. This apparently chance introduction to Australia in the early 1900s has become one of the most widely used pasture legumes in Queensland, Northern Territory, and in the north of Western Australia. It is well adapted to infertile sandy soils in seasonally dry tropical and subtropical regions. It has a good response to applied phosphorus. The ease of establishment of this species has led to its spread with minimum tillage operations, e.g. aerial seeding, after burning, or by direct sowing in roughly prepared seedbeds.

Within the species there is a range of flowering times which lead to the selection of the three cultivars to ensure that flowering would occur at the time best suited for plant survival in a range of environments. The cv. Patterson is the early flowering type and is best suited to the low rainfall regions (500–875 mm per annum). The midseason flowering cv. Lawson is suited to areas in the rainfall range 875–1100 mm per annum. The latest flowering cultivar is cv. Gordon which is well adapted to areas with a distinct dry season and an annual rainfall exceeding 1100 mm.

This species has a high proportion of hard seeds but this is overcome by storage or dehulling of the seed. However, for extensive sowing hardseededness is an advantage providing a reserve against premature germination. Compared to native pastures, Townsville stylo fertilized with superphosphate improves the carrying capacity, liveweight gain, and lengthens the period of weight gain (Bond 1974).

In mixtures with grasses Townsville stylo is often oversown with native grasses to form productive pastures with applied superphosphate. *Urochloa mosambicensis* has been shown to be a good companion grass for Townsville stylo in the Northern Territory of Australia. In Queensland good production has been obtained from Townsville stylo oversown into *Heteropogon contortus* grassland.

(iv) *Stylosanthes scabra* (shrubby stylo) cv. Seca (cowpea)

Several species forms are now known, different forms being adapted to different types of dry tropical climates (600–1700 mm rainfall). One of the forms thrives on areas generally regarded as being too dry for Townsville stylo. It is a highly persistent shrubby perennial plant (Burt and Miller 1975). Cv. Seca is a new release and little experimental work has been completed.

It forms vigorous legume dominant stands with *Brachiaria decumbens* and *Urochloa mosambicensis* in high and low rainfall environments respectively. It is readily grazed by cattle. Seca is a strong more drought-resistant perennial and is higher yielding than *S. hamata* cv. Verano in regions with low and variable annual rainfall.

(v) *Stylosanthes subsericia*

A new species that is considered to have potential in drier areas.

13. *Trifolium* spp

The genus of annuals or perennials contains about 300 species distributed mainly in temperate countries, particularly in the Mediterranean area. Two species are important in the subtropics and elevated tropics as pasture species.

(i) *Trifolium repens* (white clover) cvs. Ladino, Loussiana, Haifa, Grasslands Huia, Irrigation (specific)

A creeping perennial species which grows naturally in the temperate

Table 3.1. Summary of the agronomic adaptations of the main commercially sown tropical grasses sown in Australia

| Tropical grasses | Common name | Cultivar | Minimum * | | Sowing rate (kg ha^{-1}) |
			% purity	% germination	
Brachiaria decumbens	signal grass	Basilisk	50	15	3–6
B. mutica	para grass	–	40	15	2–5
Cenchrus ciliaris	buffel grass	Biloela, Molopo, Boorara, Nunbank Gayndah, American, West Australian	90	20	1–4
C. setigerus	birdwood grass	–	90	20	1–4
Chloris gayana	Rhodes grass	Pioneer, Katambora, Callide, Samford	90	20	1–6
Cynodon dactylon	Bermuda couch	–	97	60	1–3
Melinis minutiflora	molasses grass	–	40	30	1–4
Panicum coloratum	Makarikari grass	Bambatsi, Pollock, Burnett	80	20	2–3
P. maximum	guinea grass	Hamil, Coloniao, Makueni, Gatton, Riversdale	40	25	2–6
P. maximum	green panic var. trichoglume	Petrie	70	20	1–6
Paspalum dilatatum	paspalum, (dallis grass)	–	60	60	6–10

*The minimum purity and germination values are the standards defined in the Queensland Government Seed Regulations.

	Tolerance to		Minimum rainfall (mm)	Soil preference	Comments
Frost	Drought	Water-logging			
Poor	Fair	Good	1000	Versatile	Good ground cover, prefer to pangola and guinea in high rainfall areas
Poor	Fair	Very good	1250	Versatile. Good on heavy types	Performs well under water-logged conditions. High production on coastal lowlands
Fair	Very good	Poor	350	Light–medium textured	Drought resistant. Cultivars adapted to wide range of conditions, including drier inland environment
Fair	Very good	Poor	200	Sandy, light textured	Used in short season environment. Can be used as erosion control species
Fair	Good	Fair	650	Tolerates saline conditions	Widely used in Brigalow scrub lands. Easy to establish, giving quick cover
Fair	Good	Fair	600	Versatile	Restricted growth in dry seasons. Widely used for lawn establishment
Poor	Fair	Poor	1000	Poor, well drained	Pioneer grass in high rainfall areas. Carries fire
Good	Very good	Good	500	Self-mulching clay	Cool season, productivity good
Poor	Fair	Fair	1300	Versatile	Well adapted to high rainfall tropical lowlands. Robust, erect
Good	Good	Poor	600	Versatile	Palatable, shade tolerant. Combines well with siratro and greenleaf desmodium
Good	Fair	Good	900	Heavy texture	Best in subtropics, combines well with clovers

(continued)

Table 3.1. (*cont'd*)

| Tropical grasses | Common name | Cultivar | Minimum | | Sowing rate (kg ha^{-1}) |
			% purity	% germi- nation	
Paspalum notatum	bahia grass	–	60	60	2–5
P. plicatulum	plicatulum	Rodd's Bay Bryan, Hartley	60	40	2–4
Pennisetum clandestinum	kikuyu	Whittet, Breakwell	93	60	1–2
Setaria anceps	setaria	Nandi, Kazungula Narok	60	20	2–5
Sorghum almum	columbus grass	Crooble	97.3	65	1–10
Urochloa mosambicensis	sabi grass	Nixon	60	3	1–6

zones of the world. In the subtropics it is useful under raingrown conditions in the moist valleys receiving moderate spring and autumn rainfall. It can also be grown on the elevated tablelands of the tropics where temperatures are moderated by altitude.

Spring is the major growth period and the lush growth can cause bloat problems in stock. It is not drought tolerant and rarely grows in areas receiving less than 875 mm in the subtropics (Humphreys 1974). The high quality herbage produced by white clover has led to renewed interest in this plant for the subtropics, particularly when management and digestibility problems were delineated in the tropical species.

In mixtures white clover combines well with *Pennisetum clandestinum* and *Paspalum dilatatum*. Good management and fertilizer applications are required to maintain a balanced pasture. High levels of milk and animal production have been recorded from white-clover-based pastures.

(ii) *Trifolium semipilosum* (Kenya white clover) cv. Safari (specific)
This species grows naturally at high elevations in Yemen, Ethiopia, and tropical East Africa from 1400 m upwards (Bogdan 1977). It is adapted

| Tolerance to | | | Minimum rainfall (mm) | Soil prefer-ence | Comments |
Frost	Drought	Water-logging			
Good	Fair	Good	750	Sandy light	Poor stock acceptability. Lawn establishment – tropics
Fair	Good	Good	750	Versatile	Good legume compatibility. Coastal and subcoastal areas. Palatable
Good	Fair	Fair	850	Basaltic table-lands	Palatable, good autumn growth. Excellent for erosion control. Combines with white clover
Good	Fair	Good	750	Versatile	Adapted to coastal areas. Aggressive, early spring growth
Fair	Good	Fair	450	Scrub and Brigalow	Good pioneer species with drought and salt tolerance. Short lived. Good initial establishment
Poor	Good	Fair	400	Versatile	Combines well with townsville stylo

to areas receiving more than 800 mm of annual rainfall and is not drought tolerant. Early establishment in this species is slow and a well prepared seed bed is essential for success. This may be related to a rugose leaf virus. Kenya white clover develops a strong tap root which may impart a degree of persistence to a pasture not found in white clover pastures. It is also able to compete for light in the pasture better than white clover owing to its ascending stems.

In mixtures Kenya white clover has been tried with: *Pennisetum clandestinum*, *Setaria anceps* cv. Nandi, *Chloris gayana*, *Hyparrhenia* sp, *Paspalum plicatulum*, and may combine well with *Lotononis bainesii*. The major mixture would be with *Pennisetum clandestinum*.

Bloat may occur on Kenya-white-clover pastures but is not as big a problem as on white-clover pastures. Kenya white clover is more persistent in dry conditions and will extend production into the summer longer than *T. repens* (Jones 1973). It will be of particular importance on the elevated tablelands of the tropics.

Table 3.2. Summary of the agronomic adaptations of the main commercially sown tropical legumes in Australia

Tropical legumes	Common name	Cultivar	Minimum*		Sowing rate (kg ha^{-1})
			% Purity	% Germination	
Calopogonium mucunoides	calopo	—	93.5	50	1–3
Centrosema pubescens	centro	Common, Belalto	93.8	50	3–5
Desmodium heterophyllum	hetero	Johnstone	94.5	70	1–2
D. intortum	desmodium	Greenleaf	94.5	70	1–2
D. uncinatum	desmodium	Silverleaf	94.5	70	1–3
Glycine wightii	glycine	Tinaroo, Cooper, Clarence, Malawi	97.5	60	2–5
Lablab purpureus	lablab	Rongai Highworth	98.6	75	10–20
Leucaena leucocephala	leucaena	Peru, El Salvador	97.5	60	4–6
Lotononis bainesii	lotononis	Miles	93	50	½–1
Macroptilium atropurpureum	siratro	Siratro	97.5	70	1–3
M. lathyroides	phasey bean	Murray	98	70	1–3
Macrotyloma axillare	axillaris	Archer	97.5	60	3–5

	Tolerance to		Mini-mum rainfall (mm)	Soil prefer-ence	*Rhizobium* requirement	Comment
Frost	Drought	Water-logging				
Poor	Fair	Fair	1200	Versatile	cowpea	Pioneer legume, vigorous growth for weed control
Poor	Fair	Good	1200	Versatile	specific	Suited to tropical lowland environments
Poor	Fair	Good	1500	Versatile	specific	Wet tropical coast. Combines well with pangola and signal grass
Fair	Fair	Good	900	Versatile	specific	Combines well with grasses in coastal areas of tropics and sub-tropics
Fair	Fair	Poor	900	Versatile	specific	More persistent than Greenleaf under hardier conditions
Fair	Good	Poor	750	Well drained	cowpea	Adapted to sub-tropics and elevated tropics
Fair	Good	Fair	600	Versatile	cowpea	Cover crop, green manure, good for hay or sileage
Fair	Good	Poor	750	Well drained	specific	Perennial shrub. Combines well with signal grass
Good	Fair	Very good	900	Light, sandy, well drained	specific	Well adapted to acid soils. Good palatability
Fair	Good	Fair	600	Versatile	cowpea	Easy establishment. Prolific grower, persistent
Poor	Fair	Very good	750	Versatile	cowpea	Self-regenerating annual. Well adapted to waterlogging
Poor	Good	Fair	1000	Well drained	cowpea	Combines well with many grasses and legumes

(continued)

Table 3.2. *(Cont'd)*

| Tropical legumes | Common name | Cultivar | Minimum* | | Sowing rate (kg ha^{-1}) |
			% Purity	Germi- nation	
Macrotyloma uniflorum	uniflorus	Leichhardt	97.5	60	5–20
Pueraria phaseoloides	puero	–	93.5	50	1–3
Stylosanthes guianensis	stylo	Schofield, Cook, Endeavour	90	40	2–5
S. humilis	Townsville stylo	Common, Patterson, Lawson, Gordon	90	40	3–6
S. hamata	carribean	Verano	90	40	3–6
S. scabra	shrubby stylo	Seca	90	40	3–6
Trifolium semipilosum	Kenya white clover	Safari	93.5	75	1–3

*The minimum purity and germination values are the standards defined in the Queensland Government Seed Regulations.

14. *Vigna* spp

Annual or perennial herbs of various habits, the genus containing about 150 species distributed throughout the tropics and subtropics (Bogdan 1977). *Vigna unguiculata* (cowpea) is an important lowland tropical pulse (Rachie and Roberts 1973). One species provides a pasture plant.

(i) *Vigna luteola* (Dalrymple vigna) cv. Dalrymple (cowpea)
This species is a dense, leafy perennial summer grower with long, twining stems (Humphreys 1974). Bogdan (1977) reports that this species

Tolerance to			Mini-mum rainfall (mm)	Soil prefer-ence	*Rhizobium* requirement	Comment
Frost	Drought	Water-logging				
Fair	Good	Fair	600	Sandy	cowpea	Annual, easy establishment, good for deferred autumn grazing
Poor	Poor	Good	1250	Heavy tolerates acid soil	cowpea	Pioneer species, palatable and productive
Poor	Good	Fair	850	Versatile	cowpea	Adapted to humid tropics, even poor soils
Poor	Good	Poor	600	Versatile well drained	cowpea	Easy establishment, good reseeder
Poor	Good	Poor	600	Well drained	cowpea	Perennial under grazing. Well adapted to current Townsville stylo areas
Poor	Good	Poor	450	Versatile	cowpea	Adapted to seasonally dry tropics
Poor	Good	Fair	1000	Versatile	specific	Better adapted to subtropics and elevated tropics, than white clover, combines well with kikuyu

occurs in swampy or seasonally wet grasslands, lake shores, and wet woodlands. It is well adapted to tropical coastal conditions and produces well on seasonally wet or waterlogged soils. In this characteristic this species differs from most of the tropical legumes. It is an excellent pioneering legume which outyields most others in its first year (Humphreys 1974). It is highly palatable to stock but does suffer from insect damage.

The agronomic adaptations of the main commercially sown tropical grasses and legumes are summarized in Tables 3.1 and 3.2 respectively.

3.3 ADAPTATIONS AND TOLERANCE OF STRESS

A pasture sward can be permanent and productive only if the species sown are adapted to the climatic and edaphic conditions of the site, and are able to survive periods of stress which will occur with varying frequency. The general effects of climate and soils on pasture growth, and the agronomic characteristics of the pasture species have been discussed in previous chapters. In this chapter the adaptations of the pasture species to photoperiod, and their tolerance of extreme conditions of temperature, moisture deficit, and salinity are examined.

1. *Adaptation to photoperiod*

The response of plants to photoperiod determines whether plants will flower at a given latitude, and if so at what time of the year. While this is of particular importance for seed production, it also determines the length of the vegetative growing season, as onset of flowering and seed set in many species is associated with increasing maturity and declining nutritive value. This adaptation of the main legume and grass genera is reviewed below.

(a) *The tropical legumes*

(i) *Stylosanthes* species

The genus *Stylosanthes* with some 25 species occurs naturally over a wide geographical range and exhibits a wide range of adaptation to soil, rainfall, temperature, and daylength conditions (Burt, Edye, Williams, Grof, and Nicholson 1971). Within this genus 't Mannetje (1965) found both long-day (LD) plants and short-day (SD) plants (Table 3.3). *S. humilis* and *S. guianensis* responded as SD plants, while *S. mucronata* appeared to be indeterminate. *S. montevidensis* an introduction from Uruguay reacted as an LD plant and made little growth in short days.

Within seven selections of the annual *S. humilis* Cameron (1967) recognized early types with a critical maximum daylength of 13 hours, midseason types 12 hours, and late types 11½ hours. These results have been confirmed in a number of sowing date experiments in the field ('t Mannetje and van Bennekom 1974; Skerman and Humphreys 1975). In years of above average rainfall late-flowering types, with their longer vegetative period, tend to yield more. Short-season types are better adapted to lower rainfall environments. Even within these lower rainfall areas late-flowering types were found in depressions and roadside drains where better moisture conditions prevailed (Cameron 1967), demonstrating the wide range of adaptation in this species.

Table 3.3. Effect of photoperiod on days to flowering and growth of four
         Stylosanthes species ('t Mannetje 1965)

Photo-period (h)	*S. humilis* No. days	Dry weight*	*S. guianensis* No. days	Dry weight	*S. mucronata* No. days	Dry weight	*S. montevi-densis* No. days	Dry weight
8	33	5.3	52	9.6	51	2.8	v	0.2
10	33	4.3	52	8.8	53	2.7	v	0.2
12	44	6.2	v	12.3	54	2.2	68	1.4
14	v	10.7	v	12.1	62	4.1	64	1.5

*Dry weight (g per plant)   v — vegetative, did not flower.

In the species *S. guianensis* a range of critical photoperiods has been
demonstrated for a range of accessions (Bryant and Humphreys 1976)
(Table 3.4). Temperature effects can modify photoperiod responses,
and in this experiment where plants were grown at constant 27 or 20 °C,
one accession (CPI 33706) failed to flower at 27 °C, but flowered at
13 hours in 20 °C.

Table 3.4. Effect of photoperiod on days to first flowering in five selections of
         S. guianensis (data at 27 °C only shown) (Bryant and Humphreys 1976)

Photo-period (h)	CPI 34906*	40255	cv. Cook	cv. Endeavour	cv. Schofield
13½	117	v	—	—	—
13.0	106	110	—	—	—
12½	98	77	v	v	—
12.0	—	—	84	91	v
11½	—	—	89	72	89
11.0	—	—	—	—	86

*CPI — Commonwealth Plant Introduction number.
v — vegetative, did not flower.

Flowering in the field was ranked in similar order, CPI 40255 and
34906 at approximately 13½ hours (including civil twilight), Cook and
Endeavour at approximately 12 hours, while Schofield was last to
flower at about 11½ hours. The short critical photoperiod for Schofield
may explain its poor flowering and seed production in low latitude sites

such as Kluang, Malaysia (2°N) and Bali (8°S) (Bryant and Humphreys 1976). Another cultivar of *S. guianensis*, cv. Oxley fine stem is an LD plant (Humphreys 1974).

(ii) *Macroptilium atropurpureum*

The released cultivar siratro reacts as an SD plant with critical maximum daylength above 12 hours (Table 3.5). Lower temperatures tend to delay flowering but do not alter the basic SD response (Table 3.6) (Imrie 1973a).

Table 3.5. Effect of photoperiod on days to flowering and growth at constant 28°C in *M. atropurpureum* cv. Siratro (Whiteman, unpublished data)

Photoperiod (h)	Days to flowering	Dry weight (g per plant)
8	64	6.2
10	69	10.3
12	76	16.2
16	v	13.7
24	v	17.7

v — vegetative, did not flower.

Table 3.6. Effect of temperature and photoperiod on days to flowering in three genotypes of *M. atropurpureum* (Imrie 1973a)

Photoperiod (h)	Siratro		CPI 38591		CPI 49762	
	30/25*	24/19	30/25	24/19	30/25	24/19
8	60.5	60.7	62.3	81.7	51.0	63.0
11	65.6	81.3	53.3	82.5	61.0	73.0
16	v	v	v	v	v	v

*30/25 — 30°C day temperatures/25°C night.
v — vegetative, did not flower.

However, siratro plants given a night light-break treatment remained vegetative for up to 180 days, after which they flowered, suggesting that siratro is a quantitative SD species (Whiteman, unpublished data).

(iii) *Desmodium* sp

This is another geographically widely distributed genus, with species exhibiting a range of photoperiod adaptations. The commercial culti-vars *D. uncinatum* cv. Silverleaf and *D. intortum* cv. Greenleaf are SD plants (Bryan 1969). In southern Queensland *D. uncinatum* flowers in April at daylengths of 12–12½ hours. *D. uncinatum* did not flower under 14 hour daylengths (Chow and Crowder 1972). *D. intortum* flowers later in mid-May to June at daylengths of 11–11½ hours. Within *D. intortum* there is a range of daylength response with earlier and later flowering lines (Imrie 1973*b*).

In other *Desmodium* species Rotar, Park, Bromdep, and Urata (1967) report that *D. gyroides* is an SD plant, while *D. canum* and *D. sandwicense* flowered under 8 and 16 hour daylengths, and are indeter-minate. Both species flower continuously in Hawaii under natural conditions. However Chow and Crowder (1972) report that *D. canum* failed to flower under 11 hours, but the low temperatures imposed (24/20 °C) may have inhibited flowering.

(iv) *Glycine wightii*

All the accessions studied by Wutoh, Hutton, and Pritchard (1968) were SD plants, although critical daylength varied between varieties, Table 3.7. At higher day temperatures (33 °C) no accession flowered in 16 hour daylengths, but at lower day and night temperatures (27/16, 27/10, 21/16, and 21/10 °C) one accession (CPI 18103) flowered at 16 hours, although numbers of days to flowering was greater than at the 8-hour photoperiod.

Table 3.7. Effect of photoperiod on days to flowering in four accessions of
G. wightii grown at 27/19 °C day/night temperature (Wutoh *et al.*
1968)

Photoperiod (h)	Days to flower			
	CPI 25918	18103	cv. Cooper	cv. Tinaroo
8	58	46	47	54
10	58	46	48	57
11	58	47	47	58
12	55	56	81	v
13	83	118	127	v
14	105	v	v	v
16	130	v	v	v
18	v	v	v	v

v – vegetative, did not flower.

The critical daylengths defined above reflect closely their flowering behaviour in the field. In southern Queensland cv. Tinaroo is the latest flowering variety and CPI 25918 the earliest. The other released cultivar Clarence is a little earlier flowering than cv. Cooper, which allows seed to mature before frosting in northern New South Wales (Edye and Kiers 1966). Within *G. wightii* Ferguson (1969) has found a wide range of flowering behaviour from very early types to late types.

(v) *Calopogonium mucunoides*
Detailed studies with controlled photoperiod conditions do not appear to be available. In southern Queensland (lat. 27½°S) *Calopogonium* flowers in the shortest daylengths in June, but is usually frosted before flowering (R. J. Williams, personal communication). It is probably an SD plant.

(vi) *Pueraria phaseoloides*
Few reports are available on photoperiod responses. In Puerto Rico (lat. 18°N) flowering and seed-set occur in the short daylength period from January to March (Vicente-Chandler, Rivera-Brenes, Caro-Costas, Rodriguay, Boneta, and Gracia 1953) suggesting that it may react as an SD plant.

(vii) *Centrosema pubescens*
This species appears to be an SD plant (Wang 1961). In north Queensland (lat. 17½°S) common centro flowers in April and in October when daylength is 12½–13½ hours. The new cultivar Belalto appears to require shorter days and does not flower until June at a daylength of 12 hours (Tietzel and Burt 1976).

(viii) *Lablab purpureus*
Again appears to react as an SD species. Flowering in southern Queensland commences in May at daylengths of approximately 11–11½ hours.

(ix) *Lotononis bainesii*
This species, introduced from southern Africa (lat. 23° to 34°S) is well adapted to subtropical environments, and is markedly more cold hardy and frost tolerant than other tropical legumes (Gates 1973). Experimental data under controlled photoperiod conditions is not available, but in southern Queensland this species has a main flowering period in August to October and a smaller production of flowers in February to April, with a few flowers produced over the summer period (Fig. 3.1) (Bryan 1961).

As this species commences flowering with increasing daylength in

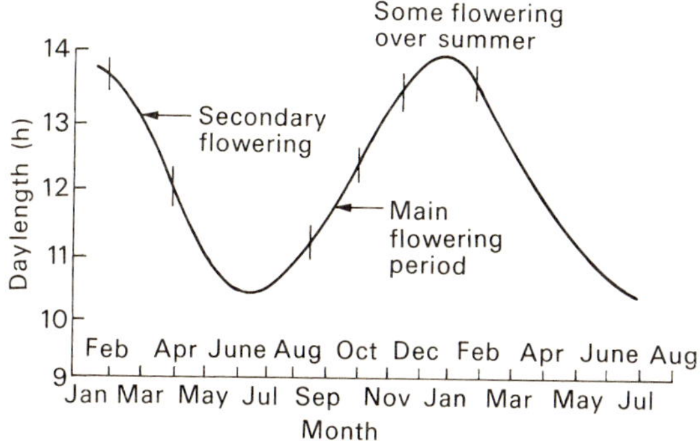

Fig. 3.1. Flowering times in *L. bainesii* in relation to daylength (sunrise to sunset) at Samford, south-east Queensland (lat. 27½° S)

spring, and will produce some flowers in daylengths of up to 14 hours in summer (see Fig. 3.1), and does not flower in the short days of winter, it is suggested that it may be an LD plant. The critical minimum photoperiod would appear to be approximately 11–11½ hours (Fig. 3.1).

(x) *Macrotyloma* spp

The two released cultivars differ in their flowering responses and data is insufficient to classify them into daylength response groups. *M. axillare* cv. Archer tends to be late flowering in the short day period. Planted in January it commences flowering in mid-April and early February plantings in early June. *M. uniflorum* cv. Leichhardt is short-season summer-growing species. Planted in January it commences flowering in mid-March at daylengths of approximately 12 hours.

(xi) *Trifolium semipilosum*

This introduction, from higher altitudes in Kenya, is adapted to the cooler environments of the subtropics. It appears to respond as an LD plant with a minimum critical daylength of 10½ to 11 hours (Table 3.8) ('t Mannetje and Pritchard 1968).

Higher night temperatures precluded flowering at 10 hours, and delayed flowering in the 12 and 14 hour daylengths.

(b) *The tropical grasses*

In the previous section, results were presented in some detail in order to demonstrate the type of data required to properly classify species in

Table 3.8. Effects of photoperiod and temperature on days to flowering in
T. semipilosum ('t Mannetje and Pritchard 1968)

Photoperiod (h)	Days to flowering at day/night temperatures	
	25/10°C	25/20°C
10	175	v
12	138	215
14	168	181

v — vegetative, did not flower.

their photoperiod responses. Generally fewer controlled photoperiod
experiments have been conducted on the tropical grasses so that only a
brief summary of the major genera is possible.

(i) *Brachiaria* spp
Probably respond as SD plants with a range of critical maximum daylengths within species. *B. decumbens* flowers as early as December at
lat. 17°S (daylength 14 hours) (Grof 1968), but continues flowering
until May (Humphreys 1974). *B. mutica* flowers in March at lat. 17°S
(daylength 13 hours) (Grof 1969). *B. ruziziensis* is later flowering in
April at daylengths of 12½ hours.

(ii) *Cenchrus* spp
Cultivars vary in time to flowering, but flower over a wide range of
daylengths, dependent more on rainfall conditions than daylength, and
are classed as indeterminate (Evers, Holt, and Bashaw 1969).

(iii) *Chloris gayana*
This species flowers over a wide range of daylengths. Flowering was
most rapid in 10 to 13 hour photoperiods, and was delayed by 15 days
in 7 to 10 hours, and by 120 days in 13–17 hours (Bogdan 1969).
However the delay in 7–10 hour photoperiods was probably due to
limitations on photosynthesis. Thus *C. gayana* might be classed as a
quantitative SD plant. Cultivars within *C. gayana* exhibit a range of
critical daylengths. In an evaluation of 24 introductions Hutton (1961)
found first flowering over a range from November to May. Released
cultivars have a range of flowering response; cv. Pioneer is early flowering and flowers throughout the growing season, cv. Katambora is mid-
season flowering, while cv. Callide and cv. Samford are late flowering.

(iv) *Melinis minutiflora*
Probably an SD plant (Humphreys 1974).

(v) *Panicum* spp

Although there is a range of flowering times within the genus, most species appear to begin flowering in the longer daylength periods of the summer growing season.

P. *coloratum* var. *makarikariense* flowers from December to April at lat. 26°S (Lloyd and Scateni 1968). In P. *maximum* at lat. 17°S cv. Common Guinea flowers November-December; cv. Hamil and cv. Coloniao flowers February-March (McCosker and Teitzel 1975), while green panic, P. *maximum* var. *trichoglume* cv. Petrie, flowers November to May. Such data do not confirm whether these species are LD, DN, or indeed SD plants where critical maximum daylength is quite long.

(vi) *Paspalum* sp

This genus appears to have species with different daylength responses. P. *dilatatum* is an LD plant with an optimum photoperiod for flowering of 14-16 hours (Knight 1955). P. *commersonii* appears to react as an LD plant, beginning flowering in January at lat. 26°S (Barnard 1967). P. *plicatulum* is an SD plant with a critical photoperiod of 13 hours (Chadhokar and Humphreys 1974).

(vii) *Setaria* spp

Flowering appears to be promoted by long days with first flowering in September-November, but flowering throughout the growing season (Humphreys 1974). Although Evans (1964) suggests that the usual photoperiod response is an SD response in panicoid grasses, there does appear to be a wide range from LD, DN, to SD plants. Obviously more research under controlled photoperiod conditions is required to properly quantify the tropical species. However for most practical applications it is probably sufficient to undertake a range of field plantings to determine onset of flowering and seed set in a particular region. It should be remembered though that the critical daylength response is determined at the time of floral initiation and not at flower emergence.

2. Adaptation to temperature extremes

Within the tropical climatic realm the boundary between the humid and semi-arid and arid regions also marks a major difference in temperature regimes. On the humid side of the boundary zone very high temperatures, in excess of 38 °C occur only rarely, and the diurnal temperature range rarely exceeds 10 °C. In the arid regions high temperatures, above 40 °C are common, and the diurnal and seasonal ranges are large. Thus regions on the margins of humid tropics where tropical pasture species are grown can experience extremes of temperature which can markedly affect the growth and survival of these species.

Areas adjacent to large inland arid regions as in Nigeria and the Queensland coast experience occasional heat waves. Heat waves are defined as three consecutive days with maximum temperature in excess of 38 °C (Skerman 1956). Heat waves are particularly devastating duing seedling emergence and establishment, and at flowering and seed formation stages. High temperatures alone, in the range 38–45 °C, may have little effect on the tropical grasses, but can be expected to reduce growth rate of the legumes. However heat wave conditions are associated with high evaporative demand, leading to rapid wilting and often death of leaves as a result of desiccation. The selected semi-arid grasses such as *Cenchrus* and *Urochloa*, and the legumes *Stylosanthes humilis* and *S. hamata* are better adapted to these conditions than most other sown species.

At the other extreme, inland areas in the tropics and the subtropical areas experience low temperature and frost conditions. In southern Queensland the most important determinant of the length of the growing season, and in many cases the survival, of tropical pasture species is the onset of low temperatures and ground frosts in the winter period (Coleman 1964). In inland Queensland occasional frosts are experienced as far north as lat. 17°S.

Onset of the cool season in subtropical and inland tropical climates is usually marked by a steady decline in night temperatures. Low night temperatures, in the region of 5 to 15 °C may be experienced while day temperatures may still be as high as 27 °C. In a number of tropical grasses Ivory (1975) found that a reduction of night temperature from 20 to 10 °C at a day temperature of 20 °C reduced growth rates by 22 to 36 per cent. In buffel (*C. ciliaris*) and *P. coloratum* a reduction to 8 °C reduced growth rate by 60 and 37 per cent, while a further reduction to 4 °C night, reduced growth rate by 90 and 80 per cent respectively. Rhodes (*C. gayana*) and kikuyu (*P. clandestinum*) were less affected and at 4 °C night were reduced by approximately 40 per cent. The reduction in growth rate was linearly related to the number of cold nights experienced, and was related to a reduced net assimilation rate, associated with a higher stomatal resistance to CO_2 uptake (Ivory 1975). The growth of tropical legumes, with the exception of *L. bainesii*, was more sensitive to low temperature than the grasses ('t Mannetje and Pritchard 1974). *C. pubescens* was most sensitive, regrowth being reduced by 85 per cent when temperature was reduced from

32/24 to 26/15 °C, compared with a reduction of 35 per cent in *M. atro-purpureum*, *G. wightii*, *P. phaseoloides*, *S. humilis*, and *S. guianensis*. At 20/6 °C, all species except *L. bainesii* made little growth, and some even died. Thus even before the onset of frosting, growth rate of tropical pasture species may be greatly reduced by cold nights.

In subtropical and inland Queensland, radiation frosts are associated with clear, cloudless, still night air conditions, allowing a rapid loss of long-wave radiation from the ground surface. Terrestrial minimum temperatures as low as –13.5 °C have been recorded at Toowoomba (elevation 300 m, lat. 26°S). In the area south from Monto (lat. 24°S) in subcoastal Queensland there is a 100 per cent probability of receiving at least one frost of –3 °C terrestrial minimum each year (Ivory 1975).

In the tropical grasses frosts cause death and senescence of the above ground foliage, although survival of crowns is not usually a problem. There is some variation in frost tolerance. The temperatures causing 50 per cent foliage death in a range of buffel (*C. ciliaris*) ecotypes was –1.8 to –2.1 °C, Rhodes grass (*C. gayana*), –2.6 to –3.5 °C, and for *Setaria anceps* ecotypes –2.1 to –4.3 °C (Ivory 1975). The setaria cultivar Narok was selected on the basis of its frost tolerance and ability to remain green over winter.

The tropical legumes are considered to be more susceptible to frost damage than the grasses, and frosting causes death of leaves and stems. Survival of plants in the field following frosting was recorded by Jones (1969) as follows: *M. atropurpureum* (siratro) 65 per cent, *D. uncinatum* (silverleaf) 56 per cent, *G. wightii* (clarence) 48 per cent, *D. intortum* (greenleaf) 36 per cent, *Lablab purpureus* 0 per cent. Within the tropical legumes *Lotononis bainesii* shows the greatest tolerance of frosting both in field observations (Bryan, Sharpe, and Haydock 1971) and in controlled environment studies (Bryan 1972; Gates 1973).

In humid tropical environments it is unlikely that temperature extremes will be a major limitation to pasture growth, but in subtropical regions and the semi arid/humid margin areas, both low and high temperature extremes can reduce pasture production and affect long-term survival of pasture species.

3. Tolerance of water stress

In most pasture regions of the world periods of soil moisture stress, either short-term or long-term (drought), are a major determinant of pasture yield. As an example, McCown, Gillard, and Edye (1974) examined the effects of long-term variation in rainfall on yield of pastures at Townsville, North Queensland. Cumulative yields of native spear grass (*Heteropogon contortus*) plus *S. humilis*, or buffel plus nitrogen, 600 kg N ha^{-1} a^{-1}, varied widely over five years, but yield was closely related to evapotranspiration. Using 69 years rainfall records and estimates of soil water storage, run off and evapotranspiration were calculated for each year 1901–72 and yields were predicted (Table 3.9).

Table 3.9. Predicted probability of yields exceeded stated yield based on 69 years rainfall record and calculated evapotranspiration (McCown *et al.* 1974)

Dry matter yield (t ha^{-1})	Probability of exceeding stated yield (%)		
	H. contortus and *S. humilis*	*C. ciliaris* and *S. humilis*	*C. ciliaris* and 600N
2	80	95	100
4	40	85	100
6	nil	50	93
8	nil	5	85
12	nil	nil	50
18	nil	nil	20

These data demonstrate the effects of rainfall variability on pasture yield, but for many environments ability of species to survive protracted periods of moisture stress will be an important determinant of pasture stability. Drought avoidance is best demonstrated by the annual pasture species which grow during the wet season and pass through the dry season as a dormant seed population. A good example is provided by Cameron's (1967) study of flowering times in *S. humilis* populations. All late-flowering collections came from sites receiving at least 1130 mm rainfall per annum. Early-flowering types were collected mainly from drier areas. However, even in these lower rainfall areas late-flowering types were found in sites of more favourable water supply, e.g. roadside drains and depressions. Thus where sufficient genetic diversity is present, environmental pressures have sorted populations into ecotypes adapted to the length of growing season. In this way populations of an annual species can adapt, through flowering response to daylength, to avoid the main period of water stress, while still making maximum use of the particular length of growing season.

Most perennials, however, must tolerate and survive periods of water stress through various mechanisms of drought resistance. Many adaptations are involved such as dry season dormancy, deep rooting, leaf shedding, and restriction of transpiration. As discussed previously, C_4 tropical grasses have a more sensitive control over stomatal resistances. Because of higher photosynthetic rates and a more sensitive control over transpiration, C_4 grasses produce much more dry matter per unit of water loss than the C_3 species (Downes 1970).

There are few critical studies which provide a physiological basis of comparison between tropical pasture species with different field drought resistance. Most information is based on the agronomic characteristics of the species in their performance under dryland conditions. The relative adaptation of the main tropical pasture species to moisture stress has been discussed in the agronomic descriptions of the species.

4. *Tolerance of waterlogging and flooding*

Many areas in the humid tropics with potential for sown pasture development are exposed to seasonal or occasional waterlogging or flooding. Waterlogging is the condition of complete saturation of the soil pores in the profile with water, while flooding involves the inundation of the plants by a free water surface. Saturation of the soil profile for extended periods causes anaerobic conditions, and while reduction in oxygen content and build up of carbon dioxide are major effects, other changes such as increased solubility of manganese, iron, and aluminium may also be important.

Pasture species vary widely in their tolerance of waterlogging and flooding. Oxygen deficiency under flooded conditions has been shown to severely limit water uptake by roots. Wilting and desiccation may rapidly ensue, and nutrient uptake is inhibited (Bergman 1959). Kramer (1951) suggested that downward translocation of carbohydrates and auxins to the roots is inhibited and that their accumulation in the lower stem may account for the production of adventitious roots in some species.

More recent work has compared the respiratory pathways and associated enzymes in flood tolerant and intolerant plants (Crawford 1969; McManmon and Crawford 1971). In the roots of species intolerant of flooding oxygen deprivation leads to accelerated glycolysis and the end-product is ethanol which is toxic to plant cells. In the tolerant species, flooding induces a partial blocking of the normal respiratory pathway, glycolysis to ethanol is not accelerated, and oxaloacetate is converted to malate. As malic enzyme is not present, malate accumulates and is non-toxic.

Tolerance of flooding in the field is affected by many factors including: age of plants, whether actively growing or dormant, previous defoliation management, duration and depth of flooding, temperature and oxygen content of water, whether flowing or stagnant, and silt load of the water. Depth of flooding is particularly important. Effects were less severe in tropical grasses (Anderson 1970) and in white clover (*T. repens*) (Davis and Martin 1949) where part of the plant was above water level than when completely covered.

Assessment of comparative tolerance is valuable for agronomic selection of species for flooded situations. Francis and Poole (1973) have shown a wide range of tolerance of flooding in annual temperate legume species, where reductions in yield due to 16 days flooding varied from only 13 per cent in *Trifolium subterraneum* to 51 per cent in *Medicago truncatula* to 79 per cent in *M. minima*. Similarly Anderson (1970) has shown considerable differences between tropical grass species, and even within species, e.g. cultivars of *Cenchrus ciliaris* had different tolerances of flooding. Less data are available on the flooding tolerances of tropical legumes, but many of these species appear to be less well adapted than the tolerant grasses. *Macroptilium lathyroides* has been shown to be much less affected by 21 days flooding than *M. atropurpureum, Centrosema pubescens* and *S. guianensis* (Seitleko and Whiteman, unpublished). *M. lathyroides* maintained top and root growth at similar rates to unflooded plants and even maintained root nodules, while in the other species root and top growth was markedly inhibited and nodules decayed.

The relative tolerance of flooding of a range of tropical grasses and legumes are summarized in Table 3.10.

Within a species cultivars vary in their flooding tolerance. Anderson (1974) has shown that flooding tolerance in the tall rhizomatous forms of *Cenchrus ciliaris* was greater than in the short non-rhizomatous types. Mean tolerance of cultivars with up to 30 days flooding expressed as percentage survival were ranked cv. Tarewinnibar 84, Nunbank 80, Boorara 77, Biloela 74, Molopo 68, Gayndah 35, American 20, the latter two cultivars being short non-rhizomatous types.

5. *Tolerance of salinity and acid soil conditions*

High levels of soluble salts in the soil solution, particularly in sodic soils in drier regions, may limit pasture growth. While NaCl toxicity is most common, other ions such as Mg^{2+}, HCO_3^-, and SO_4^- may be involved. In some humid tropical soils with low pH, excess soluble manganese

Table 3.10. Relative tolerance of some tropical grasses and legumes to flooding.
Species listed in approximate descending order of tolerance.
(Compiled from Anderson 1970, 1974; Humphreys 1974;
Sietleko and Whiteman, unpublished)

Level of tolerance	Tropical grasses	Tropical legumes
Good	*Brachiaria mutica*	*Macroptilium lathyroides*
	Panicum coloratum var. *makarikariense*	*Lotononis bainesii*
	cv. Bambatsi	*Desmodium heterophyllum*
	cv. Kabulubula	*Pueraria phaseoloides*
	Paspalum plicatulum	*Desmodium intortum*
	Paspalum dilatatum	
	Digitaria decumbens	
	Setaria anceps	
Moderate	*Pennisetum clandestinum*	*Calopogonium mucunoides*
	Pennisetum purpureum	*Centrosema pubescens*
	Brachiaria decumbens	
	Cenchrus ciliaris	*Macrotyloma axillare*
	Chloris gayana	*Desmodium uncinatum*
	Panicum maximum (guinea)	*Stylosanthes guianensis*
Poor	*Brachiaria ruziziensis*	
	Panicum maximum (green panic)	*Macroptilium atropurpureum*
	Panicum antidotale	*Glycine wightii*
	Melinis minutiflora	*Stylosanthes humilis*
	Urochloa mosambicensis	*Lablab purpureus*
		Cajanus cajan
		Leucaena leucocephala

and aluminium may have toxic effects. Salinity leads to reduced water-uptake owing to effects on the osmotic potential of the soil solution, while high levels of uptake of specific ions may cause toxic effects in the plant. More importantly in the legumes salinity may reduce nodulation and inhibit nitrogen fixation.

Most studies have sought simply to compare species or cultivar tolerance to salinity. However scientists have generally recognized osmotic and water potential effects as separate from specific ion toxicities. Some studies have been extended to establishing aspects of tolerance through comparisons of uptake and concentration of particular ions in the roots and tops.

The effects of NaCl have been most widely studied. Hutton (1971) has reported on comparative levels of tolerance of 14 legume species (Table 3.11), while Russell (1976) compared tropical and temperate legumes and tropical grasses (Table 3.12). In Hutton's study the ratio of yield with salt added to yield without added salt was used as an index

Table 3.11. Comparative salt tolerance of 14 legume species grown at 3500 p.p.m. NaCl (Hutton 1971)

Species	Yield without salt A (g)	Yield with salt B (g)	B/A
Medicago sativa cv. Hunter River	3.8	2.2	58
M. atropurpureum cv. Siratro	13	7	51
C. pubescens	11	4	40
L. bainesii	4.2	1.6	38
V. luteola	27	9	35
D. sandwicense	15	5	33
D. uncinatum	11	3	33
S. humilis	5.5	1.6	29
S. guinanensis cv. Schofield	4.3	1.2	28
M. lathyroides	27	6	23
G. wightii cv. Clarence	3.8	0.9	23
G. wightii cv. Cooper	3.2	0.4	12
D. intortum	14	1	7
G. wightii cv. Tinaroo	4.3	0.2	5

of salt tolerance. Russell (1976) grew plants at six levels of added salt in a heavy clay soil and computed salt levels giving zero yield and half yield from the response curves. Both studies ranked *M. sativa* and *M. atropurpureum* as the most tolerant legumes, but there was less agreement between the middle and low tolerance rankings. In the temperate legumes the *Trifolium* species, except for *T. alexandrinum*, were generally less tolerant than the *Medicago* species. The most tolerant grasses were *C. gayana*, *P. coloratum*, *P. clandestinum*, and *S. almum*.

Within species there are differences in tolerance between cultivars so that selection for salt tolerance appears to be quite feasible. Gates (1974) found that within 22 lines of *Glycine* a range in salt tolerance was evident. The most vigorous *Glycine* lines in the absence of salt were also more resistant to salinity stress, and had a higher tendency to exclude sodium and to a lesser extent chlorine from the plant saps. They also maintained higher nitrogen, phosphorus, and potassium contents in both the control and saline treatments. Gates (1974) suggests that the higher resistance to salinity stress was related to the ability of some *Glycine* lines to maintain higher levels of enzymatic efficiency at all salinity levels.

Table 3.12. Comparison of salt tolerance over a range of salinity levels in some tropical legumes, temperate legumes, and tropical grasses (Russell 1976)

Species	Soil salinity at zero plant yield EC* siemens m^{-1} at 25°C	Rank	Soil salinity at half plant yield EC siemens m^{-1} at 25°C	Rank
Tropical legumes				
M. sativa	1.88	1	1.02	1
M. lathyroides	1.75	2	0.95	3
M. atropurpureum	1.74	3	0.99	2
V. unguiculata	1.29	4	0.72	6
G. wightii	1.23	5	0.69	7
L. bainesii	0.96	6	0.66	8
L. purpureus	0.95	7	0.55	9
M. uniflorum	0.85	8	0.78	5
D. intortum	0.84	9	0.79	4
S. humilis	0.84	10	0.51	10
D. uncinatum	0.82	11	0.49	11
Temperate legumes				
M. sativa	1.61	1	0.88	1
T. alexandrinum	1.51	2	0.83	2
M. scutellata	1.48	3	0.82	3
M. truncatula				
cv. Jemalong	1.41	4	0.78	5
cv. Cyprus	1.38	5	0.77	6
M. littoralis	1.38	6	0.77	7
T. fragiferum	1.14	7	0.65	8
T. repens	1.09	8	0.62	9
T. hirtum	0.89	9	0.81	4
T. semipilosum	0.67	10	0.42	10
Tropical grasses				
P. coloratum	3.25	1	1.70	3
P. clandestinum	2.94	2	2.15	2
C. gayana	2.90	3	2.32	1
S. almum	2.78	4	1.46	5
D. decumbens	2.77	5	1.47	4
C. ciliaris				
cv. Nunbank	1.61	6	1.42	6
U. mosambicensis	1.55	7	1.40	7
P. maximum	1.53	8	0.98	8
P. dilatatum	1.28	9	0.72	9
S. anceps	1.05	10	0.60	10

*Electrical conductivity of soil saturation extract.

Similarly in 24 varieties of *Sorghum bicolor* Gates (1974) found that differences between sorghums in growth and dry-matter yield at various salinity levels were a reflection of differences at the control level. Within other grass genera rhodes grass (*Chloris gayana*) has been shown to tolerate salinity remarkably well (Bogdan 1969). Rhodes grass was found to be tolerant not only of NaCl, but also of $NaHCO_3$, $CaCl_2$ and Na_2SO_4, but was very sensitive to $MgCl_2$. Rhodes grass accumulates sodium in high sodium substrates, up to 86–121 milliequivalents sodium per g dry matter without visible ill effect; under similar conditions *Paspalum dilatatum* accumulated only 16–31 milliequivalents sodium per g dry matter. Playne (1970) has also shown that *C. gayana*, and *P. maximum* and *Cenchrus setigerus* were 'high sodium' species (> 0.4 per cent Na); *Urochloa mosambicensis* was a 'medium sodium species', and *C. ciliaris*, *H. contortus*, and the legumes *S. humilis* and *M. atropurpureum* were 'low sodium species' (< 0.10 per cent Na).

Within buffel grass cultivars, although overall salinity tolerance is low, a range of tolerance to salinity has been shown (Graham and Humphreys 1970). Again tolerance to salinity was related to growth and vigour in non-saline conditions. Cultivar Biloela maintained highest dry-matter yields at the control and intermediate salinity levels, but there was no difference between cultivars at the highest salinity level (160 milliequivalents per litre). The higher tolerance in Biloela did not appear to be related to a greater ability to exclude sodium or chloride ions from uptake.

The comparative tolerance to manganese excess of eight tropical legumes was studied by Andrew and Hegarty (1969). The relative ranking of tolerance, percentage reduction in yield at the highest manganese level compared with control, and toxicity threshold manganese levels in top dry matter are shown in Table 3.13. *M. atropurpureum* and *G. wightii* are shown to be intolerant of high manganese levels.

Aluminium is another element which can be at toxic levels in acid soils in the tropics and subtropics. The effects of soluble aluminium on the growth and chemical composition of some tropical and temperate pasture legumes was studied by Andrew, Johnson, and Sandland (1973), at four levels of aluminium (0, 0.5, 1.0, 2.00 p.p.m.). Addition of aluminium to the substrate had no significant depressing effect on the growth of *M. lathyroides*, *L. bainesii*, and *S. humilis*. The growth of *D. uncinatum* was depressed only at the 2.0 p.p.m. level, whereas growth of *G. wightii* and *M. sativa* was severely depressed by all additions

Table 3.13. Relative tolerance to manganese excess, percentage reduction in dry
matter yield at the highest manganese level (40 p.p.m. in solution)
and toxicity threshold levels of manganese in plant top dry matter
(Andrew and Hegarty 1969)

Species	Rank	Reduction in yield (% of control)	Toxicity threshold (p.p.m. manganese)
C. pubescens	1	70	1600
S. humilis	2	50	1140
L. bainesii	3	42	1320
M. lathyroides	4	29	840
L. leucocephala	5	23	550
D. uncinatum	6	24	1160
M. sativa	7	18	380
G. wightii	8	11	560
M. atropurpureum	9	9	810

of aluminium. In the temperate group *Medicago* spp were more severely
depressed than the *Trifolium* species. The tropical grasses also show a
range in tolerance to aluminium excess, from *C. ciliaris* which is very
sensitive to *Brachiaria decumbens* which showed negligible effects at all
levels of aluminium (Andrew, personal communication).

Tolerance of aluminium and manganese excess is only one factor in
adaptation to acid soils. Apart from direct pH effects, calcium and
molybdenum deficiencies are commonly associated with acid soils.
However the tropical legumes are more efficient at extracting calcium
from calcium deficient soils than are the temperate legumes (Andrew
and Norris 1961). The relative yields of dry matter in the absence of
added calcium compared with maximum yields with added calcium in a
range of tropical and temperate legumes were as follows — tropical
legumes, *M. lathyroides* 68, *S. guianensis* 64, *C. pubescens* 52, *Indigofera spicata* 40, *D. uncinatum* 25 per cent; temperate legumes, *T.
repens* 6.5, *T. fragiferum* 2.0, *M. sativa* 1.5, *M. tribuloides* 0.2 per cent.
Norris (1967) suggests that the better adaptation of tropical legumes to
acid soils is perhaps not surprising as they have evolved under tropical
acid soil conditions. There are exceptions such as *Leucaena leucocephala* which is adapted to higher pH soils in the tropics and responds
readily to additions of lime (Diatloff 1973).

Adaptation to one factor does not necessarily imply tolerance of other toxic factors. *Medicago sativa* was ranked first in tolerance to salinity but ranked very poorly in tolerance of excess manganese or aluminium, or deficiency of calcium. *Centrosema pubescens, Stylosanthes humilis*, and *Lotononis bainesii* appear to be well adapted to adverse soil conditions, while *Glycine wightii* is poorly adapted to these conditions. *Macroptilium atropurpureum*, while well adapted to salinity, was severely depressed in the presence of excess manganese. The relative tolerance of some pasture legumes to adverse conditions associated with acid soils is summarized in Table 3.14. This shows the wide range of responses and demonstrates that some species are more demanding in their soil requirements than others. Thus it is important to understand the limiting soil conditions at particular sites and the range of physiological adaptation of the selected species being tested.

Table 3.14. Relative responses of some pasture legumes to certain acid soil factors (Andrew 1978)

Species	Low pH	Low calcium	Excess aluminium	Excess manganese	Low molybdenum
L. bainesii	5	5	5	5	4
S. humilis	5	5	5	4	4
C. pubescens	4	4	5	5	–
M. lathyroides	5	4	5	2	3
M. atropurpureum	5	4	5	1	3
D. intortum	3	4	3	3	3
D. uncinatum	3	4	4	4	2
L. leucocephala	3	3	–	3	–
G. wightii	2	1	1	1	1
M. sativa	1	1	1	1	3
T. semipilosum	3	4	4	–	–
T. rueppellianum	3	4	4	–	–

*Rating scale 1 (least tolerant) to 5 (most tolerant).

REFERENCES

Anderson, E. A. (1970). Effect of flooding on tropical grasses. *Proc. 11th Int. Grassl. Cong.*, Australia, p. 591.

— (1974). The reaction of seven *Cenchrus ciliaris* L. cultivars to flooding. *Trop. Grassl.* 8, 33.

Anderson, E. W. (1967). The rotation of deferred grazing. *J. Range Mgmt* **20**, 5.

Andrew, C. S. (1978). Mineral characterization of tropical forage legumes. In *Mineral nutrition of legumes in tropical and sub-tropical soils* (eds C. S. Andrew and E. J. Kamprath). CSIRO, Melbourne.

— and Hegarty, N. P. (1969). Comparative responses to manganese excess of eight tropical and four temperate pasture legume species. *Aust. J. agric. Res.* **20**, 687.

— and Norris, D. O. (1961). Comparative responses to calcium of five tropical and four temperate pasture legume species. *Aust. J. agric. Res.* **12**, 40.

— Johnson, A. D., and Sandland, R. L. (1973). Effect of aluminium on the growth and chemical composition of some tropical and temperate pasture legumes. *Aust. J. agric. Res.* **24**, 325.

Austin, J. D. A. (1970). Townsville stylo; looking for companion grasses. *Turnoff* **2**, 28–34.

Barnard, C. (1967). *Australian herbage plant register.* CSIRO Division of Plant Industry, Canberra.

— (1972). *Register of Australian herbage plant cultivars.* CSIRO Division of Plant Industry, Melbourne.

Beale, I. F. (1973). Tree density effects on yields of herbage and tree components in south west Queensland mulga (*Acacia aneura* F. muell) scrub. *Trop. Grassl.* **7**, 135.

Begg, J. (1963). Comparative responses of indigenous, naturalised and commercial legumes to phosphorus and sulphur. *Aust. J. exp. Agric. Anim. Husb.* **3**, 17.

Bergman, H. F. (1959). Oxygen deficiency as a cause of disease in plants. *Bot. Rev.* **25**, 417.

Bogdan, A. V. (1969). Rhodes grass – a review. *Herb. Absr.* **39**, 1.

— (1977). *Tropical pasture and fodder plants.* Longman, London.

Bond, J. H. (1974). Cattle performance and management on Townsville stylo pastures. *Trop. Grassl.* **8**, 60.

Bryan, W. W. (1961). *Lotononis bainesii* Baker – a legume for sub-tropical pastures. *Aust. J. exp. Agric. Anim. Husb.* **1**, 4.

— (1968). Grazing trials for the wallum of south eastern Queensland. 1. A comparison of four pastures. *Aust. J. exp. Agric. Anim. Husb.* **8**, 512–20.

— (1969). *Desmodium intortum* and *Desmodium uncinatum.* Herb. Absr. **39**, 183.

— (1972). Growth and chemical composition of lotononis (*Lotononis bainesii*) under simulated winter conditions. *Aust. J. exp. Agric. Anim. Husb.* **12**, 396.

— Sharpe, J. P., and Haydock, K. P. (1971). Some factors affecting the growth of *Lotononis bainesii. Aust. J. exp. Agric. Anim. Husb.* **11**, 29.

Bryant, P. M. and Humphreys, L. R. (1976). Photoperiod and temperature effects on the flowering of *Stylosanthes guyanensis. Aust. J. exp. Agric. Anim. Husb.* **16**, 506.

Burt, R. L. and Miller, C. P. (1975). *Stylosanthes* – a source of pasture legumes. *Trop. Grassl.* **9**, 117–23.

— Edye, L. A., Williams, W. T., Grof, B., and Nicholson, C. H. L. (1971). Numerical analysis of variation patterns in the genus *Stylosanthes* as an aid to plant introduction and assessment. *Aust. J. agric. Res.* **22**, 737.

Cameron, D. F. (1967). Flowering time and the natural distribution and dry matter production of Townsville lucerne (*Stylosanthes humilis*) populations. *Aust. J. exp. Agric. Anim. Husb.* **7**, 501.

Chadhokar, P. A. and Humphreys, L. R. (1974). Short day and plant age effects on flowering of *Paspalum plicatulum*. *J. Aust. Inst. agric. Sci.* **40**, 75.

Chow, K. and Crowder, L. V. (1972). Hybridization of *Desmodium canum* (Gmel.) Schin. and Thell. and *D. uncinatum* (Jacq.) DC. *Crop Sci.* **12**, 784.

Coleman, R. G. (1964). Frosts and low night temperatures as limitations to pasture development in subtropical eastern Australia. CSIRO Division of Tropical Pastures, Tech. Paper No. 3.

Coleman, R. L. (1975). Growing kikuyu in association with legumes. *Agric. Gaz. N.S.W.* **86**, 14.

Crawford, R. M. M. (1969). The physiological basis of flooding tolerance. *Ber. dt. bot. Ges.* **82**, 111.

Davis, A. G. and Martin, B. F. (1949). Observations on the effects of artificial flooding on certain herbage plants. *J. Br. Grassl. Soc.* **4**, 63.

Diatloff, A. (1973). *Leucaena* needs inoculation. *Qld agric. J.* **99**, 642.

Donald, C. M. (1963). Competition among crop and pasture plants. *Adv. Agron.* **15**, 1.

Downes, R. W. (1970). Differences between tropical and temperate grasses in rates of photosynthesis and transpiration. *Proc. 11th Int. Grassl. Cong.*, Australia, p. 527.

Ebersohn, J. P. (1969). A reconnaissance collection in four homoclimates for herbage plants with potential in semi-arid northeastern Australia. *Trop. Grassl.* **3**, 1.

Edye, L. A. and Kiers, H. J. (1966). Variation in maturity, stolon development and frost resistance in *Glycine javanica*. *Aust. J. exp. Agric. Anim. Husb.* **6**, 380.

Evans, L. T. (1964). Reproduction. In *Grasses and grasslands*. Macmillan, London.

Evans, T. R. (1967). Primary evaluation of grasses and legumes for the northern wallum of south east Queensland. *Trop. Grassl.* **1**, 143.

Evers, G. W., Holt, F. C., and Bashaw, E. C. (1969). Seed production characteristics and photoperiodic responses in buffel grass, *Cenchrus ciliaris* L. *Crop Sci.* **3**, 309.

Ferguson, J. E. (1969). Characterization of introductions of *Glycine javanica* L. *Qld J. Agric. Anim. Sci.* **26**, 517.

Francis, C. M. and Poole, M. L. (1973). Effect of waterlogging on the growth of annual *Medicago* species. *Aust. J. exp. Agric. Anim. Husb.* **13**, 711.

Gates, C. T. (1973). Comparative efficiency of development under cold stress of the tropical legumes *Lotononis bainesii* Baker and *Stylosanthes humilis* H.B.K. *Aust. J. biol. Sci.* **26**, 693.

— (1974). Water shortage and agriculture: some responses. *J. Aust. Inst. agric. Sci.* **40**, 121.

Graham, T. W. G. and Humphreys, L. R. (1970). Salinity responses of cultivars of buffel grass (*Cenchrus ciliaris*). *Aust. J. exp. Agric. Anim. Husb.* **10**, 725.

Grof, B. (1968). Viability of seed of *Brachiaria decumbens*. *Qld J. Agric. Anim. Sci.* **25**, 149.

— (1969). Viability of para grass (*Brachiaria mutica*) seed and the effect of fertilizer nitrogen on seed yield. *Qld J. Agric. Anim. Sci.* **26**, 271.

— and Harding, W. A. T. (1970*a*). Yield attributes of some species and ecotypes of *Centrosema* in north Queensland. *Qld J. Agric. Anim. Sci.* **27**, 237–43.

— — (1970*b*). Dry matter yields and animal production of Guinea grass (*Panicum maximum*) on the humid coast of north Queensland. *Trop. Grassl.* **4**, 85–95.

Gutteridge, R. C., Whiteman, P. C., and Watson, S. E. (1976). Final Report of Pasture Specics Evaluation and Soil Fertility Assessment 1973–1976. Solomon Islands Pasture Research Project, South Pacific Aid Programme, Australian Development Assistance Agency.

Hall, R. L. (1971). The influence of potassium supply on competition between *Nandi setaria* and Greenleaf desmodium. *Aust. J. exp. Agric. Anim. Husb.* **11**, 415.

Hartley, W. (1960). Plant introduction for the arid zone of Australia – its scope and limitations. *Proc. Arid Zone Tech. Conf.*, Vol. 1, p. 1. CSIRO, Melbourne.

— and Williams, R. J. (1956). Centres of distribution of cultivated pasture grasses and their significance for plant introduction. *Proc. 7th Int. Grassl. Cong.*, pp. 190–201.

Hopkinson, J. M. (1977). Siratro seed production. *Trop. Grassl.* **11**, 33–9.

Humphreys, L. R. (1974). *A guide to better pastures for the tropics and sub-tropics*, 3rd edn. Wright Stephenson, Victoria, Australia.

Hunter, R. A., Mcintyre, B. L., and McIlroy, R. J. (1970). Water-soluble carbohydrates of tropical pasture grasses and legumes. *J. Sci. Fd Agric.* **21**, 400–4.

Hutton, E. M. (1971). Variation in salt response between tropical pasture legumes. *Soc. Advanc. Breed. Res. Asia Oceania Newsl.* **3**, 75.

— and Beall, L. B. (1977). Breeding of *Macroptilium atropurpureum*. *Trop. Grassl.* **11**, 15–31.

Hutton, F. M. (1961). Inter-variety in Rhodes grass (*Chloris gayana* Kunth). *J. Br. Grassl. Soc.* **16**, 23.

Imrie, B. C. (1973*a*). Flowering response of *Macroptilium atropurpureum* to daylength temperature and nitrogen levels. *Soc. Advanc. Breed. Res. Asia Oceania Newsl.* **5**, 95.

Imrie, B. C. (1973*b*). Variation in *Desmodium intortum*: a preliminary study. *Trop. Grassl.* **7**, 305.

Ivory, D. A. (1975). The effects of temperature on the growth of tropical pasture grasses. Ph.D. Thesis, University of Queensland.

Jones, R. J. (1973). The effect of cutting management on the yield, chemical composition and *in vitro* digestibility of *Trifolium semipilosum* grown with *Paspalum dilatatum* in a sub-tropical environment. *Trop. Grassl.* **7**, 277–84.

— (1974*a*). Effect of previous cutting interval and of leaf area remaining after cutting on regrowth of *Macroptilium atropurpureum* cv. Siratro. *Aust. J. exp. Agric. Anim. Husb.* **14**, 343.

— (1974*b*). The relation of animal and pasture production to stocking rate on legume based and nitrogen fertilized subtropical pastures. *Proc. Aust. Soc. Anim. Prod.* **10**, 340–3.

— Blunt, C. G., and Holmes, J. H. G. (1976). Enlarged thyroid glands in cattle grazing *Leucaena* pastures. *Trop. Grassl.* **10**, 113–16.

Jones, R. M. (1969). Mortality of some tropical grasses and legumes following frosting in the first winter after sowing. *Trop. Grassl.* **3**, 57.

Knight, W. E. (1955). The influence of photoperiod and temperature on growth, flowering and seed production of Dallis grass, *Paspalum dilatatum* Poir. *Agron. J.* **45**, 268.

Kramer, P. J. (1951). Causes of injury to plants resulting from flooding of the soil. *Pl. Physiol., Lancaster* **26**, 722.

Lange, R. T. (1969). The piosphere: sheep track and dung patterns. *J. Range Mgmt* **22**, 396.

Lloyd, D. L. and Scateni, W. (1968). Makarikari grasses for heavy soils. *Qld agric. J.* **94**, 721.

Loch, D. S. (1976). *Paspalum plicatulum* cultivars. A case of incorrect labelling. *Trop. Grassl.* **10**, 219–20.

McCosker, T. H. and Teitzel, J. K. (1975). A review of Guinea grass for the wet tropics of Australia. *Trop. Grassl.* **9**, 177.

McCown, R. L., Gillard, P., and Edye, L. A. (1974). The annual variation in yield of pastures in the seasonally dry tropics of Queensland. *Aust. J. exp. Agric. Anim. Husb.* **14**, 328.

McIlvain, E. H. and Shoop, M. C. (1971). Shade for improving cattle gains and rangeland use. *J. Range Mgmt* **24**, 181.

McKay, J. H. E. (1973). Register of Australian Herbage Plant Cultivars, *Desmodium heterophyllum* cv. Johnstone. *J. Aust. Inst. agric. Sci.* **39**, 147.

— (1974*a*). Register of Australian Herbage Plant Cultivars, *Urochloa mosambicensis* cv. Nixon. *J. Aust. Inst. agric. Sci.* **40**, 89–91.

— (1974*b*). Register of Australian Herbage Plant Cultivars, *Brachiaria decumbens* cv. Basilisk. *J. Aust. Inst. agric. Sci.* **40**, 91–3.

McManmon, M. and Crawford, R. M. M. (1971). A metabolic theory of flooding tolerance. The significance of enzyme distribution and behaviour. *New Phytol.* **70**, 299.

't Mannetje, L. (1964). The use of some African clovers as pasture legumes in Queensland. *Aust. J. exp. Agric. Anim. Husb.* **4**, 22.

—— (1965). The effects of photoperiod on flowering growth habit, and dry matter production in four species of the genus *Stylosanthes*. *Aust. J. agric. Res.* **16**, 767.

—— (1972). The effects of some management practices on pasture production. *Trop. Grassl.* **6**, 260–3.

—— and Pritchard, A. J. (1968). The effects of photoperiod and night temperature on flowering and growth in some African *Trifolium* species. *New Phytol.* **67**, 257.

—— —— (1974). The effect of daylength and temperature on introduced legumes for the tropics and subtropics of coastal Australia. I. Dry matter production, tillering and leaf area. *Aust. J. exp. Agric. Anim. Husb.* **14**, 173.

—— and van Bennekom, K. H. L. (1974). Effect of time of sowing on flowering and growth of Townsville stylo (*Stylosanthes humilis*). *Aust. J. exp. Agric. Anim. Husb.* **14**, 182.

—— Jones, R. J., and Stobbs, T. H. (1976). Pasture evaluation by grazing experiments. In *Tropical pasture research: principles and methods* (eds N. H. Shaw and W. W. Bryan) pp. 194–234. Comm. Agric. Bur. Bulletin 51.

Mears, P. (1975). Beef cattle on the north coast. *Agric. Gaz. N.S.W.* **86**, 24.

Nitis, I. M., Rika, K., Supardjata, M., Norbudhi, K. D., and Humphreys, L. R. (1976). *Productivity of improved pastures grazed by Bali cattle under coconuts: a preliminary report*. F.K.H.P., Universitas Udayana, Denpasar, Bali, Indonesia.

Norris, D. O. (1967). The intelligent use of inoculants and lime pelleting for tropical legumes. *Trop. Grassl.* **1**, 107.

O'Reilly, M. V. (1975). *Better pastures for the tropics*. Arthur Yates, Revesby, New South Wales.

Peake, D. C. I., Strickland, R. W., and Hacker, J. B. (1975). CSIRO Tropical Crops and Pastures Divisional Report 1975–76.

Playne, M. J. (1970). The sodium concentration in some tropical pasture species with reference to animal requirements. *Aust. J. exp. Agric. Anim. Husb.* **10**, 32.

Rachie, K. O. and Roberts, L. M. (1973). Grain legumes of the lowland tropics. *Adv. Agron.* **26**, 1–132.

Reid, R. (1973). A numerical classification of sown tropical pasture regions based on the performance of sown pasture species. *Trop. Grassl.* **7**, 331.

Rotar, P. P., Park, S. J., Bromdep, A., and Urata, U. (1967). Crossing and flowering behaviour in Spanish clover *Desmodium sandwicense* E. Mey., and other *Desmodium* species. Technical Progress Report No. 164. Hawaii Agricultural Experimental Station, University of Hawaii.

Russell, J. S. (1976). Comparative self tolerance of some tropical and temperate legumes and tropical grasses. *Aust. J. exp. Agric. Anim. Husb.* **16**, 103.

—— and Moore, A. W. (1970). Detection of homoclimates by numerical analysis with reference to the brigalow region (eastern Australia). *Agric. Meteor.* 7, 455.

Russell, J. S. and Moore, A. W. (1976). Classification of climate by pattern analysis. *Agric. Meteor.* **16**, 45–70.

— and Tothill, J. C. (1974). CSIRO Tropical Agronomy Divisional Report, 1974–75.

Santhirasegaram, K. and Ferdnandez, D. E. F. (1967). Yield and competitive relationship between two species of *Brachiaria* in association. *Trop. Agric., Trinidad* **44**, 229–34.

Shaw, N. H. and Whiteman, P. C. (1977). Siratro — a success story in plant breeding. *Trop. Grassl.* **11**, 7–14.

— Ehlich, T. W., Haydock, K. P., and Waite, R. B. (1965). A comparison of seventeen introductions of *Paspalum* species and naturalized *P. dilatatum* under cutting at Samford, south-eastern Queensland. *Aust. J. exp. Agric. Anim. Husb.* **5**, 22.

Skerman, P. J. (1956). Heatwaves and their significance in Queensland's primary industries. Proc. UNESCO Arid Zone Research Conf. *Climatology and microclimatology*, Canberra, Australia, p. 195.

— and Humphreys, L. R. (1975). Flowering and seed formation of *Stylosanthes humilis* as influenced by time of sowing. *Aust. J. exp. Agric. Anim. Husb.* **15**, 74.

Strickland, R. W. (1971). CSIRO Division of Tropical Pastures, Annual Report 1971–72.

Teitzel, J. K. and Burt, R. L. (1976). *Centrosema pubescens* in Australia. *Trop. Grassl.* **10**, 5.

Vicente-Chandler, J., Rivera-Brenes, L., Caro-Costas, R. R., Rodriguay, J. P., Boneta, E., and Gracia, W. (1953). The management and utilization of the forage crops of Puerto Rico. *Agric. Expt. Sta. Bull.* 116.

Vicente-Chandler, R., Silva, S., and Figarella, J. (1959). Effect of nitrogen fertilization and frequency of cutting on yield and composition of three tropical grasses. *Agron. J.* **51**, 202–6.

Walker, B. (1977). Productivity of *Macroptilium atropurpureum* cv. Siratro pastures. *Trop. Grassl.* **11**, 79–86.

Walker, J., Moore, R. M., and Robertson, J. A. (1972). Herbage response to tree and shrub thinning in *Eucalyptus populnea* shrub woodlands. *Aust. J. agric. Res.* **23**, 405.

Wang, C. C. (1961). Growth, flowering, and forage production of some grasses and legumes in response to different photoperiods. *Herb. Absr.* **33**, 864.

Whiteman, P. C. (1968). The effects of temperature on the vegetative growth of six tropical legumes pastures. *Aust. J. exp. Agric. Anim. Husb.* **8**, 528–32.

— and Gillard, P. (1971). Species of *Urochloa* as pasture plants. *Herb. Absr.* **41**, 351–7.

— Humphreys, L. R., and Monteith, N. H. (1974). *A course manual in tropical pasture science.* Australian Vice-Chancellors' Committee.

Williams, R. J., Burt, R. L., and Strickland, R. W. (1976). Plant introduction. In *Tropical pasture research — principles and methods* (eds

N. H. Shaw and W. W. Bryan). Commonw. Agric. Bur. Bull. No. 51, pp. 77–100.

Wilson, G. P. M., Dale, A. B., and Spurrs, A. B. (1975). Kikuyu seed production. *Agric. Gaz. N.S.W.* **86**, 9.

Wutoh, J. G., Hutton, E. M., and Pritchard, A. J. (1968). The effects of photoperiod and temperature on flowering in *Glycine javanica*. *Aust. J. exp. Agric. Anim. Husb.* **8**, 544.

Younge, O. R. and Pluncknett, D. L. (1966). Beef production with heavy phosphorus fertilization in fertile wetlands of Hawaii. *Proc. 9th Int. Grassl. Cong.*, p. 959.

4

Management of tropical pastures

4.1 PASTURE ESTABLISHMENT

Tropical pasture species are sown under a wide range of conditions varying from fully cultivated prepared seed beds to oversowing of seed into undisturbed grasslands. Regardless of seed-bed conditions, development of the sown sward may be divided into three phases — germination and emergence, establishment, and the consolidation phase. The germination phase is the period from imbibition of the seed and emergence of the plumule and radicle up to the time when the seed reserves are exhausted and the seedling becomes autotrophic. The establishment phase is more difficult to define but can be considered in two parts. Firstly, the seedling establishment period which in the legumes covers the period of tap-root growth and the development of secondary roots, and in the grasses the period in which fibrous roots are formed up to the onset of tillering. The second stage could be termed the plant establishment stage covering the period of rapid growth during which tillers or secondary branches are developed and stolons are spreading, up to the time of first flowering. The sward consolidation phase may begin in the second growing season when new seedlings emerge or the interplant spaces from the originally established plants are occupied by stolons of the sown species. This last stage is commonly termed 'thickening-up'. In each stage of pasture development different factors exert a major influence.

Germination

Detailed accounts of the processes and factors involved in germination may be found in reviews, e.g. Koller (1972). This section is mainly concerned with those factors affecting field sown seed germination which are influenced by agronomic management.

Moisture conditions

The most important factors in control of germination are the weather conditions prior to and after sowing which influence soil moisture conditions. Temperature is unlikely to limit germination in the tropics, although in the subtropics low soil temperatures in winter sowing cause delayed and uneven germination in *Cajanus cajan* (Wallis, personal communication) and rotting of imbibed seed of *Sorghum bicolor* (Swanson and Hunter 1936). Successful germination requires that the rate of water movement from soil to seed meets the requirements for imbibition, and that the rate of water loss from the seed is not excessive. Rate of transfer of water to the seed is very much a function of soil–seed contact. As soil–seed contact becomes poorer the rate of water absorption and rate of germination becomes slower (Collis-George and Hector 1966). Thus rate and evenness of germination is improved when seed is sown into fine seed beds and covered, which also reduces the rate of water loss from the seed (Campbell and Swain 1973). Light rolling of the seed bed is often employed to improve soil–seed contact. Another advantage of sowing into prepared seed beds is that seed may be placed in the moist lower layers of the soil depending on depth and seed size. The depth from which seed may emerge is largely a function of seed size, and most tropical pasture seeds are planted at 0.5 to 1.5 cm (Stonard 1969; Smith 1967).

The main factor which will influence soil moisture conditions at sowing is the choice of sowing time. Analysis of climatic data will assist in determining the periods of highest rainfall probability adequate for germination and establishment (Slatyer 1960). As an example in northern Australia, Winkworth (1969) calculated that there is on average four times a year when surface moisture conditions are suitable for germination and establishment of surface sown *S. humilis*. These periods coincided with the entry of tropical cyclones or major low pressure troughs.

Seed quality and seed treatments

If conditions are suitable for germination the potential number of plants per unit area at a given sowing rate will be a function of seed size (number of seeds per kg) and the quality of the seed. Standard procedures for measuring these are defined by the International Seed Testing Association (Anon. 1966; Prodonoff 1965). The seed-purity analysis determines the amount (by weight) of pure whole seed apart from the trash, broken seeds, weed seeds, and other contaminants in the sample. Seed viability is determined by standard germination testing procedures on pure whole seed. A single index of seed quality is given by the pure live seed percentage (PLS), determined by:

$$\text{PLS \%} = \frac{\text{\% (by weight) of pure seed} \times \text{\% germination (by number) of pure seed}}{100}$$

Table 4.1. Recommended standards for germination and purity in Queensland

Species	Percentage germinable seeds		Percentage allowable by weight			
	Minimum % germinable seeds	Maximum % hard seeds	Pure seed minimum	Weed seed maximum	Other crop seed maximum	Inert matter maximum
Calopogonium mucunoides	50	–	93.5	0.2	5.0	1.5
Centrosema pubescens	50	–	93.8	0.2	5.0	1.2
Desmodium spp	70	–	94.5	0.5	0.5	5.0
Glycine wightii	60	–	97.5	0.5	0.5	2.0
Lablab purpureus	75	10	97.5	0.5	0.5	2.0
Leucaena leucocephala	60	–	97.5	0.5	0.5	2.0
Lotononis bainesii	50	45	93.0	0.5	5.0	2.0
Macroptilium atropurpureum	70	–	97.5	0.5	0.5	2.0
M. lathyroides	70	–	98.0	0.5	0.5	1.5
Macrotyloma axillare	60	10	97.5	0.5	0.5	2.0
M. uniflorum	60	–	97.5	0.5	0.5	2.0
Medicago sativa	80	30	98.0	0.5	0.5	1.5
Pueraria phaseoloides	50	–	93.5	0.2	5.0	1.5
Stylosanthes guianensis	40	20	96.5	0.5	0.5	3.0
S. humilis	40	20	90.0	0.5	0.5	9.5
Trifolium repens	75	20	93.5	0.5	5.0	1.5

Vigna luteola	70	—	98.8	0.2	—	1.2
Brachiaria spp	15	—	40.0	0.2	0.5	59.5
Cenchrus spp	20	—	90.0	1.0	1.0	9.0
Chloris gayana	20	—	90.0	1.0	3.0	7.0
Cynodon dactylon	60	—	97.0	0.5	1.0	2.0
Melinis minutiflora	30	—	40.0	0.2	0.2	59.8
Panicum antidotale	50	—	80.0	0.2	0.5	19.5
P. coloratum	20	—	80.0	0.2	0.5	19.5
P. maximum	20	—	70.0	0.2	0.5	29.5
Paspalum commersonii	40	—	93.0	1.0	2.0	5.0
P. dilatatum	60	—	60.0	0.2	1.0	39.0
P. notatum	60	—	60.0	0.2	1.0	39.0
P. plicatulum	40	—	60.0	0.2	3.0	37.0
Setaria anceps	20	—	60.0	0.2	1.0	39.0
Sorghum almum	65	—	97.3	0.2	0.5	2.2
Urochloa mosambicensis	15	—	60.0	0.2	1.0	39.0

Compiled from The Agricultural Standards (Seeds) Regulations of 1969: *Qld Govt. Gaz.* **CCXXXII** (9), 217.

Minimum standards for seed quality are laid down, by legislation in many countries, as shown in Table 4.1 for Queensland. A line of seed which fails to comply with these standards may not be offered for sale. Determination of PLS% is very important as it allows the determination of a comparative price for different lines of seed, and also allows for adjustment of sowing rate for seed lots with low PLS%.

Failure of germination of intact seed under conducive conditions may be due to a number of intrinsic seed factors which have evolved to prevent premature germination of seed to ensure the long-term survival of species. These conditions include — seed coats impermeable to entry of water or to oxygen and CO_2 exchange; germination inhibitors present in seed coats; embryo dormancy requiring a period of 'after ripening'; intrinsic germination inhibitors within the seed requiring certain sequences of conditions for breakdown.

Many of the tropical legumes and some of the grasses have impermeable seed coats or 'hardseededness'. In extensive oversowing for example, *S. humilis* in regions of unreliable rainfall this may be an advantage ensuring that all seeds will not germinate at the same time in response to insufficient rainfall for subsequent establishment. However for sowings into cultivated seed beds rapid, even germination is usually required. Hardseededness is evident in nearly all the commercially available tropical legumes except *Lablab purpureus* and *Macrotyloma uniflorum*. It is most marked in freshly harvested seed, particularly in hand-harvested samples, but declines with time in storage. In the grasses impermeable seed coats are found in *Brachiaria decumbens* (Grof 1968), *B. ruziziensis* (McLean and Grof 1968), *Panicum coloratum* (Roe 1970), and *Urochloa mosambicensis* (Harty 1972) and again germination increases with time in storage up to a maximum, after which it declines.

Various scarification treatments are employed prior to sowing to abrade the seed coat and improve permeability. These methods include:

Mechanical scarification
The seed coat is abraded by passing over abrasive surfaces. Simple rubbing between emery paper covered boards is widely used for laboratory germination testing. For commercial quantities of seed, revolving-drum scarifiers lined with abrading surfaces are available. Another type has a stationary drum and the seed is agitated by compressed air.

Acid scarification
For laboratory germination testing, immersion of seed for 15 to 25

minutes, depending on species, in concentrated sulphuric acid is widely used. Seed must be thoroughly washed after treatment in running water (Gray 1968). This method is difficult to apply to large quantities of seed, but Grof (1968) describes a method for larger quantities of *B. decumbens*.

Hot-water treatment
Immersion of seed in water at 80 °C for ten minutes is effective for a range of species, particularly *Leucaena leucocephala* (Gray 1968).

Dry heat
Jones (1969) found that heating *M. atropurpureum* at 80 °C for 12 hours reduced hardseededness and increased germination by 80 per cent. Heat treatment was also effective with *Panicum coloratum*. Scarified seed should be sown within reasonable time after treatment as scarified seed of many species tends to lose viability in storage more rapidly than unscarified seed.

Other seed treatments are imposed prior to sowing not to improve germination, but to aid better establishment.

Rhizobium inoculation
Inoculation of legumes prior to sowing is strongly recommended, particularly when introducing new species into new areas to ensure that the species are nodulated by the most effective and efficient *Rhizobium* strains. Bacterial cultures for inoculation are available in three main forms: (a) on agar slopes; (b) mixed into powdered peat, or, sometimes, (c) freeze-dried in small vials. The agar cultures are used by washing the bacterial growth from the surface of the agar and applying the suspension to the seed. Skim milk or a 10 per cent sugar solution may be used to stick the bacteria to the seed.

The peat culture is now widely used as the bacteria have a long storage life in this medium, up to one year under cool storage conditions. Peat cultures are usually applied by making up a slurry of peat and water, or with a sticking solution such as 10 per cent sugar, or 45 per cent (w/v) gum arabic solution, or a methyl cellulose solution. The aim is to provide about 50 000 bacterial cells per seed or even higher under difficult conditions.

After inoculation seed should be dried in the shade as direct sunlight is lethal to the *Rhizobium*. Seed should be sown as soon as possible after inoculation. High temperatures and humidity cause rapid loss of bacterial viability. Inoculated seed should not be sown in contact with superphosphate as the low pH (3.0) is lethal to *Rhizobium*.

Table 4.2. A guide to inoculum and pelleting requirement of legumes used in tropical pastures (Diatloff 1971)

Common name	Botanical name	Inoculum requirement	Pelleting material L = lime RP = rock phosphate
Calopo	*Calopogonium mucunoides*	cowpea	RP
Pigeon pea	*Cajanus cajan*	cowpea	RP
Centro	*Centrosema pubescens*	specific	RP
Greenleaf desmodium	*Desmodium intortum*	desmodium	RP
Silverleaf desmodium	*D. uncinatum*	desmodium	RP
Archer axillaris	*Macrotyloma axillare*	cowpea	RP
Leichhardt uniflorus	*M. uniflorum*	cowpea	RP
Lablab	*Lablab purpureus*	cowpea	RP
Cooper, Clarence, or Tinaroo	*Glycine wightii*	cowpea	RP
Peruvian leucaena	*Leucaena leucocephala*	specific	L
Miles lotononis	*Lotononis bainesii*	specific	RP
Lucerne	*Medicago sativa*	lucerne	L
Siratro	*Macroptilium atropurpureum*	cowpea	RP
Green gram, mungbean	*Phaseolus aureus*	cowpea	RP
Phasey bean	*Macroptilium lathyroides*	cowpea	RP
Black gram	*Phaseolus mungo*	cowpea	RP
Tropical kudzu	*Pueraria phaseoloides*	cowpea	RP
Velvet bean	*Stizolobium deeringianum*	cowpea	RP
Schofield stylo	*Stylosanthes guyanensis*	cowpea	RP
Oxley fine stem style	*S. guyanensis*	specific	RP
Townsville stylo	*S. humilis*	cowpea	RP
White clover	*Trifolium repens*	clover	L
Kenya white clover	*T. semipilosum*	specific	L
Dalrymple vigna	*Vigna luteola*	cowpea	RP
Cowpea	*V. sinensis*	cowpea	RP

If seed is to be sown mixed with superphosphate, and this is often required in aerial sowing, then it should be pelleted after inoculation to protect the seed. For most tropical legumes, except *Leucaena*, the best pelleting agent is finely sieved *rock* phosphate. This is used as the 'cowpea' type of *Rhizobium* associated with most tropical legumes produces an alkaline reaction. Hence pelleting with lime may be harmful to these strains (Norris 1967). A guide to inoculation and pelleting requirements is given in Table 4.2 (Diatloff 1971). Full details regarding preparation of cultures and inoculation are given by Norris (1964) and Vincent (1970).

Chemical treatments

Various fungicides may be applied to seeds to prevent seed and seedling rots such as 'damping off' (*Pythium* spp). Usually such treatments are confined to small scale or experimental sowing. Fungicides should not be used in combination with inoculation.

A common problem with surface sown seeding is harvesting by ants. This is common in sowing of *S. humilis* or buffel grass (*Cenchrus ciliaris*). Dusting of seed with a BHC preparation such as 'Lindane' is often used to prevent this. Pelleting of seed with lime or rock phosphate also reduces seed loss (Jones 1975*b*).

Another problem sometimes encountered in sowing of *Macroptilium* species and *Vigna marina* is attack by bean fly (*Melanagromyza phaseoli*). The larvae eat through the stem base and can cause almost 100 per cent loss of plants. Soaking seed before planting in cyclodiene insecticide (aldrin, dieldrin, or endrin) at 2 to 4 g of active ingredient per kg of seed effectively controls bean fly during establishment. Seed is dried after wetting. The presence of the seed dressing had little effect on *Rhizobium* viability and subsequent nodulation (Jones 1965).

Seedling establishment

Once the germination processes are under way, rate of seedling root development is often very important in conditions of rapid soil desiccation (McWilliam, Clements, and Dowling 1970). Seedling vigour and rate of radicle elongation can be related to seed size (Campbell and Swain 1973), both between species (Whiteman 1968; Tudsri and Whiteman 1977*a*) and within species (Imrie 1972). Also in legumes a larger seed size may overcome some of the problems arising from slow nodulation. Diatloff (1974) showed in *Glycine wightii* cv. Tinaroo that nitrogen reserves in the seed were depleted before nitrogen fixation began. A larger seed nitrogen reserve maintained seedling growth until nodulation was initiated.

When seed reserves are exhausted the seedling becomes dependent on soil nutrients, and deficiencies can depress establishment. Also excessive applications of fertilizer may depress or delay establishment. Low moisture levels and high rates of urea application at sowing reduced emergence of *Medicago sativa* by 93 per cent (Mar and Nielsen 1970). Application of high levels of potassium chloride (100–200 kg KCl ha^{-1}) caused seedling mortality in *D. intortum* (Jones 1973). Establishment of *Stylosanthes humilis* was also shown to be reduced by increased levels of chloride ion (Hall 1971).

More usually nutrient deficiencies will be a major factor affecting establishment. Seedlings of *Trifolium subterraneum* require exogenous nutrients within seven days for calcium, ten days for phosphorus, 14 days for nitrogen and magnesium, and 21 days for potassium (Krigel 1967). However, McWilliam *et al.* (1970) have shown that clover seedlings respond to phosphorus as early as four days after imbibition. In the legumes phosphorus has a greater effect than nitrogen or potassium in stimulating the growth of seedlings. The effects of increased availability of phosphorus on the field establishment of tropical legumes have been widely reported in Queensland (Blunt and Humphreys 1970), in Indonesia (Steel and Humphreys 1974), in Kenya (Keya, Olsen, and Holliday 1971), and in Uganda (Olsen and Moe 1971). The marked effects of increased phosphorus levels in low phosphorus soils on growth rates and nodulation of *S. humilis* has been described in detail by Gates (1974). He showed that nodules were first detected at 11 days in high phosphorus plants, but not until day 14 in low phosphorus. Nodule growth rates in high phosphorus plants were twice those of low phosphorus. Phosphorus promoted the formation of an effective symbiosis and nitrogen fixation. Assimilation of nitrogen by the whole plant was increased from 17 mg per g nodule dry weight per day at low to 53 mg per g nodule dry weight per day with high phosphorus over days 23 to 26. Sulphur caused an increase in dry weight, but response to sulphur mainly occurred late in development and was smaller than the response to phosphorus in young seedlings.

Pathogens and pests may have important effects in early seedling stages. Treatments for 'damping off' fungi (*Pythium* spp) and bean fly have been previously discussed. Seedling mortality due to termite (*Macrotermes* spp) damage has been noted in *Stolysanthes* spp in Malawi (Thomas 1976), northern Australia (Wallace 1970), and Thailand (personal observation). With *Trifolium semipilosum* 'rugose leaf curl virus' causes heavy plant mortality in the first year, but the stand usually recovers to form a productive sward (Jones and Date 1975).

Plant establishment and consolidation

As the developing seedlings grow to fill the space available, competition for growth factors in limiting supply will then ensue. Competition or interference may be with other plants of the same species, with other species sown at the same time, with existing plants in an oversowing situation, or with volunteer weed species. Plants growing together in the same space modify the environment of adjacent plants. In some cases this may be beneficial as in the case of the legume which fixes nitrogen, part of which ultimately becomes available to adjacent grass plants. On the other hand one plant may reduce the availability of some factor such as light, water, or nutrients to an adjacent plant, causing suppression of that plant and dominance of the other.

One of the major roles of pasture management is to modify the interactions of species to maintain a desired balance. This is of particular significance during the establishment and consolidation phase. Application of fertilizers may markedly alter the interaction. When *D. intortum* and *Setaria anceps* were grown together in a soil of low potassium status, the *D. intortum* was suppressed, yet growth was satisfactory when grown alone (Hall 1974). When potassium fertilizer was applied growth rate in *D. intortum* with *S. anceps* was restored to near its original level. Hall (1974) thus demonstrated that responses to applied nutrients are modified when mixtures are grown compared with pure stands.

Another important nutrient interaction in plant establishment is between phosphorus and nitrogen. In soils low in nitrogen, application of phosphorus usually enhances legume establishment, but in soils high in nitrogen, application of phosphorus may increase grass dominance and reduce legume establishment (Blunt and Humphreys 1970). Jones (1975*a*) has also shown that defoliation treatments to reduce grass and weed shading effects on legume establishment had greater effects in high fertility than in low-fertility soils. This experiment also showed that survival of weaker sown species, in this case *D. intortum* was improved by early topping or grazing. Cut material should not be left on plots to shade seedlings beneath.

Where weeds present a particular problem weedicides may be used, but the legumes vary in their sensitivity. Siratro is very sensitive to 2,4D, the *Desmodiums* had good tolerance, *Glycine wightii* intermediate tolerance, while stylo is very resistant (Bailey 1969). An important weed problem in many humid tropical pasture sowings is *Mimosa invisa*.

Although one or two mower slashings in the first year will control this weed, this is expensive and equally good control has been achieved by dragging a large roller behind a tractor to crush and break the weed. The pasture soon grows over the weeds.

Management in the consolidation phase aims to limit weed development or excessive grass growth by judicious cutting or grazing, or even hand slashing or hoeing or spot spraying of woody weeds. With adapted pasture species given adequate nutrition, weeds should be only a minor problem by the second year (Jones 1975).

Land clearing

Methods used for pasture establishment will depend on the intensity and scale of development. When an introduced species is sown some modification of the environment is usually required, which may range from complete replacement of existing vegetation to oversowing accompanied by small changes in grazing management.

Where cultivation for seed-bed preparation is undertaken, then tree clearing is obviously required. Also dense forest must be cleared to allow pasture development. In savannah and open woodlands aerial oversowing of legumes, particularly *S. humilis*, has been undertaken without tree clearing, but yields of pasture were closely related to the degree of thinning or removal of trees (Gillard, 1970; Walker, Moore, and Robertson 1972), and levels of animal production were also related to tree density in oversown grasslands (Williams and Gillard 1971).

The control of woody species on grazing lands may present at least three separate problems — killing the original trees and shrubs, killing vegetative regrowth following original clearing, and preventing regeneration from seed. Clearing may be undertaken by hand methods — felling, ring barking, basal bark spraying or basal injection with herbicides such as 2,4D, 2,4,5T, and Picloram or mixtures of these marketed as 'Tordon' (Robertson and Young 1976). Basal spraying is applicable to multistem species, while injection has been highly successful with a range of *Eucalyptus* species.

For large-scale clearing of open *Eucalyptus* forest, *Acacia harpophylla* scrub, *Acacia cambagei* scrub, and scrub heath-lands in Australia a method of pulling the forest down with a large chain mounted between two crawler tractors has been widely employed. Up to 40 to 50 ha per day can be cleared by this method at an approximate cost of US$6.50 to 12.50 per ha (Coaldrake 1970). With dense scrub, after a suitable period of drying, the material can be burnt *in situ*, and pasture seed sown into the ash, which provides a good weed-free seed bed (Purcell 1964). Approximately 2 million ha have been developed in this way in Queensland, and sown to Rhodes grass (*C. gayana*), green panic (*P. maximum* var. *trichoglume*), or buffel grass (*C. ciliaris*). Where the original vegetation was not dense enough, it had to be pushed up and raked into windrows before burning at an added cost of some US$25–40 per ha.

Clearing humid tropical rainforest is more difficult and expensive and should be carefully planned before starting. The value of the existing forest in terms of timber production and watershed protection should be properly considered compared with potential returns from conversion to pasture land. Important land-use questions and economic comparisons of rainforest clearing are discussed in the South American context by Nelson (1973). Many projects based on large-scale clearing of virgin rainforest have failed because of limitations of soil fertility, lack of markets, distance from markets, and poor technical and administrative ability of the developers. Nelson (1973) also provides data on the costs of clearing for a number of projects, based on large-scale machinery — bulldozers with blades and tree pushers, root rakes, tree crushers, and giant disc ploughs. Costs per hectare varied from US$80 to US$670 (Table 4.3).

In some areas, e.g. in the Solomon Islands, pasture development follows hand clearing of rainforest. First undergrowth and small trees are hand cut and when possible burnt. Large trees are felled with chain saws and millable commercial timber is removed. Other trees are burnt if possible or left to rot. Stumps are left and vegetative cuttings of para grass (*Brachiaria mutica*), koronivia grass (*B. dictyoneura*), or batiki blue grass (*Ischaemum aristatum*) and legume seed (*C. pubescens, P. phaseoloides*, or *G. wightii*) are planted by hand. Regular hand slashing of any woody weed-growth in the first year leads to rapid and successful pasture establishment.

Oversowing

Low-cost pasture improvement has been achieved in many areas of the world by sowing a legume into natural tropical grasslands. The most important factors affecting germination of surface-sown seed will be suitable moisture conditions for a sufficient period of time, surface soil characteristics to allow penetration of the radicle, and competition from the existing species. Miller and Perry (1968) and Winkworth 1969) have shown that at least 100 hours of low soil moisture tension

Table 4.3. Capacity of equipment and cost of mechanized forest clearing and land preparation in the humid tropics (Nelson 1973)

Machinery in the clearing unit	Area of project	Type of forest[c]	Annual rainfall (mm)[d]	Land-clearing and preparation capacity[a] Ha per hour[e]	Equipment investment[b] (US $) Total	Cost per ha (US $) Total (excluding profit)	Owner's profit[f]	Equivalent contract price
(A) 4 270-hp tractors with shearing blades[g]	Tingo	Medium	2400	0.60	400 000	240	35	275
1 160-hp tractor with tree-pusher bar	Maria-Tocache, Peru							
(B) 1 180-hp tractor with shearing blade	Cariari, Costa Rica	Medium-dense	4000	0.26	160 000	170	25	195
1 180-hp tractor with tree-pusher bar								
1 95-hp tractor with bulldozer blade								
2 giant discs								
(C) 2 235-hp tractors with shearing blades	Peru	Medium	—	0.21	230 000	250[h]	40	290
1 235-hp tractor with angle-dozer blade								
2 tree-pusher bars								

2 root rakes								
1 100-metre chain								
1 giant disc								
(D) 2 90-hp tractors with angle-dozer blades	Peru	Light	–	0.22	90 000	100[i]	15	115
2 tree-pusher bars								
2 root rakes								
1 100-metre chain								
1 giant disc								
(E) 1 475-hp tree-crusher	Ivory Coast	Medium	–	0.80	450 000	–	–	–
1 tree-crusher tender								
3 240-hp tractors with bulldozer blades								
2 giant discs								
(F) 1 475-hp tree-crusher	Pucallpa, Peru	Medium	1800	0.85[j]	400 000	95[k]	–	210[l]
1 960-hp tree-crusher								
1 tree-crusher tender								
1 960-hp diesel–electric rubber-tyred tractor								
1 635-hp diesel–electric rubber-tyred tractor								

(continued)

Table 4.3. (*cont'd*)

Machinery in the clearing unit	Area of project	Type of forest[c]	Annual rainfall (mm)[d]	Land-clearing and preparation capacity[a] Ha per hour[e]	Equipment investment[b] (US $) Total	Cost per ha (US $)		
						Total (excluding profit)	Owner's profit[f]	Equivalent contract price
1 root rake								
2 giant discs								
(G) 1 960-hp tree-crusher with root rake	British Honduras	Medium	1300	0.90[m]	–	60[n]	20[o]	80
1 170-hp tractor with root rake								
1 giant disc								
(H) 2 960-hp tree-crushers	(Theoretical)	Medium	–	2.50	680 000	–	–	–
2 960-hp diesel–electric tractors with root rakes								
2 635-hp diesel–electric tractors with root rakes								
2 270-hp tractors with bulldozer blades								
2 giant discs								

(I)	Combination of 230–270-hp, 150–170-hp, and 93–115-hp tractors	Ivory Coast	Dense medium light	— — —	— — —	— — —	— — —	— — —	450–570h 290–450p 180–290p
(J)	4 180-hp tractors with bulldozer blades	La Chontalpa	Medium-light	—	—	—	125	—	250q
	2 giant discs								
	1 100-metre chain								

Sources: Enrique Ferrayros y Cía, S. A., *Operación Tocache* (Lima, February 1968), p. 68; Manuel Abastos Gómez, *Inventario y evaluación de la concesión Tournavista*, R. G. Le Tourneau, Inc. (Lima, July 1967); files of the Comisión de Grijalva, Cárdenas, Tabasco, Mexico; files of UNDP Huallaga Central Project, Tarapoto, Peru; and *Oleagineaux*, November 1966.

Note: Dashes indicate 'not available'.

[a]Land clearing and preparation includes felling, windrowing (either after burning in place or without burning) with 50- or 100-metre centres, reburning in the windrow, and two giant discings. [b]Investment is based on the price of machinery at the port of entry, excluding taxes and including 10 per cent allowance for inventory of replacement parts, with costs of transport to the site added. [c]Dense forest is taken as that with 600–1000 trees per hectare with over 100 being more than 60 cm DBH. Medium is defined as an average of 50–99 trees per ha of more than 60 cm DBH. Light is second growth or dry land forest of the xerophytic, caatinga, or cerrado type. [d]The volume and distribution of rainfall together with the consistency and drying rate of the soil will have a significant effect on idle time of machinery and consequently on unit costs. [e]Based on the assumption that machinery is working at 100 per cent efficiency. [f]Unless otherwise stated, owner's profit is assumed at 15 per cent of total costs. [g]Shearing blades, such as the Rome Plough Company's 4-ton model K-G, are bulldozer blades equipped with a horizontal knife edge along the base and a vertical stinger, or wedge, on one edge used for splitting standing trees. [h]Based on 14.5 tractor (235-hp) hours per ha at $20 per hour. [i]Based on 9 tractor (90-hp) hours per ha at $11 per hour. [j]Based on approximately 1000 hours by treecrushers to clear 850 ha. [k]Based on crushing and burning in place at $40 per ha, windrowing at $30, and discing at $25. [l]Approximate contract price for similar unit in the Pucallpa region. Price for crushing, burning in place, windrowing and reburning – $185 per ha. [m]Crushing rate alone, 1.4 ha per hour. [n]Assumed cost of root raking after windrowing, reburning, and two giant discings – $25 per ha. [o]Based on a margin of 30 per cent for crushing, removal of stumps, windrowing, and burning. [p]Contract quotations. [q]Contract price on the La Chontalpa project, not necessarily with the same equipment.

in the top 2.6 cm of soil is required for successful establishment of *S. humilis* oversown into natural grasslands. Surface moisture conditions may be improved by the presence of vegetation or litter and the rate of germination increased compared with bare burnt surface (Miller and Perry 1968; Tudsri and Whiteman 1977*b*).

However, pretreatments are usually applied to reduce competition from existing species. A single light cultivation improved establishment of *S. humilis* under conditions when weather was less favourable (Norman 1961). Heavy grazing prior to sowing has also improved establishment (Norman 1961). Burning has the advantage of a low cost rapid pretreatment. Stocker and Sturtz (1966) found that an early wet season burn, which destroyed young seedlings in an annual *Sorghum intrans* grassland was most effective in reducing competition in the establishment of oversown *S. humilis*. Burning of *Imperata cylindrica* prior to oversowing with addition of 30 kg P ha^{-1} plus 100 g molybdenum ha^{-1}, allowed the successful establishment of *M. atropurpureum* cv. Siratro, *Stylosanthes guyanensis* cv. Schofield, and *Glycine wightii* cv. Tinaroo. Good establishment and yield of Siratro in Fiji (Partridge 1975) and of *S. guyanensis* in the Solomon Islands (Gutteridge and Whiteman 1977) was obtained after burning and oversowing of *Pennisetum polystachyon* grasslands. Burning prior to oversowing of legumes has been widely used and other examples are noted from East Africa (Keya, Olsen, and Holliday 1972), West Africa (Haggar, de Leeuw, and Agishi 1971), and the Philippines (Magatan, Janier, and Madamba 1974).

Chemical spraying with herbicides has been used with varying success. In Hawaii aerial application of the herbicide 'Silvex' to control mixed shrub and tree vegetation including *Melastoma* sp and *Psidium guajava* prior to aerial oversowing of *D. intortum*, *Stylosanthes*, and *P. maximum* led to successful pasture establishment in steep topography (Motooka, Plucknett, Saiki, and Younge 1967). Keya *et al.* (1972) obtained successful establishment of *D. uncinatum* after spraying *Hyparrhenia* although the heavy litter build up tended to trap seed and smother developing seedlings.

While considerable success has been achieved in oversowing legumes into natural grasslands, attempts to re-establish legumes into vigorous stands of sown grasses have been less successful. Pretreatments of cutting, burning, or spraying with 'Paraquat' in *Setaria anceps* swards did not sufficiently reduce tiller density to allow oversown Siratro to establish (Tudsri and Whiteman 1977*b*). Light cultivation was also unsuccessful (Middleton 1973). Only rigorous cultivation of the Setaria allowed successful establishment, and this was not improved by any treatments prior to cultivation (Tudsri and Whiteman 1977*b*). *Lotononis bainesii* was successfully oversown into *Setaria anceps* swards (Tudsri and Whiteman 1977*a*) and *Digitaria decumbens* swards (L. R. Humphreys, personal communication) following heavy grazing and cultivation, and by sowing at the start of the cool season when *L. bainesii* is able to grow and the grass species are relatively dormant.

Management following oversowing varies. With *S. humilis* heavy stocking following sowing is recommended to reduce grass competition (Shaw 1961). *S. humilis* is less utilized during the vegetative growth stages. Stock should be removed prior to flowering and seed set, after which they can be returned to utilize the dry plant material which is preferred. Because of its low growing habit, *Lotononis bainesii* is also favoured by reasonable levels of grazing after seedling establishment. With the twining tropical legumes grazing should be deferred until the legumes are well established and have set seed, after which grazing pressure may be increased to open up the stand and allow establishment from the current seed crop (Walker and Potere 1974).

4.2 THE ROLE OF THE LEGUME IN TROPICAL PASTURE PRODUCTION

The importance of nitrogen in pasture productivity

In many tropical areas nitrogen deficiency is a major limitation to pasture productivity. This is demonstrated by the large number of simple nitrogen rate trials which show marked responses in pasture dry matter production to increasing inputs of fertilizer nitrogen. One example is from the work of Henzell (1963), where nitrogen inputs up to 448 kg ha^{-1} a^{-1} increased dry-matter yields of a Rhodes grass pasture from 2000 kg ha^{-1} at 0, to 20 000 kg ha^{-1} at 448 kg N ha^{-1}. There are many similar examples.

Deficiencies of other macro- and micro-elements can be overcome by basal and maintenance applications of appropriate fertilizers at a moderate cost. Furthermore, in the case of phosphorus and potassium some soil store of these elements can be built up, so that in time, maintenance applications may be reduced. With micro-elements such as copper, zinc, molybdenum, etc., applications are required only at longer intervals of three to seven years. However with nitrogen, a soil store is not built up, and similar amounts must be applied each year, as shown in Table 4.4, making nitrogen far more expensive than other fertilizer costs.

Table 4.4. Relative costs per hectare of applying macro-elements

Nutrient	Approximate price per kg element (US $ kg^{-1} (1974))	Cost ha^{-1} (US $)	
		Basal	Maintenance
N: urea, at 400 kg N ha^{-1}	0.20	80.00	80.00
*P: superphosphate	0.48	19.20	9.60
†K: KCl	0.20	20.40	10.20

*P as 400 kg super basal, and 200 kg ha^{-1} a^{-1} maintenance.
†K as 200 kg KCl basal, and 100 kg ha^{-1} a^{-1} maintenance.

Nitrogen is important in pasture productivity in two major aspects; firstly, in maintaining dry-matter yields which are often linearly related to nitrogen availability up to certain levels, and secondly in maintaining an important aspect of pasture quality, that is the protein content. Since stocking rate is very much a function of dry-matter yield, and production per head is affected by pasture quality, it can be seen that nitrogen plays a key role in the productivity of livestock at pasture.

Sources of nitrogen for pasture production

Before considering the role of the legume in tropical pastures, the sources of nitrogen for pastures are briefly assessed. These can be listed as:

(a) soil nitrogen – nitrogen built up in the soil from accumulation of organic matter, plus the contribution of free-living nitrogen fixing bacteria, blue–green algae, and nitrogen compounds in rainfall;
(b) nitrogen from industrial synthesis – fertilizer nitrogen;
(c) nitrogen from fixation by the legume – *Rhizobium* symbiosis; and
(d) nitrogen fixation by organisms associated with some tropical grasses.

(a) Soil nitrogen

Under extensive grazing conditions and in the absence of a significant legume component, the soil stored nitrogen is the main source of nitrogen for vast areas of natural pastures. Under many tropical situations soil nitrogen status is low, leading to severe limitations in grassland production and herbage protein content, even though in these soils the total amount of nitrogen in the root zone may appear to be high. Hubble and Martin (1962) calculated total nitrogen in the profiles of a range of Australian soils. Tropical red–yellow podzolics contained approximately 4000 kg N ha^{-1}, while a red krasnozem after 60 years pasture following clearing from rainforest contained 10 000 kg N ha^{-1}.

However, the limitation to pasture production is not the total amount of nitrogen in the profile, but the rate at which this nitrogen is mineralized to available forms for uptake by plants. Henzell (1968) has shown that the rate of mineralization under a grassland or sward cover is usually less than 1 per cent per annum. Under an undisturbed sward cover the rate of mineralization is inhibited by factors not yet fully understood. Thus in a profile containing 5000 kg total N ha^{-1}, less than 50 kg N ha^{-1} a^{-1} is made available for uptake. Norman (1966) found in a sandy-loam soil profile under natural pastures at Katherine containing 3000 kg N ha^{-1}, that only 6 kg N ha^{-1} a^{-1} became available for uptake. These amounts of nitrogen are obviously too low to support the potential production of tropical pasture swards.

(b) *Nitrogen from industrial synthesis*

Fertilizer nitrogen has been widely used for intensive pasture production on temperate pastures in Europe and the United States. In the Netherlands, for example, an average rate of approximately 100 kg N ha^{-1} a^{-1} is applied over all pasture lands, while in England an average usage in the grassland sector is about 70 kg ha^{-1} a^{-1} (Crozier 1976). However, production of nitrogenous fertilizers requires a high energy input, and in Britain nitrogenous fertilizers account for 26 per cent of total energy consumed in British agriculture. With the increasing costs of energy this is of considerable concern, but it is unlikely that the use of nitrogen fertilizers in intensive temperate grassland farming can be replaced by legume nitrogen, at least in the short term.

Under the more extensive grazing conditions in the tropics the use of nitrogen fertilizers, especially for beef production, is unlikely to be economic. Tropical grasses under optimum conditions with high nitrogen inputs are capable of very high dry-matter yields, up to 86 000 kg ha^{-1} a^{-1} have been recorded (Vicente-Chandler, Silva, and Figarella 1961). However, under practical field conditions other limitations of feed quality, rainfall, deficiencies of other nutrients, and genetic potential of the livestock, combined with the high cost of nitrogen, tend to limit the economic potential of high nitrogen, high production systems in the tropics. Nevertheless there are roles for nitrogen-fertilized pastures in tropics and these are discussed in Section 5.3.

(c) *Nitrogen from the legume–Rhizobium symbiosis*

Because of the generally low levels of available soil nitrogen under pastures, and the costs of synthetic nitrogen fertilizers, pasture improvement based on biological nitrogen fixation will assume increasing importance, as it has already done in some countries, e.g. New Zealand and Australia. Furthermore biological nitrogen fixation avoids many of the problems of industrial synthesis, e.g. high capital costs, and pollution problems in the manufacturing process and in nitrate leaching in the field.

The *Rhizobium*-legume symbiosis is seen as giving mutual benefit to both partners, the legume supplying sugars to the *Rhizobium* in the nodules, while the *Rhizobium* in turn supplies fixed nitrogen to the host. However, Dixon (1969) points out that the bacteria which infect the host plant become bacteroids and are no longer capable of division and thus are not benefited by the association. The major benefit that the *Rhizobium* gains from the symbiotic association arises from the rhizosphere of the legume. The advantage they give to the host by nitrogen fixation is returned by the stronger and more adequately nourished plant excreting larger quantities of substances into the rhizosphere, thus benefiting the rhizosphere *Rhizobium*.

(d) *Nitrogen fixation by organisms associated with tropical grasses*

Nitrogen fixation by micro-organisms associated with the root rhizosphere of tropical grasses holds potential for increased production of quality herbage and addition of nitrogen to the soil reservoir through recycling by grazing animals. Species such as *Andropogon, Brachiaria, Cynodon, Digitaria, Hyparrhenia, Melinis, Panicum, Paspalum, Pennisetum, Saccharum*, and *Zea* have this capability (Dobereiner and Day 1974). They all possess a carbon pathway (C_4 dicarboxylic acid) in photosynthesis which is more efficient than the C_3 pathway in temperate grasses. Wide variation occurs among grasses in nitrogen fixed by the associative organisms, but estimates of about 1.5 kg ha^{-1} day^{-1} have been made for *Digitaria decumbens* and *Paspalum notatum* (Dobereiner and Day 1974).

The *Paspalum notatum–Azobacter paspali* association, described in Brazil, has been the most intensely studied (Day, Neves, and Dobereiner 1975). Activity of this organism occurred largely in the rhizosphere of grass roots, showing greatest activity during the growing season.

Accumulation of nitrogen persisted into the dry season and enhanced growth and nitrogen content of herbage. Ecotypes of *P. notatum* and *Pennisetum purpureum* differed significantly in nitrogenase activity on their roots; native forms of the grasses were more efficient nitrogen-fixers. Nitrogen fixation also varied with different soils from different localities. This suggested that commercial grass cultivars may have been selected for response to higher mineral nitrogen levels. Such variability among grass types poses a challenge to plant breeders to promote this characteristic through breeding and selection (Day *et al.* 1975).

The fate of fixed nitrogen was ascertained by studies of *P. notatum* growing in vermiculite with nitrogen-free solution. At no time did the grass plants display signs of nitrogen deficiency. After two months, the quantity of nitrogen had increased several-fold as plants developed. Plants contained a high percentage of the nitrogen but a significant quantity remained in the vermiculite, indicating that a portion would be made available to associated vegetation.

In Brazil, seasonal variation in nitrogenase activity was observed in *Pennisetum purpureum* and *D. decumbens* grown in the field, with and without applied nitrogen. Highest rates of nitrogen fixation occurred during the rainy season and when warm temperatures prevailed. Additions of 20 kg ha^{-1} of fertilizer nitrogen every two weeks did not interfere with nitrogenase activity, even after application of 160 kg ha^{-1}. Further studies with bahia grass growing in pots showed a decline in organism activity with the use of an ammonium form of nitrogen but not with a nitrate form (Dobereiner and Day 1974; Day *et al.* 1975).

The discovery of nitrogen fixation within the roots of tropical grasses has important implications in the nitrogen economy of tropical grasslands (Day and Dobereiner 1976). The organism *Spirillum lipoferum* was found concentrated in the cortex cells of *D. decumbens*. This organism appears to be associated with the C_4 grasses which have malate as one of the primary photosynthetic products. *S. lipoferum* utilizes malate as an energy source and in association with grasses may be dependent upon them for this product as a substrate. Similar strains of *Spirillum* were isolated from guinea grass (*P. maximum*) and corn (*Zea mays*) (Albrecht, Okon, Lonnquist, and Burrie 1976). Nitrogen fixation rates of up to 1 kg N ha^{-1} day^{-1} have been estimated in *Digitaria* spp in the field (Schank, Day, and de Lucas 1977). These findings suggested a symbiotic relationship as well as a specific *Spirillum* relationship with grass species. Just as with *Rhizobium* of legumes,

it should then be possible to improve the effectiveness of host plant–
bacteria symbiosis.

The role of the legume

In a mixed pasture sward the legume is an integral component of the
feed on offer to the grazing animal, and makes an important contribution to the nutritive value of the pasture, apart from its role in providing a nitrogen input into the pasture system. Thus two aspects of the
role of the legume are examined.

(a) *The legume as a component in animal nutrition*

The nutritive value of a pasture component is best assessed in terms of
its dry-matter digestibility (DMD), the voluntary intake (VI), and the
efficiency of utilization of the digested material. Two other factors to
be taken into account when the pasture is the sole source of animal
feed are the protein content and mineral content.

(i) Protein content

Because the legume is independent of the soil nitrogen and obtains its
nitrogen through biological nitrogen fixation, the protein content of
the legume is usually higher than that of the associated grasses at similar
ages or stages of growth. This is true in both temperate and tropical
pastures, but is much more important in tropical pastures where the
grasses tend to decline more rapidly in protein content with increasing
maturity (Milford and Haydock 1965) (Fig. 4.1).

Deficiencies of dietary protein depress intake of dry matter. When
the crude protein (CP percentage) content of temperate hays falls below
8.5 per cent CP (Blaxter and Wilson 1963), and in tropical grasses
below 7.0 per cent CP (Milford and Minson 1965), dry-matter intake is
depressed. Thus the legume in the pasture makes an important contribution in maintaining adequate protein levels in the feed on offer. This
legume can act as supplement, and by maintaining CP intake above the
critical levels allows increased intake of the associated grass. This has
been shown to be a major factor in the value of *Stylosanthes humilis* in
increasing animal production on mature tropical grass swards in the dry
season (Norman 1970; Playne 1969*a*).

(ii) Nutritive value

Animal production is a function of the daily intake of digestible dry
matter and therefore depends on both the quantity of food eaten and
the digestibility of the feed (Holmes, Franklin, and Lambourne 1966).
As pastures mature, the fibre content increases leading to a reduction

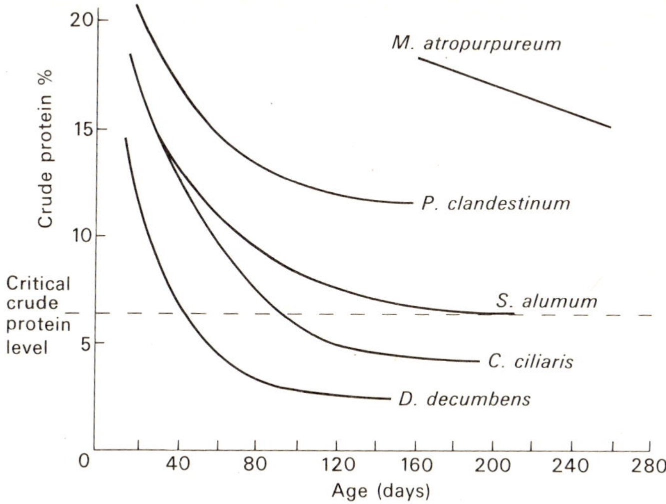

Fig. 4.1. Relationship between crude protein content and plant age in four grasses and the legume *M. atropurpureum* (after Milford and Haydock 1965).

in digestibility and voluntary intake. As in the case of protein content, this decline is more marked in tropical grasses than in the legumes. Playne (personal communication) has shown that in mature spear grass (*H. contortus*) the cell wall constituents (CWC) attained values up to 78 per cent with DMD values as low as 27 per cent, while *Stylosanthes humilis* in the same pasture attained maximum CWC values of 68 per cent, and minimum DMD of 43 per cent. Since intake is a function of digestibility and passage time in the gut, such differences can have marked effects on voluntary intake of digestible energy and hence animal production, apart from the important effects of maintaining crude protein intake.

Dry-matter digestibility and voluntary intake of legumes is generally higher than the grasses except at the earliest stages of growth where they may be similar. This is true of the temperate legumes, e.g. white clover (*T. repens*) (Bland 1968) and generally true also for the tropical legumes, with some exceptions (Table 4.5).

Although the values in Table 4.5 are difficult to compare directly as different species have been grown separately and cut at differing intervals, general trends are evident. While DMD values do not differ widely between the two groups these small differences can have marked effects

Table 4.5. Comparison of dry matter digestibility (DMD) and voluntary intake (VI) values for tropical grasses and legumes from indoor feeding trials with sheep

Species	Age of regrowth (days)	DMD%	VI g kg^{-1} body weight$^{0.75}$	Reference
Grasses				
Chloris gayana	28	61–4	50–2	Milford and Minson (1968*a*)
C. gayana	112	49–51	39–45	Milford and Minson (1968*a*)
Digitaria decumbens	7	64	48	Minson and Milford (1968)
D. decumbens	84	48–54	–	Minson and Milford (1967)
Pennisetum clandestinum	52	62	46–50	Minson and Milford (1968)
Setaria anceps	52	56	48	Minson and Milford (1968)
S. anceps	102	48	28	Minson and Milford (1968)
Cenchrus ciliaris	56	56	45	Minson and Milford (1968)
C. ciliaris	112	46	34	Minson and Milford (1968)
C. ciliaris	140	41	29	Minson and Milford (1968)
Legumes				
Lablab purpureus	70	59	64–6	Milford and Minson (1968*b*)
L. purpureus	130	51	64–6	Milford and Minson (1968*b*)
Vigna sinensis	70	64	76	Milford and Minson (1968*b*)
Macroptilium atropurpureum	112	50	38	Milford (1967)
Desmodium uncinatum	112	54	56	Milford (1967)
Lotononis bainesii	182	60	59	Milford (1967)
Medicago sativa	60	70	97	Milford (1967)

on nutritive value. Differences in VI are much more clearly shown in favour of the legumes, except in the case of *M. atropurpureum* (Siratro), where VI is low. The other important point to note in Table 4.5 is the very much higher values for DMD and VI for the temperate legume, *M. sativa* (alfalfa).

A more direct comparison of nutritive value is provided by Playne (1969*b*), where *H. contortus* and *S. humilis* were cut and fed from the same pasture at the same age. Values for *H. contortus* were DMD% – 40, VI – 43, and for *S. humilis* DMD% – 59, VI – 61.

These comparisons indicate the importance of the legume in providing a pasture component with a higher feeding value than the associated grasses. In fact some scientists consider that this is the most important role of the legume, particularly in dry season environments where the legume becomes a major source of available energy for the animal.

(iii) Mineral content

Apart from a higher nitrogen content, tropical legumes generally maintain higher sulphur and calcium concentrations in plant tops than the companion tropical grasses. Andrew and Robins (1969, 1971) give values of calcium concentration for seven tropical legumes and nine tropical grasses grown in nutrient cultures. Values for the legumes ranged between 1.13 per cent (*D. intortum*) to 1.93 per cent (*S. humilis*) and for the grasses between 0.17 per cent (*D. decumbens*) to 0.41 per cent (*C. gayana*). A similar range of calcium concentration (0.17–0.40 per cent) in five tropical grasses in East Africa was found by Marshall and Long (1971). Furthermore the tropical legumes are more efficient at extracting calcium from soils low in calcium than are the temperate legumes (Andrew and Norris 1961).

Sulphur deficiency in tropical grasses can markedly reduce VI and DMD. Rees, Minson and Smith (1974) have shown that addition of sulphur fertilizer to pangola grass (*D. decumbens*) increased sulphur content from 0.09 to 0.15 per cent, increased VI from 44.4 to 64.1 g per kg body weight, 0.75 per day, and DMD from 55.2 to 60.2 per cent. Thus sulphur deficiency can have marked effects on pasture quality.

As legumes have a higher protein content a higher sulphur content is generally expected, but direct comparisons of sulphur content of grasses and legumes have been difficult to find. Playne (1972) gives comparative values for a spear grass–TS pasture where spear grass (*H. contortus*) sulphur content ranged from 0.07 to 0.15 per cent, while *S. humilis* was higher, 0.14 to 0.20 per cent sulphur. Jones and Quagliato (1973) give some useful values of sulphur content in tropical legumes over a range of sulphur fertilization rates:

Sulphur fertilizer rate (kg ha^{-1})	0	20	60
Sulphur content (%) in *S. guyanensis*	0.096	0.130	0.209
Sulphur content (%) in *C. pubescens*	0.162	0.174	0.195
Sulphur content (%) in *G. wightii*	0.072	0.093	0.145
Sulphur content (%) in *M. atropurpureum*	0.071	0.097	0.129
Sulphur content (%) in *M. sativa* (alfalfa)	0.057	0.067	0.075

For the temperate legume bur clover (*M. hispida*), Jones, Ruckman, and Lauder (1972) suggested critical levels of plant sulphur as 0.225 per cent in leaf blades and 0.08 per cent sulphur in mid and lower stems. Furthermore sulphur deficiency in legumes reduces nitrogen content, so that sulphur is an important element in legume nutrition, and tends to be higher in the legumes than in the grass.

Comparisons of phosphorus concentration are more variable, depending on species and plant age. For example, Whiteman (1969) reported a mean value for four tropical legumes under grazing and cutting of 0.22 per cent phosphorus, while Rhodes grass was 0.36 per cent phosphorus. Jones (1968) gives values for *Urochloa mosambicensis* and *S. humilis* grown in a mixed sward. Phosphorus content of tops for *Urochloa* ranged between 0.08 to 0.13 per cent phosphorus without applied phosphorus, up to 0.39 per cent phosphorus at applications of 200 kg P ha^{-1} over three years, while *S. humilis* values were 0.10 to 0.15 per cent up to 0.37 per cent at the highest phosphorus level.

Norman (1965) compared phosphorus concentrations in a *Cenchrus setigerus-S. humilis* sward at a range of superphosphate application rates:

Superphosphate rate (kg P ha^{-1})	10	20	40
Per cent phosphorus in *Cenchrus*	0.035	0.045	0.050
Per cent phosphorus in *S. humilis*	0.054	0.060	0.084

In this case phosphorus content of legume, harvested shortly before maturity while forage was still green, was consistently higher than the grass. Similarly Havilah and others (personal communication) give values for phosphorus content for an *Axonopus*-white clover sward, where *Axonopus* phosphorus contents were 0.20 to 0.27 per cent, and white clover 0.29 to 0.36 per cent over a range of superphosphate maintenance fertilizer rates.

We can conclude that at a given stage of pasture maturity, the legume component will usually have a higher DMD, VI, protein content, and a higher content of some minerals, calcium, sulphur, or possibly phosphorus. Thus the legume is a most important sward component in the nutrition of the grazing animal. However, it must be recognized also that the legumes in the pasture can lead to problems such as bloat and pasture oestrogens. The tropical legumes are little problem in this regard, but many tropical legumes such as *Indigofera*, *Crotalaria*, *Leucaena*, and others do contain toxins which can affect animal health and production.

(b) *Nitrogen fixation in the pasture system*
The second role of the legume is, of course, through symbiotic nitrogen fixation, and the cycling of this nitrogen into the pasture system. This input can be thought of as occurring in two ways. Firstly as the legume–*Rhizobium* symbiosis provides the legume nitrogen requirement, the legume is independent of soil nitrogen supply. It has been clearly demonstrated that when a legume and grass are growing in association, the legume is a very poor competitor for soil nitrogen and most of the available nitrogen is taken up by the grass component. Henzell (1968) provides the following examples:

Pasture	% Legume in pasture	% of the soil nitrogen taken up by the legume
M. atropurpureum–C. gayana	7–17	1.2–9.5
M. atropurpureum–S. anceps	30	<10
S. humilis–H. contortus	up to 34	<10
T. repens–Lolium sp	–	5–6
M. atropurpureum–C. gayana	(in pots)	2–4

More recent studies (Vallis 1978) have shown that some legume species are more competitive for soil nitrogen in mixtures with grass. *Lotononis bainesii* and *T. repens* were more competitive for soil nitrogen in mixture with *C. gayana* than *M. atropurpureum* and *S. guianensis*. At 50 per cent legume content, *L. bainesii* and *T. repens* obtained approximately 22 per cent of total nitrogen uptake, while *M. atropurpureum* and *S. guianensis* took up about 15 and 12 per cent respectively.

Ability to compete for soil nitrogen in the seedling stage before the plants are fully nodulated and actively fixing nitrogen may be important in determining establishment. In this regard *M. atropurpureum* appeared better able to compete with *Panicum maximum* for nitrogen than was *D. intortum* (Vallis 1978). Also time of year can have an influence. *T. repens* and *L. bainesii* appear to take up a larger share of available nitrogen in spring than the associated grasses. Thus part of a legumes ability to compete with grasses and maintain itself at an adequate level in the sward may be related to its ability to obtain nitrogen from the soil when required. However, under most conditions the grass takes up the highest proportion of soil nitrogen, and the legume makes little demand on the soil nitrogen store. In terms of nitrogen economy the legume yield might be regarded as a bonus.

The nitrogen fixed in the legume dry-matter is made available to the associated grasses, after mineralization, by decay of legume plant materials, particularly of roots, nodules, and stubbles, and by return through urine and faeces of grazing animals and soil fauna. The amount of nitrogen made available to the pasture system varies widely and depends on many factors, but is mainly a function of legume yield in the pasture (Jones 1972). These factors are discussed in Section 4.3.

There are many examples of the beneficial effects of tropical legumes on pasture productivity and animal production. Effects on pasture productivity are measured in terms of dry matter and nitrogen yields, and are related to the nitrogen-fixation role, while effects on animal production, also involve nutritive value aspects previously discussed. As an example of effects on pasture production the data of Balachandran and Whiteman (1975) demonstrate the main principles. These data are from the second year of a trial in which effects of fertilizer nitrogen were compared in pure and mixed swards of *Setaria anceps* and *Desmodium intortum*. Only the data of the control (No) treatment are presented (Table 4.6).

Table 4.6. The effect of a legume on dry matter yield, nitrogen content, and nitrogen yield of pure and mixed swards of *Setaria anceps* and *Desmodium intortum* (Balachandran and Whiteman 1975)

Sward and components	DM yield (kg ha^{-1} a^{-1})	Nitrogen content (% N)	Nitrogen yield (kg ha^{-1} a^{-1})
1. Pure Desmodium			
Desmodium	10 150	3.28	354
Weeds	1910	–	26
Total	12 060	–	380
2. Pure Setaria			
Setaria	2690	1.23	35
Weeds	470	–	5
Total	3060	–	40
3. Mixed sward			
Desmodium	6130	2.94	193
Setaria	3790	1.45	55
Weeds	470	–	5
Total	10 390	–	253

The major effects of the legume on this red–yellow podzolic soil, extremely deficient in nitrogen are clearly demonstrated. The pure legume sward gave four times the dry-matter yield and almost ten times the nitrogen yield of grass alone. When grass and legume are combined,

Setaria dry-matter yield was increased by 41 per cent, nitrogen content of *Setaria* by 18 per cent, and nitrogen yield of *Setaria* by 57 per cent. Clearly even by the second year of the trial nitrogen was becoming available to the grass component. In terms of forage and nitrogen available to the grazing animal, the combination of grass and legume increased total dry-matter yield three-fold, and nitrogen yield by a factor of six.

Obviously legume effects are more evident with increasing legume content in the sward, but even relatively low legume contents can have marked effects on animal production, as is discussed in Chapter 5. The aim of pasture management must be to maintain an adequate amount of legume in the sward on which the productivity of the pasture will depend. In grazed tropical pastures legume content seldom exceeds 30–40 per cent, and except in *S. humilis* based pastures, tends to decline with time (Firth, Evans, and Bryan 1975), particularly at higher stocking rates. Under field conditions over a wide range of environments Henzell (1968) suggests that the following levels of nitrogen increment might be expected in legume-based tropical pastures.

$$\text{kg ha}^{-1}\,\text{a}^{-1}$$

(a) high quality, high yielding pastures, adequate
  maintenance fertilizer   280–400
(b) good pastures, well grown, well utilized   170–280
(c) average pasture, strongly grazed for a number of years   55–170

This range of values for annual nitrogen increments is similar to that recorded for temperate pastures (Whitehead 1970). Even at the highest levels of recorded nitrogen fixation, temperate and tropical pastures are comparable. For temperate pastures the highest nitrogen fixation rate was recorded in New Zealand over a three year period, with an average of 645 kg N ha^{-1} a^{-1}, while for tropical species the highest estimated rate of nitrogen yield was 560 kg N ha^{-1} a^{-1} for stand of *Leucaena leucocephala* at Samford, South Queensland (Hutton and Bonner 1960).

Thus the tropical pasture legumes have a potential for nitrogen fixation similar to the temperate legumes, although rates of nitrogen accumulation could be slower in the tropics as nitrogen losses may be higher under the influence of higher rainfall and temperature. Nevertheless, the tropical legumes have a similar potential to increase pasture productivity as the temperate legumes have already demonstrated in temperate pasture production.

4.3 FACTORS AFFECTING NITROGEN FIXATION
AND LEGUME CONTENT

In the previous Section(4.2) the role of the legume in the nitrogen economy and in the animal nutrition components of the pasture system was discussed. The fulfillment of this role depends on the efficiency of the nitrogen-fixation process and in the long term, on the maintenance of an adequate legume content in the pasture. Hence these two topics are discussed together.

Legumes have a particularly important role in the tropics. Following forest clearing, soil fertility, and particularly the nitrogen and organic matter components, decline rapidly in cropped soils. Under legume-based pastures this decline is greatly reduced, as shown by Bruce (1965). Over a 16-year period following forest removal, a grazed *Centrosema-pubescens*-based pasture added 100 kg N ha^{-1} in a wet tropical environment. This addition just balanced nitrogen losses from the system. In pure grass pastures yield declines with time as available soil nitrogen is immobilized. In Kenya, Thairu (1972) showed that the dry matter and nitrogen yield of a pure *Setaria* sward declined over four years to approximately 25 per cent of the first-year yields, whereas in a *Desmodium uncinatum–Setaria* sward, dry-matter yields were twice, and nitrogen yields three times those of the pure sward after four years. Similar results are demonstrated in Table 4.6. Also in the tropics nitrogen fixed during a pasture phase may be utilized in a cropping system, as discussed in Section 4.4. Finally nitrogen fixation by the legumes is extremely important in their ability to compete with grasses and maintain their position in a mixed pasture sward.

The legume–Rhizobium symbiosis

Norris (1956) has suggested that the Leguminoseae originated in the tropics, and that they evolved into more specialized species as they developed into temperate forms. The *Rhizobium* associated with the legumes followed similar evolutionary pathways to more specialized types. The basic type of *Rhizobium* which developed under acid tropical soil conditions and is associated with many present-day tropical legumes is the 'cowpea type', which is slow growing and produces an alkaline reaction in culture. As the legumes became more specialized and moved into generally alkaline temperate soils, so the associated *Rhizobium* changed to the 'clover type' which is fast growing and acid-producing in culture.

Within these two broad groups are a wide variety of strains associated with different species of legumes. Some strains can infect and form nodules over a wide range of species while others are extremely specific to one legume species or even variety.

These *Rhizobium* strains can be classified in relation to any one host into:

(a) *Effective* strains which are able to infect the host, forming viable nodules capable of fixing nitrogen. They are usually characterized as large nodules, containing the red pigment leg-haemoglobin.

(b) *Ineffective* strains which are able to infect the host and may form nodules, but are unable to form an effective nitrogen-fixing symbiosis. Usually small white nodules without leg-haemoglobin develop on the roots, often in large numbers.

(c) *Non-nodulating* strains which are unable to infect a given host or form nodules.

Thus the first requirement is for an effective *Rhizobium* strain which is able to infect the host and form nitrogen-fixing nodules. The specificity of *Rhizobium*-host relationships has been shown in Table 4.2, where inoculation for pasture establishment is discussed. The processes of infection, nodulation, and nitrogen fixation are briefly discussed below.

Infection and nodulation

Recent studies on the processes of root infection and nodulation by *Rhizobium* give a better insight into the factors affecting these processes and the possibilities for improved management to allow better nodulation and nitrogen fixation. This topic is well reviewed by Dixon (1969).

An important area of development has been in the study of the rhizosphere effect, the effect of the root system on the soil microbial population. Legumes appear to excrete specific substances which stimulate *Rhizobium* in the rhizosphere more than other micro-organisms (Rovira 1961) but there is good evidence that rhizobia survive and grow in non-legume rhizospheres. Although the stimulation is less than in the case of legumes, this may be essential for the maintenance of some rhizobia in soils in the absence of legume hosts.

This leads to some redefinition of the *Rhizobium*-host symbiosis. The major benefit that the rhizobia gain from the symbiotic association arises from the rhizosphere of the legume. Bacteria that infect the host plant and become bacteroids are no longer capable of division and are thus not benefited by the association. The advantage that they give to the host by nitrogen that is fixed is returned by the stronger and more adequately nourished plant excreting larger quantities of substances which benefit the rhizosphere and soil population (Dixon 1969).

The physiology of infection is not yet fully understood. It is known that indoleacetic acid (IAA) is involved and is demonstrated by the curling of the root hair which precedes infection. IAA may be produced by the infective rhizobia in the rhizosphere from tryptophan excreted by the legume root. However, the infection process also requires a softening and breakdown of cell-wall substances. IAA appears to play a role in the regulation of production of pectinase enzymes, and this IAA may be produced within the host cells. Also pectinase activity appears to be induced in the host plant by the presence of the polysaccharide coat of *Rhizobium* bacterium capable of infecting the host (Dixon 1969).

Once the bacteria has penetrated the cell wall an infection thread is produced which passes down the root hair and into the cortex of the root, where it stimulates the division of the cells. In many species it is tetraploid cells which are stimulated to divide giving a nodule containing large cells. After nodule initiation a nodule meristem may be formed which goes on dividing to produce the outer walls and tissues of the nodule. The rod-shaped bacteria in the infection thread are released into the cell cytoplasm. The bacteria are encased in a membrane envelope, probably derived from the plasmalemma. The bacteria may increase in size within the envelope and change shape to form large rectangular bacteroids. Division may also occur, giving groups of bacteroids within an envelope. The development of bacteroids, plasmalemma membranes, production of leg-haemoglobin, and interactions with the host cell nucleus are all closely related in the development of the nitrogen-fixation process (Dixon 1969).

The nitrogen-fixation process

In many pasture systems the fixation of nitrogen ranks in importance with CO_2 fixation in determining the overall productivity. Recent research into the biochemistry of this process has increased our understanding of the system. The development of techniques to study the process in cell free systems has contributed greatly to an elucidation of the biochemical steps.

It now seems fairly well established that the first key product is ammonia, and that this is produced by a series of reductions of N_2 which is bound to the enzyme nitrogenase, and only released from the enzyme when the reductive steps are completed, as shown in Fig. 4.2.

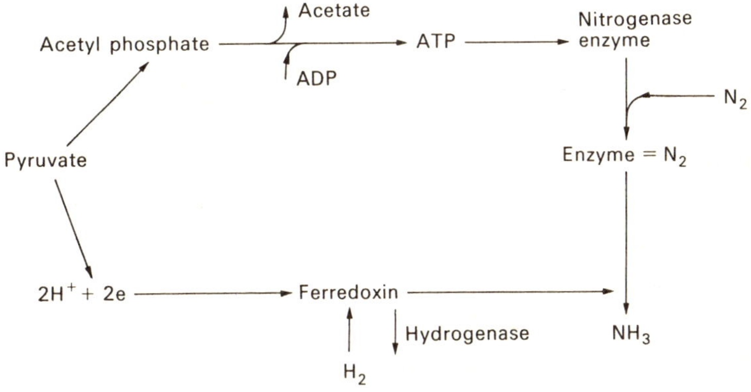

Fig. 4.2. Simplified scheme of electron and energy transfer in nitrogen fixation (after Epstein 1972).

The enzyme nitrogenase is thought to be a complex of two proteins, one a Mo–Fe protein, the other an Fe protein. The iron molecules in the enzyme appear to be sites of the initial binding of the nitrogen diatomic molecule, while the molybdenum sites appear to function in weakening the bond between the two nitrogen atoms, which renders them susceptible to reduction by the reducing agent (Epstein 1972). Nitrogenase is also able to reduce acetylene (C_2H_2) to ethylene (C_2H_4) which has provided a rapid and accurate technique for assay. While these studies have increased our understanding of the nitrogen-fixing system, we can do little to influence the process directly. However, the importance of iron, molybdenum, and cobalt has been realized in the maintenance of an efficient nitrogen-fixing system.

Efficiency of nitrogen fixation

While a range of different *Rhizobium* strains may form an effective symbiosis with a particular legume cultivar, different associations may differ widely in their efficiency of nitrogen fixation as shown in Table 4.7. Plants are grown in nitrogen-free media and the efficiency of a *Rhizobium* strain is clearly reflected in the dry-matter yield of the legume (Norris 1967).

Thus legume-seed inoculation prior to sowing is important to ensure that the most effective and efficient strains are available for infection and nodulation. Choice of *Rhizobium* strain CB81 is recommended for *Leucaena leucocephala* on acid soils while strain NGR8 is recommended for neutral to alkaline soils (Norris 1973).

Table 4.7. *Rhizobium* strain by variety interaction on dry-matter yield of *Glycine wightii* (Norris 1967)

	Glycine variety		
Rhizobium strain	Tinaroo	Cooper	Clarence
	(dry-matter yield relative to yield in the highest yielding treatment)		
Control (not inoculated)	4	6	8
CB439	34	37	100
CB453	23	80	85
CB466	100	30	92
CB756	85	82	82
QA922	77	100	89

Factors affecting nitrogen fixation

In this section those factors more directly affecting nodulation and nitrogen fixation at the single plant level will be considered. Total nitrogen input into a pasture system is a function of the growth and yield of the total legume component and these factors will be discussed below.

If a plant has been infected with an effective efficient *Rhizobium* strain, the amount of nitrogen fixation will be affected by (a) the amount of nodule tissue formed; (b) the time over which the nodule tissue is active in nitrogen-fixation; and (c) external and environmental factors which may affect directly the mechanisms of nodulation or nitrogen fixation or cause loss of nodules from the plant.

(a) The amount of nodule tissue

The total weight of nodules on the root system is a function of the number of nodules and the size of each nodule. It might be thought that the number of nodules formed is related to the number of infections through the root, but Nutman (1956) has shown that this varies widely with species. In *Medicago sativa* only 5 per cent of root hairs infected form nodules; also in *Pisum sativum* there is an excess of infections over nodules formed, whereas in *Trifolium fragiferum* the number of infections is closely related to the number of nodules formed. Nutman (1956) also suggests that species with more abundant lateral root formation tend to form more nodules than sparsely rooted species.

More important than total weight of nodules per plant is the volume of nitrogen-fixing bacterial tissue. This is closely regulated by the host to match its nitrogen requirement, and nitrogen content in the plant is maintained despite great variations in nodule size and number in a species, as shown in Fig. 4.3.

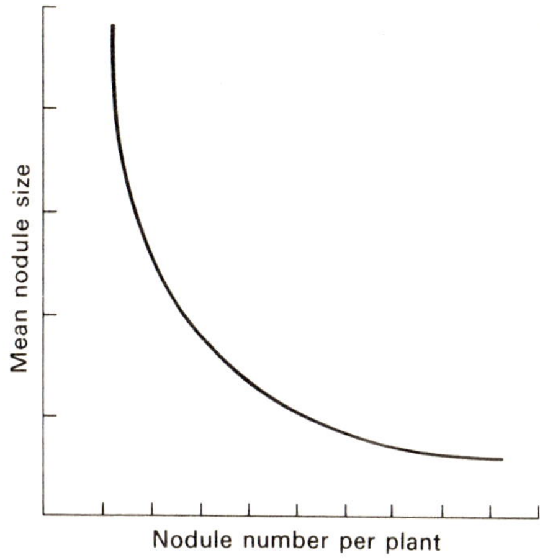

Fig. 4.3. Relationship between nodule number and size of nodules in *Trifolium subterraneum* (after Nutman 1956).

This self regulation of nodule volume is suggested to arise from two mechanisms. Formation of nodules differentially inhibits the formation of new nodules, since excision of growing nodules can cause an immediate increase in the rate of nodule formation. Nutman (1956) believes that the inhibitory effect on new nodule formation may be related to the amount and activity of nodule meristem tissue, and mediated through release of the growth regulating hormone IAA from the meristematic tissue. Secondly competition for available nutrients and photosynthates between nodule sites will regulate nodule growth and will be related to the growth and vigour of the plant and hence to its demand for nitrogen.

Thus we should expect some relationship between plant size and nodule weight per plant, and the amount of nitrogen fixed should be related to nodule volume. These relationships have been shown for clover nodules by Bergersen (1961). This relationship has been further developed by Dobereiner (1966) and Whiteman (1970a) for field-grown tropical legumes. Dobereiner (1966) examined the regression of nitrogen yield per plant against nodule weight per plant in the form ($y = a + bx$): (Fig. 4.4).

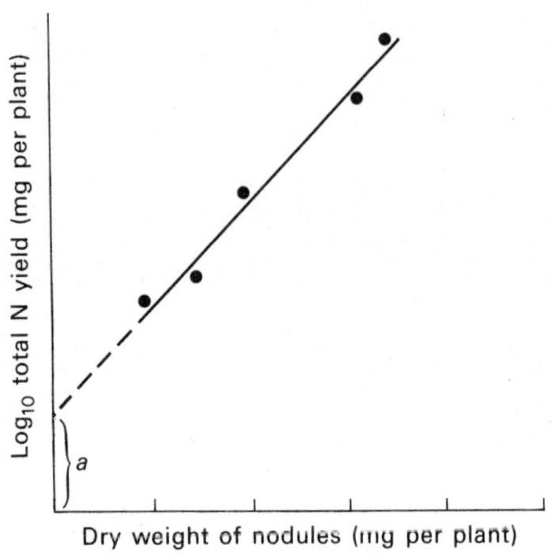

Fig. 4.4. Relationship between dry weight of nodules and nitrogen yield per plant in soyabean (Dobereiner 1966).

By extrapolating to the theoretical zero nodulation point, the intercept a is a measure of the amount of nitrogen taken up by the plant from the soil and seed. The constant, b is a measure of the efficiency of nitrogen fixation. Whiteman (1970a) showed that this relationship only holds in perennial species over the period when the plant is actively growing and both plant weight and nodule weight are increasing. Thomas and Whiteman (1971) used this analysis to compare effects of different soils on efficiency of nitrogen fixation in *Glycine wightii*.

It is sometimes suggested that tropical legumes nodulate less well, and fix less nitrogen than their temperate counterparts. However, the data in Table 4.8 (Whiteman 1970b) show that the tropical legumes may reach high nodule weights per plant and these are similar to the temperate species as a proportion of plant weight.

Table 4.8. Comparison of maximum recorded nodule weights per plant
(Whiteman 1970*a*)

Species	Place	No. of nodules	Nodule fresh weight (g per plant)	Nodule weight as proportion of plant weight (%)
Temperate				
Vicia faba	UK		4.70	2.24
Phaseolus vulgaris	UK		4.47	1.88
Pisum sativum	UK		1.16	2.60
Tropical				
Peanut (*Arachis hypogaea*)	Sudan	1000	3.4	2.3
Peanut (*Arachis hypogaea*)	Israel	413	5.3	
Cowpea (*Vigna sinensis*)		100	4.0	
Soyabean (*Glycine max*)	Malaya	256	3.3	1.3
Psophocarpus tetragonolobus	Malaya	440	21.0	
	Glasshouse UK	406	24.0	
Phaseolus atropurpureus	Queensland	1720	42.0	2.3
Desmodium uncinatum	Queensland	1460	37.7	2.9

(b) Duration of active nitrogen fixation

The amount of nitrogen fixed will also be affected by the length of
time over which a high nodule volume is maintained and active nitrogen
fixation proceeds. This will be a function of the pasture species, whether
an annual or perennial, and by the length of the growing season.

In an annual species such as *Vicia sativa* the nodule volume increases
in harmony with increasing plant weight, reaches a short plateau, and
then declines with the onset of flowering and seed formation (Fig. 4.5(a))
(Pate 1958). In the perennial tropical pasture legumes these changes do
not appear to be as closely related to plant ontogeny. In *Desmodium
uncinatum* there is a rapid increase in nodule weight, followed by a
decline, well before maximum plant dry weight is reached, and before
the onset of flowering (Fig. 4.5(b)) (Whiteman and Lulham 1970).
Over the cool dry season the nodule population is very low. A similar
pattern was shown for *Centrosema pubescens* by Bowen (1959). Thus
in both annual and perennial legumes the length of time over which

maximum nodule weights and active nitrogen fixation are maintained is relatively short.

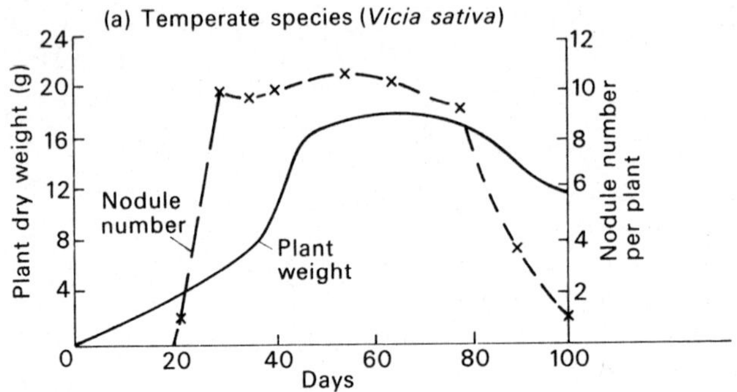

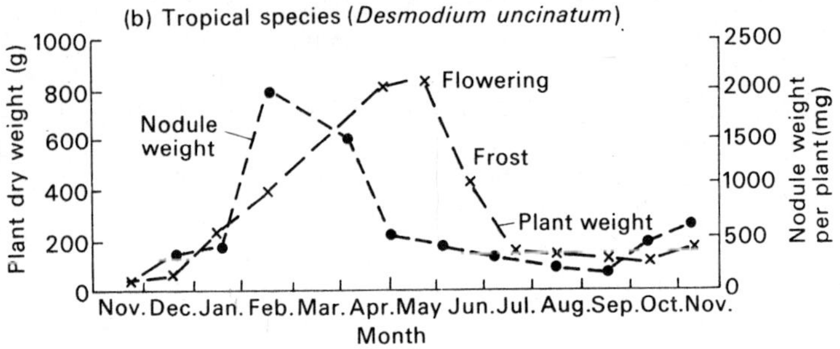

Fig. 4.5. Changes in nodulation and plant weight with time in (a) annual, *Vicia sativa* (after Pate 1958); (b) a perennial, *Desmodium uncinatum* (after Whiteman and Lulham 1970).

(c) External and environmental factors

Many external factors influence the amount of nitrogen fixed through effects on the process of nodulation, effects on the nitrogen fixation process, or by causing shedding or loss of part or all of the nodule population.

Temperature

In a temperate species such as *Trifolium repens*, rate of nitrogen fixation reached a maximum at root temperatures of 20-3 °C, and was negligible at 5 °C (Crofts 1969). In the tropical species *G. wightii*, *D. intortum*, *D. uncinatum*, *S. humilis*, and *M. atropurpureum* nodulation and nitrogen fixation were optimal at 30 °C when compared over the range of 18–36 °C (Gibson 1971). Souto and Dobereiner (1968, 1970) found a similar optimum temperature range for *Glycine*, *Pueraria*, and *Stylosanthes*, although some varieties of *Stylosanthes* and *Pueraria* were stimulated by higher soil temperatures up to 34 °C.

Soil moisture conditions

This factor affects nodulation in a number of ways. Firstly when inoculated seed is sown into dry hot soils, high mortality of *Rhizobium* may occur, leading to poor nodulation when germination does eventually occur. When plants are established soil moisture stress may cause shedding of nodules from nodulated plants. Excess moisture leading to waterlogging can cause reduced nitrogen fixation due to poor aeration, and also the shedding of nodules.

Photosynthate supply

A number of factors appear to operate through the ability of the plant to provide sufficient photosynthate to maintain the nodule population. Defoliation has been shown to cause loss of older nodules and lateral roots in red clover (*T. pratense*) and *Lotus corniculatus* (Butler and Bathurst 1956). This is followed by regrowth of laterals and new nodule formation, so that with cycles of grazing, there is a turnover of nodules. In the tropical species also, defoliation affects the nodule population, as shown in Table 4.9 (Whiteman 1970*c*).

Table 4.9. Effects of defoliation treatments on mean plant dry weight and mean nodule weight in *D. intortum* and *M. atropurpureum* (Whiteman 1970*c*)

Treatment	Plant dry weight		Nodule dry weight	
	(g)	% of control	(mg per plant)	% of control
Control	68	–	405	–
Cut at 7.5 cm	7	11	142	35
All leaves removed	11	16	135	34
Half leaves removed				
young leaves	42	61	214	52
old leaves	45	65	340	84

These data show that nodule loss was greater with more severe defoliation, and was greater when young photosynthetically active leaves were removed than when older less active leaves were removed. The effects of reduction in photosynthate supply are also demonstrated by the loss of nodules when plants are shaded (Butler, Greenwood, and Soper 1959), or with the onset of flowering and seed formation (Pate 1958).

Mineral nutrition

Although the deficiency of any essential element may lead to reduced growth and yield, and hence to reduced total nitrogen yield in the pasture, in this discussion we will be mainly concerned with those elements which are known to have more direct effects on nodulation or nitrogen fixation. Because of specific requirements for certain elements for nitrogen fixation, the legume may be more sensitive to some nutrient deficiencies than the associated grasses.

Phosphorus   Phosphorus deficiency is an important limitation to legume growth in many tropical soils. It is important for plant growth, nodule formation, and the nitrogen fixation process, and deficiencies of phosphorus are evident early in the life of the plant. Gates (1974) showed that *S. humilis* plants adequately supplied with phosphorus initiated nodules at day 11 after sowing, but not until day 14 in low phosphorus. Furthermore increasing rates of phosphorus lead to increasing nodule number, nodule weight, rate of nitrogen fixation, and nitrogen yield, as shown in Table 4.10.

Table 4.10. Effect of phosphorus level on nodulation and nitrogen fixation in
Stylosanthes humilis at 26 days (Gates 1974)

Phosphorus level (kg P ha^{-1})	0	3	6	12.5	25
Number nodules per plant	10	15	40	55	148
Nodule weight (mg per plant)	0.1	0.4	1.1	3.3	8.5
Nitrogen fixation rate (mg N per g nodule wt per day)	17	8	20	39	53
Nitrogen yield (mg per plant)	6	8.5	10.5	12	14

Souto and Dobereiner (1968) also demonstrated that phosphorus had marked effects on nodulation and nitrogen fixation in *G. wightii* at early stages of growth. In the field, legume responses in dry matter yield to added phosphate, have been widely reported for many tropical soils. Not only dry matter is increased, but also the rate of nitrogen accumulation as shown by Henzell, Fergus, and Martin (1966) in Table 4.11.

In the clover-based pastures of southern Australia, Donald (1960) estimated that nitrogen increment in pasture soils was linearly related

Table 4.11. Effect of total amount of phosphorus applied on the rate of nitrogen accumulation in a *Desmodium uncinatum* pasture (Henzell *et al.* 1966)

Total phosphorus applied since sowing (kg P ha^{-1})	54	104	215	420
Rate of nitrogen accumulation (kg ha^{-1} a^{-1})	71	98	110	126

to the amount of phosphorus applied, and that approximately 84 kg N per ha was fixed per 100 kg of superphosphate applied. Thus in low phosphorus status soils, maintenance of legume content and rate of nitrogen input into the pasture system is closely related to the rate of phosphorus application.

Sulphur   In many areas pasture responses to sulphur have been masked by the application of superphosphate, which supplies both phosphorus and sulphur. With the wider use of high analysis triple-superphosphate, in which sulphur content is very low, sulphur-deficiency symptoms have become more apparent.

Sulphur deficiency in legumes leads to stunted chlorotic growth, giving symptoms similar to nitrogen deficiency. In sulphur-deficient plants, protein content is often low, although nitrogen may accumulate in non-protein forms. Sulphur does not appear to be important in nodulation as Gates (1974) showed that responses to sulphur in *S. humilis* were much smaller than to phosphorus in the early seedling stage. Sulphur response seems to be more important in the nitrogen fixation and protein synthesis stages. Janssen and Vitosh (1974) showed that addition of sulphur to sulphur-deficient *Phaseolus vulgaris* increased nitrogen fixation (by acetylene reduction assay) by 60 per cent.

These effects of sulphur on nitrogen fixation are reflected in higher nitrogen content in the plant. Oke (1969) applied sulphur to *Cajanus cajan* and guar (*Cyamopis tertragonolobus*) and found that nitrogen content in the nodules and the plant was significantly increased. Also in soyabean (*Glycine max*) Wooding, Poulsen, and Murphy (1970) showed that seedlings grown in nitrogen-free solution with adequate sulphur had fixed 2.5 times more nitrogen than those without sulphur.

In many soils an interaction between phosphorus and sulphur is evident. This was shown by Shaw, Gates, and Wilson (1966), where either phosphorus or sulphur applied alone to *S. humilis* had little effect on nitrogen content, but applied together markedly increased nitrogen content as shown in Table 4.12.

Table 4.12. Effect of phosphorus and sulphur on the nitrogen content of *S. humilis* (Shaw *et al.* 1966)

Phosphorus rate (kg ha^{-1})	Sulphur rate (kg ha^{-1})	Nitrogen content (% of DM)
0	0	2.5
0	100	2.5
60	0	2.6
20	100	3.2
60	100	3.3

Calcium and magnesium   Plant growth and nodule formation are inhibited by calcium and magnesium deficiencies, and both elements are required for *Rhizobium* function, but the requirement for calcium by the bacteria is very much less than for magnesium (Norris 1970). Application of calcium in the carbonate form as lime affects both the soil calcium status and the soil pH. The temperate legumes are, in general, more sensitive to both calcium status and soil pH than are the tropical legumes.

In an elegant series of experiments, Munns (1968) demonstrated the interactions of pH and calcium on nodulation of *Medicago sativa*. At pH below 5.5 nodule number was reduced, and pH below 4.5 prevented nodulation. Low pH prevented roots from curling and becoming infected. Raising pH from 4.4 to 5.4 allowed root-hair curling. After curling and infection pH could be lowered again to pH 4.4 without hindering normal completion of nodulation. This stage of infection while being the most acid-sensitive stage, was also the most calcium demanding. Calcium concentration needed for nodulation increased with reduction in pH. Below pH 4.8 no nodulation occurred even at the highest calcium concentration. At calcium concentrations below 0.2 m nodulation was inhibited regardless of pH. Munns (1968) postulates that as a pectinase is released soon after *M. sativa* is inoculated with a compatible *Rhizobium*, the pH requirements for its attack on pectin may determine the pH requirement for infection, and also the requirement for calcium. However, it is known that legume species differ in their tolerance of low pH for nodulation, some *Trifolium* species being more tolerant than *Medicago* spp.

The tropical legumes are more efficient at extracting calcium from low calcium soils, and also appear to be able to nodulate at lower pH values than the temperate species. As previously discussed (p. 175)

Andrew and Norris (1961) compared the growth of a range of temperate and tropical legumes in calcium-deficient soil of pH 5.2, with and without added calcium. The tropical species were clearly better able to grow than the temperate species. Also nodulation of *S. humilis* has been observed at pH as low as 4.0 to 4.5, and in *L. bainesii* at pH 4.0.

Thus in tropical pasture development, lime is usually applied at rates sufficient to correct soil calcium deficiency rather than to raise soil pH, as is often required in temperate pasture situations. In some cases sufficient calcium is added in superphosphate to meet plant calcium requirement.

Molybdenum   In the previous discussion of the nitrogen-fixation process it was shown that molybdenum is a constituent of nitrogenase and functions in the fixation of nitrogen by weakening the dinitrogen bond. Jensen and Betty (1943) showed that legume root nodules contained 5- to 15-times more molybdenum than other plant parts. In the absence of molybdenum legumes fix very little nitrogen (Hagstrom and Berger 1963).

Thus the symptoms of molybdenum deficiency are very similar to those of nitrogen deficiency, and can be corrected by the application of nitrogen. However, application of molybdenum leads to longer-term solution to the problem and an increase in nitrogen content as shown in Table 4.13 ('t Mannetje, Shaw, and Elich 1963). The molybdenum treatment also increased nodule size and effectiveness.

Table 4.13. Effect of molybdenum on yield and nitrogen content of *M. lathyroides* and *M. sativa* in a clay loam prairie soil ('t Mannetje *et al.* 1963)

Treatment	*M. lathyroides* yield (g per pot)	N%	*M. sativa* yield (g per pot)	N%
+ Mo (as Na_2MoO_4 at 0.28 kg ha^{-1})	2.67	2.62	4.38	3.35
− Mo	2.01	1.94	2.33	2.20

The usual application rates are between 0.25 and 0.50 kg ha^{-1} of sodium molybdate. In higher rainfall tropical areas it is suggested that molybdenum should be re-applied every three to four years (Swain 1959; 't Mannetje *et al.* 1963). Molybdenum is commonly applied in mixtures with superphosphate, but can also be applied in a seed-pellet coating at planting.

Cobalt　Cobalt deficiency commonly occurs in (a) acid, highly leached sandy soils; (b) soils derived from granite; (c) some highly calcareous soils, and (d) some peat soils (Stephens and Donald 1958). Animal responses to cobalt deficiency are well documented, but there is little clear evidence for a cobalt requirement for the growth of higher plants, except in the legume–*Rhizobium* symbiosis.

The cobalt containing vitamin B_{12} is synthesized by *Rhizobium* and concentrates in effective nitrogen-fixing nodules containing haemoglobin. The concentration of vitamin B_{12} in effective pink nodules was found to be four times that in non-effective white nodules (Levin, Funk, and Tendler 1954). Since nitrogen fixation is related to haemoglobin content and to vitamin B_{12} cobalt deficiency leads to impaired nitrogen fixation. Cobalt deficiency gives symptoms of nitrogen deficiency, which can be corrected by application of nitrate.

Application of 0.01 p.p.m. cobalt to lucerne resulted in earlier nodulation, a threefold increase in nodule weight, a sevenfold increase in nitrogen fixation, and a sixfold increase in dry weight compared with cobalt deficient plants (Reisenauer 1960). Field responses to cobalt application have been obtained. Powrie (1960) obtained a 30 per cent yield response with subterranean clover and lucerne on application of 0.45 kg ha^{-1} cobalt sulphate on a low pH sandy soil. Similarly Ozanne, Greenwood, and Shay (1963) increased dry-matter production by 30 per cent and nitrogen content of clover on application of 0.12 and 0.57 kg cobalt sulphate ha^{-1}. When clover tissue fell below 0.04 p.p.m. cobalt growth was sharply reduced.

The amount of nitrogen fixed

The estimation of total nitrogen fixation in the field is difficult as the increments in nitrogen yield in all components of the system must be measured as follows:

Total nitrogen fixation = [(DM yield tops, roots, nodules × nitrogen content of tops, roots, and nodules) + (increment in soil nitrogen, from t_1 to t_2) + (nitrogen losses due to leaching or volatilization) + (nitrogen removal in crop or animal products)] − [nitrogen uptake from the soil].

In most studies only net nitrogen fixation is measured because of the difficulties in estimating nitrogen losses from the system. Also it is difficult to measure directly the amount of legume nitrogen derived from uptake from the soil, although techniques using ^{15}N isotope are available (Vallis, Haydock, Ross, and Henzell 1967). Thus most estimates of nitrogen fixation are based on indirect estimates of the contribution of soil nitrogen to total nitrogen yield. These methods are summarized below.

(a) Direct measurement of nitrogen yield

Although nitrogen yield of the pasture components can be determined from dry matter yield and nitrogen content, estimation of changes in soil nitrogen are much more difficult. This is so because measurements have to be made against a large and variable background of soil nitrogen. Coefficients of variation for individual samples in the field are typically as high as 25–30 per cent (Beckett and Webster 1971). Also in most short-term experiments changes of the order of 50–100 kg N ha^{-1} a^{-1} are being sought against a background of 1000–5000 kg N ha^{-1} in the soil profile, i.e. changes of the order 1–10 per cent. Vallis (1973) has shown that to detect changes within ± 50 kg N ha^{-1} requires extremely detailed sampling. Thus it is difficult to make accurate estimates of nitrogen fixation by direct measurement in the short term.

(b) Measurement of soil nitrogen changes

Estimates of legume contribution through nitrogen fixation have been made by measuring increment in soil nitrogen. However, as discussed above reasonable estimates can only be achieved in long-term experiments. Ideally this technique requires a long-term experiment, with high levels of fixation in soils of initially low nitrogen content. Vallis (1972) has summarized results of a number of estimates based on this technique and provided estimates for the pastures in Queensland (Table 4.14).

(c) By comparison with pure grass swards

One of the most widely used techniques of estimating the legume nitrogen contribution is to compare the nitrogen yield of a legume-based pasture with a pure grass sward. The nitrogen yield of the grass sward gives an estimate of the amount of nitrogen coming from soil uptake. In comparisons with legume–grass pastures this is a valid assumption as the grass component usually takes up more than 90 per cent of available soil nitrogen. Then

Net nitrogen fixation =   [(nitrogen yield in grass + legume sward) + (Increment in soil nitrogen)] – [nitrogen yield in pure grass sward].

Table 4.14. Annual increment in soil nitrogen under tropical legume-based
pastures measured by changes in soil nitrogen (Vallis 1972)

Legume and location	Nitrogen increment (kg N ha^{-1} a^{-1})	Duration (years)	Original soil N%	Comments
S. humilis				
1. Katherine	+16	7	0.07	Grazing not stated
	+106	3	–	Grazed
2. Rodd's Bay	−29	3	0.06	Grazed. Dry years, below average rainfall
	+30	3	0.06	Grazed. Two wet years
M. atropurpureum				
Samford, Queensland	+129	4.3	0.09	Heavily grazed three times a year
Samford, Queensland	+44	4	0.1·1	Grazed at 1.67 head ha^{-1} all year
D. uncinatum				
Beerwah, Queensland	34–86	5	0.05	Intermittent grazing. Nitrogen input increased with increasing rate of phosphate
D. uncinatum, D. intortum, L. bainesii, T. repens				
Beerwah, Queensland	81	4	0.05	Grazed at 1.23 to 2.47 head ha^{-1}
C. pubescens				
Ibadan, Nigeria	280	2.2	0.07	Lightly grazed
Utchie Ck, Queensland	103	16	0.38	Regularly grazed
South Johnstone, Queensland	122	2.6	0.13	Periodically grazed
P. phaseoloides				
South Johnstone, Queensland	130	2.6	0.13	Periodically grazed
G. wightii				
Kenya	167	9		

A variation of this method is to compare the nitrogen yield of the
legume–grass pasture with pure grass swards receiving a range of nitrogen
fertilizer applications (Fig. 4.6). The nitrogen yield of the legume-based
pasture can then be equated with an amount of nitrogen fertilizer
required to give the same nitrogen yield.

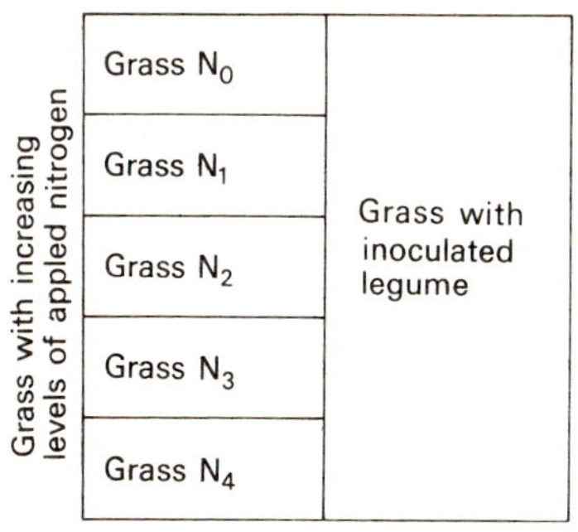

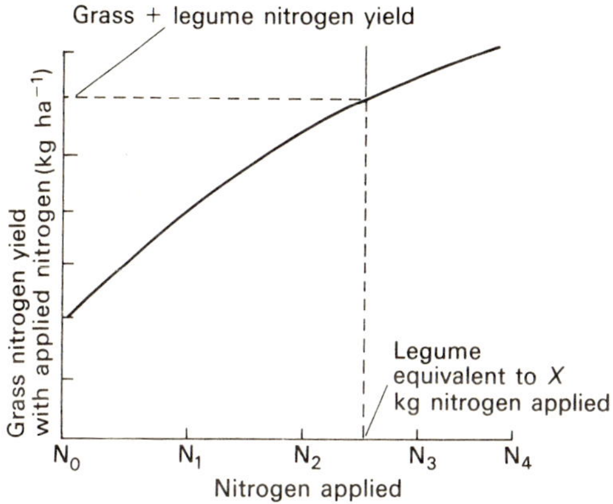

Fig. 4.6. Diagrammatic representation of estimation of legume contribution by nitrogen fixation.

In Section 4.2 (Table 4.6), the application of this technique was demonstrated where the contribution of *D. intortum* in mixed swards and alone was compared with *S. anceps* swards receiving nitrogen fertilizer (Balachandran and Whiteman 1975).

The nitrogen contribution of the legume component varies widely, depending on the effects of the many factors discussed above and on grazing management. These effects will be reflected in the total yield and vigour of the legume component in the sward, and thus a range of values is apparent from a review of published results.

The important point is that both temperate and tropical legumes fix similar quantities of nitrogen with similar efficiencies even up to the highest levels. The highest nitrogen fixation for a temperate pasture was recorded at Palmerston North, New Zealand in a sward of sown grasses with red and white clovers over a three year period. Nitrogen yields plus soil increment gave an overall fixation rate of 645 kg ha^{-1} a^{-1}. For tropical legume species the highest estimated rate is for *Leucaena leucocephala* at Samford, Queensland of approx. 560 kg N ha^{-1} a^{-1}. Under ideal controlled environment conditions, Gates (1970) has measured an annual fixation rate equivalent to 1500 kg ha^{-1} a^{-1} for *Stylosanthes humilis* grown with nutrient solutions, which demonstrates the potential of the symbiotic fixation systems.

Under field conditions the following nitrogen increments under legume-based pastures might be expected (Henzell 1968):

(a) high quality, high yielding pasture   280–400 kg N ha^{-1} a^{-1};
(b) good pasture, growing well   170–280 kg N ha^{-1} a^{-1};
(c) average pasture, strongly grazed   55–170 kg N ha^{-1} a^{-1}.

Nitrogen transfer and losses

Nitrogen fixed by legume symbiosis is incorporated within the plant and must be released before it is available within the pasture system. Plant material may be returned to the soil by ingestion through grazing animals, including insects, or by senescence. At the soil surface this material is further broken-down by soil fauna, fungi, and bacteria. The process by which complex nitrogen compounds are reduced to the forms available for plant uptake, viz. the NO_3^- or NH_4^+ ion forms, is termed mineralization. Under certain conditions, particularly in water-logged soils, the process of denitrification may occur leading to losses of nitrogen in the gaseous form.

There are a number of mechanisms by which legume plant material may enter the soil mineralization process, as follows:

(i) Direct excretion of nitrogenous compounds

Although excretion of nitrogenous compounds from legume roots can be demonstrated in the laboratory, the amounts are extremely small and are negligible on a field scale (Whitney and Kanehiro 1967).

(ii) Underground transfer by sloughing of roots and nodules

As previously described, various factors such as defoliation, shading, water stress, and flowering can cause loss of lateral roots and nodules. Butler *et al.* (1959), suggested that this may allow transfer of up to 70 kg N ha^{-1} a^{-1} in clover-based pastures, but experiments in which grasses and legumes were grown together demonstrate that this mechanism allows only limited nitrogen transfer while the legume component is actively growing (Dilz and Mulder 1962; Simpson 1965). However, with annual legume species, on the death of the plant, a large proportion of the nitrogen which becomes available may be derived from the underground plant parts. Watson and Lapins (1969) demonstrated that much of the nitrogen in the senescent plant material above ground can, under certain conditions, be lost (see Fig. 4.7).

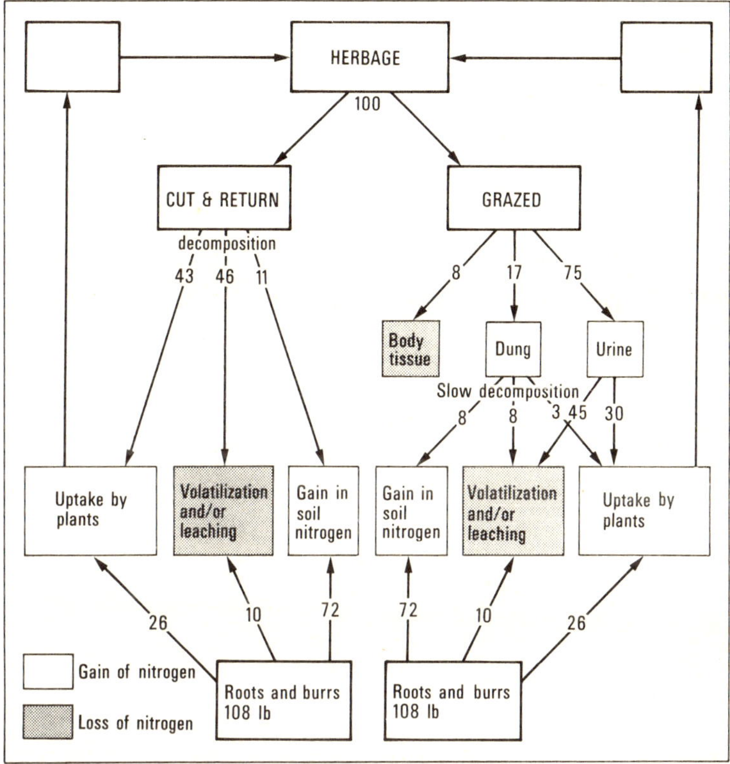

Fig. 4.7. Nitrogen transfer pathways under grazing and cutting. The quantitative estimates are based on the gains and losses of 100 lb of herbage nitrogen in a *Trifolium subterraneum*-based pasture (after Watson and Lapins 1969).

(iii) Senescence of stems and leaves

In the absence of the grazing animal, the incorporation of plant material into the soil surface becomes the major pathway of nitrogen transfer. In southern Queensland the onset of frosts at the end of the growing season causes up to 85 per cent loss of plant dry-matter, which falls to the ground and is available for breakdown. Whiteman (1970*b*) has estimated the amounts of nitrogen derived from the main plant components following frosting (Table 4.15).

Table 4.15. Return of nitrogen after frosting in *D. uncinatum* and *M. atropurpureum* swards in southern Queensland (Whiteman 1970*b*)

Plant component	Nitrogen return (kg ha^{-1})		As a percentage of total release	
	D. uncinatum	*M. atropurpureum*	*D. uncinatum*	*M. atropurpureum*
Tops	85	119	71	78
Roots	18	18	15	12
Nodules	17	15	14	9
Total	120	152		

This shows that 20–30 per cent of the total nitrogen can be released through loss of roots and nodules. Much less of this nitrogen would be lost compared with the material on the soil surface. It has been shown also that the rate of mineralization of leaf material varies between legume species, and was very much faster for *M. atropurpureum* than for *D. intortum* (Vallis and Jones 1973). It is suggested that the higher concentrations of polyphenols in the leaves of *D. intortum* may protect the proteins from rapid decomposition.

(iv) Transfer through the grazing animal

In pastures, the grazing animals are the most important source of nutrient cycling and mineral turnover. Grazing animals include also insects and other microfauna, which have been shown to increase in the pasture, as the stocking rate of domestic animals is reduced (Hutchinson and King 1970).

Depending on the type of animal and the nitrogen content of the pasture, domestic livestock generally retain only between 7 and 15 per cent of the nitrogen ingested, the balance being returned mainly in the urine fractions as shown in Fig. 4.7. The presence of the grazing animal leads to uneven return of nutrients, and to areas of high nitrogen content in urine patches. This can lead to high losses of nitrogen due to the breakdown of urea by the enzyme *urease* which is present in most soils.

$$\underset{\text{Urea}}{\underset{\underset{O}{\overset{\|}{C}}}{\overset{NH_2 \quad NH_2}{\diagdown \diagup}}} \quad \xrightarrow{\text{Urease}} \quad \underset{\text{Ammonia and carbon dioxide}}{NH_3 + CO_2}$$

Under conditions of high soil temperature with some moisture, losses up to 80 per cent of nitrogen have been recorded from the volatilization process in urine patches (Watson and Lapins 1969).

The grazing animal is also responsible for redistribution of nutrients by transferring nutrients to areas of high concentration such as to watering points, shaded rest areas, and stock yards. Thus the grazing animal is responsible for much of the transfer and turnover of nutrients, and also for important nitrogen losses from the pasture system.

Factors affecting legume content

The botanical composition of a grazed pasture can vary widely with time depending upon many edaphic, environmental, competition, and management factors. The important aim in pasture management is to control changes in composition, and particularly the legume–grass balance, while producing the maximum output of forage of adequate quality. Maintenance of an adequate legume content in the pasture is important in meeting these requirements.

Defining what is an adequate legume content for maximum animal production is difficult. In a review of the role of white clover (*T. repens*) in temperate pastures, Martin (1960) suggests that the maximum amount of forage is obtained when clover content is of the order of 30–50 per cent of total dry matter. In a tropical pasture based on pangola grass (*D. decumbens*) with *Desmodium intortum*, *D. uncinatum*, *L. bainesii* and *T. repens*, Evans (1970) has shown that liveweight gain of beef cattle was linearly related to legume content of the pasture (Fig. 4.8).

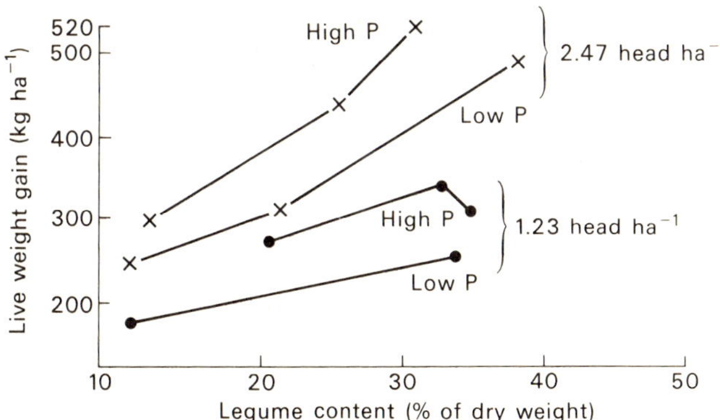

Fig. 4.8. Live weight gain per ha and the legume content of pasture at two stocking rates and two levels of phosphate (Evans 1970).

In northern Australia, Norman (1970) has also demonstrated that liveweight gain of beef cattle grazing natural pastures was linearly related to the number of days in a grazing cycle that cattle were allowed access to *S. humilis* pasture. Also there was a linear relationship between liveweight gain and the proportion of *S. humilis* up to 75 per cent legume content.

While no general optimum level of legume content can be recommended, pasture management should aim to maintain a balance between the grass and legume components in order to achieve adequate levels of total yield. The main factors affecting this balance are discussed below.

Species compatability

The first requirement in selection of pasture species for a particular area is that they are adapted to the environmental and edaphic conditions. Because of the complex interactions between species under any particular set of conditions and grazing management, it is often difficult to predict the level of compatability between legume and grass species. However, the usual problem has been in maintaining the legume component.

Many of the tropical grasses have been selected for high yield and vigour in pure stands with nitrogen fertilization. These have then been combined with legumes which have performed well in a region, usually to the detriment of the sown legume. Realizing the importance of a high legume content it has been suggested that it may be better to select far less vigorous grasses (Hutton 1968). Selection for high yields with high nitrogen fertilization is of less importance in legume-based pastures as it is unlikely that the legume will supply sufficient nitrogen to meet the grasses' potential for maximum yield. The other important principle is that if pasture development is to be based on legume pastures then selection of species must be by field testing of mixtures and not an evaluation in plots of single species.

Sowing rate

Increasing the sowing rate of one of the species in a mixture will have marked short-term effects on increasing the density of established seedlings, and reducing the density of the other species (Middleton 1970). However, in the long term, environmental and management effects are of overriding importance, and different ratios of grass and legume seed may have limited effects on pasture composition by the second year (Jones 1975). Recommended sowing rates for tropical pasture are given in Humphreys (1974) and Anon (1970). Because of cost of legume seed, the tendency is often to sow at too low a rate, which can lead to poor legume content, or to an increase in the time required to reach a satisfactory legume content.

Fertilization

The effects of nutrients on legume growth and nitrogen fixation has been discussed, but here the importance of fertilizer inputs as a management tool in the manipulation of botanical composition is emphasized.

Application of nitrogen fertilizers to grass–legume pastures has dramatic effects on the legume component. Presence of high levels of NO_3^- or NH_4^+ will inhibit nodulation and in plants already nodulated, reduces rate of nitrogen fixation. When a legume is growing with a grass, as shown in Table 4.16, the grass is a stronger competitor for available nitrogen, and takes up most of that applied. This will lead to an increased rate of growth, leaf expansion, and tillering in the grasses, often leading to suppression of the legume owing to shading. Also grass root development may be enhanced leading to greater competition for other nutrients and water. Thus application of nitrogen to legume–grass swards usually leads to reduction in legume content (Linehan and Lowe 1960) (Table 4.16).

Table 4.16. Effect of rate of application of fertilizer nitrogen on clover content, dry matter yield, and protein yield (Linehan and Lowe 1960)

	Nitrogen application (kg ha^{-1} a^{-1})					
	0	40	80	160	240	400
% Clover	36	28	21	7	1.3	0.3
Total field (kg ha^{-1} × 1000)	7.4	7.5	7.8	8.5	9.8	11.6
Protein yield (kg ha^{-1} × 1000)	1.26	1.24	1.16	1.26	1.42	2.09

<table>
<tr><td></td><td colspan="4">Replacement of clover nitrogen</td><td colspan="2">Net increase in protein</td></tr>
</table>

The rate of decline in legume content depends on the rate of nitrogen applied, and the time over which nitrogen is applied, as shown in Tables 4.17 and 4.18 (Jones 1967, 1970).

As there was little effect of application of 100 kg N ha^{-1} applied in one year only on legume content, yet grass yield was increased by 30–50 per cent, Jones (1970) suggests that 100 kg N ha^{-1} could be applied to different areas of pasture in one year only to stimulate grass production early in the growing season.

Table 4.17. Effect of rate of application of nitrogen on yield of Siratro in a Siratro–Rhodes grass pasture (Jones 1967)

Nitrogen level (kg ha^{-1})	Dry matter yield (kg ha^{-1})			% Reduction in legume
	Siratro	Other spp	Total	
0	3700	6000	9700	–
75	3120	7600	10 720	16
225	2500	9400	11 900	33

Table 4.18. Effect of rate and duration of nitrogen application on yield of Siratro with Rhodes grass (Jones 1970)

Year	Yields of siratro (kg ha^{-1}) at N rate (kg ha^{-1} a^{-1})		
	0	100	300
1962–3	434	470	172
1963–4	1920	1020	150
1964–5	560	380	99
1965–6	1650	500	0

Other nutrients are also important. Phosphorus must be applied to deficient soils in order to establish legumes and long-term maintenance of legume content may be related to the input of phosphorus fertilization, as shown by Evans (1970) (Table 4:19).

Table 4.19. Effect of maintenance rate of superphosphate on legume content and animal performance in a *D. uncinatum*, *L. bainesii*, *T. repens*, and *D. decumbens* pasture (Evans 1970)

Maintenance rate of superphosphate (kg ha^{-1})	Legume content (%)	Live weight gain (kg ha^{-1})
125	18	315
250	24	395
500	30	505

However, application of phosphate may not always result in increased legume content. Both Norman (1965) and Jones (1968) found that much of the response to added phosphorus in *S. humilis*-grass pastures was due to increased yield in the grass component. Similarly Tietzel (1969) found little increase in *S. guyanensis* yield on poor wet coastal soils, but a marked increase in grass yield. These responses may result from the ability of *Stylosanthes* to extract phosphorus from low phosphorus soils (Andrew 1966; Andrew and Robins 1969).

In detailed studies of competition for phosphorus and potassium in *P. maximum* var. *trichloglume*–*M. atropurpureum* mixtures in a soil low in phosphorus and potassium, Wildin and Hall (1975) demonstrated that the grass component was favoured by low potassium and high phosphorus while the legume appeared to be the stronger competitor under conditions of low phosphorus and high potassium. These relationships were also evident in field studies. Thus application of macro-elements can be a useful tool to manipulate species composition in a pasture.

Defoliation and grazing management

This is perhaps the most important factor under the control of the pasture manager. Cutting trials have shown that the erect types such as *S. guyanensis*, and the twining climbing types of tropical legumes, such as *M. atropurpureum*, *D. intortum*, and *G. wightii* are sensitive to the frequency and height of defoliation, and that severe defoliation can lead to rapid depletion of legume content (Jones 1967; Whiteman 1969), as shown in Table 4.20.

Table 4.20. The effect of cutting interval on dry matter yield of *M. atropurpureum* cv. Siratro sown with grasses (Jones 1967)

Cutting interval (weeks)	Dry matter yield (kg ha^{-1})		
	Siratro	Grass and weeds	Total
4	1600	8700	10 300
8	2370	8650	11 020
12	5400	5600	11 000
16	7350	3760	11 110

In contrast the more prostrate legume types such as *Trifolium semi-pilosum* and *S. humilis* are favoured by a more intensive defoliation, which reduces the shading effect of the associated grasses, as shown in

Table 4.21 (Jones 1973). This table shows that *T. semipilosum* yields were almost doubled by the lower cutting height, but were less affected by cutting interval.

Table 4.21. Mean annual yield of *Trifolium semipilosum–Paspalum dilatatum* pasture as influenced by cutting interval and cutting height over four years (Jones 1973)

| Cutting interval (weeks) | T. semipilosum yield Cutting height | | | Mean total yield (grass and legume) |
	3.8	7.5	Mean	
4	2160	1210	1690	5360
8	2430	1260	1850	6130

The different reactions to defoliation of the prostrate temperate type legumes and the twining climbing tropical legumes may be related, in part, to the differences in morphology. When the tropical twining legumes are cut or grazed, the upper younger leaf and stem tissue is removed. This is the most active photosynthetic tissue and thus there tends to be a longer 'recovery phase' after defoliation while new leaves and shoots are produced from lower meristems. In contrast, with temperate species such as clovers, it is the older leaves which are held higher in the canopy and which are removed by grazing, allowing the younger leaves to be exposed to radiation. Thus high rates of photosynthesis are established leading to rapid recovery after defoliation.

Although the general principles established in cutting trials may still apply, these may be modified under grazing. Whereas in cutting trials defoliation is uniform, the grazing animal is selective both in terms of species preference and plant parts. Also uneven return of dung and urine and effects of trampling and fouling of pastures can influence species response to defoliation.

Selective grazing can lead to the reduction of the selected species even though overall grazing pressure may be quite low. Furthermore utilization of species within a pasture may change with time. With both *M. atropurpureum* and *S. humilis* selection and intake is higher at the end of the growing season (Stobbs 1975).

Regardless of the method of grazing, the intensity of defoliation is primarily a function of the stocking rate. Botanical composition of mixed pasture swards grazed at a range of stocking rates can be very different, as is illustrated by the following examples. As in the case of the cutting trials, effects of grazing intensity on the legume component differ with different legume types.

In a grazing trial under coconuts in Bali, Indonesia, with a *Brachiaria ruziziensis*, *M. atropurpureum*, *C. pubescens* pasture, Nitis, Rika, Supardjata, Nurbudhi, and Humphreys (1976) found after three years grazing that legume content at the four stocking rates were: 1.8 Au ha^{-1} siratro 50 per cent, centro 14 per cent; at 2.4 Au ha^{-1} 16 per cent and 30 per cent; and at 3.1 Au ha^{-1} siratro was only 1 per cent and centro 16 per cent.* Siratro was less tolerant of higher stocking rates than Centro, and both species declined at the highest stocking rate.

Walker (1975) has shown that changes in legume content with stocking rate occur within one year of imposing treatments. In a *Setaria anceps-M. atropurpureum-S. guyanensis* pasture, *S. guyanensis* declined to low yields very quickly, while Siratro yields in July, after one year's grazing were – 1.2 steers ha^{-1}, 900 kg ha^{-1}; 1.7 steers ha^{-1}, 600 kg ha^{-1}; 2.0 steers ha^{-1}, 200 kg ha^{-1}; 2.5 steers ha^{-1}, 150 kg ha^{-1}; 3.3 steers ha^{-1}, 100 kg ha^{-1}; 5.0 steers ha^{-1}, 25 kg ha^{-1}. At the highest stocking rates siratro had almost disappeared from the pasture.

The effect of stocking rate on siratro plant populations has been demonstrated by Bisset and Marlow (1974). After four years grazing at stocking rates of 1.24, 0.82, and 0.62 steers ha^{-1}, siratro densities were 12 000, 33 000 and 39 000 crowns ha^{-1}, respectively.

The tropical trailing and twining or erect species are considered to be susceptible to overgrazing and legume yield will decline as stocking pressure increases. However the prostrate legumes such as *Lotononis bainesii* and *Desmodium heterophyllum* are more tolerant of heavy grazing pressure, while the content of the annual legume *S. humilis* is increased by heavy grazing. Austin (1970) compared three stocking rates on a mixture of *Urochloa mosambicensis* and *S. humilis*, originally in a ratio of 75 per cent and 25 per cent respectively. After two years grazing botanical composition and animal production were as follows:

Stocking rate (steers ha^{-1})	0.6	1.25	2.5
S. humilis content (%)	8	35	75
Live weight gain (kg head^{-1})	123	124	147
Live weight gain (kg ha^{-1})	74	156	368

These data clearly demonstrate the contrast in *S. humilis* response compared with the trailing tropical legumes. As grazing pressure increased, so legume content increased, which led to improved animal production.

* 1 Au = 1 steer of 400 kg liveweight.

Thus *S. humilis*, *L. bainesii*, and *D. heterophyllum* are similar to the clover-based temperate pastures and require moderate to heavy grazing pressure to maintain the legume content. With the other legume species such as siratro, glycine, *D. intortum*, *D. uncinatum*, and perennial stylo, reaction to stocking rate is very different from that of the clover-based pastures. Research data from rainfall environments around 1250 mm per annum suggests that stocking rate should not exceed 1.7 mature steers ha^{-1}, and at higher stocking rates rapid decline in legume content may be expected. Many of the failures of tropical legumes in sown pastures can be related to over-optimistic assessments of the year-long carrying capacity.

4.4 INTEGRATION OF PASTURES IN CROPPING SYSTEMS

In many tropical countries, higher fertility areas devoted to arable cropping tend to be highly utilized and individual farm size tends to be small. Pasture and forage production is seldom a priority in these areas. The extensive grazing areas, and major areas of improved or sown tropical pastures are usually found on land systems less suited to arable crop production.

Ruminant animals are important in many cropping areas as draught animals and could play a more important role in the overall production system providing not only draught, but milk, meat and valuable manures. These animals provide a means whereby the waste products of arable cropping such as vegetable materials (e.g. kenaf tops, cassava tops), stubbles, weeds, and crop by-products such as rice bran, diseased grains, etc., can be converted into animal products. However, in cropping areas there are periods of severe forage deficit. In the wet rice-cropping areas one important period is during the growth of the main rice crop before stubbles are available for grazing after the grain is harvested.

If ruminant populations are to be maintained or increased in the cropping areas, then it is important to increase the forage resources, particularly during the deficit periods. This may be achieved by introducing a short-term, or ley pastures into the cropping system. A ley pasture is defined as a temporary pasture grown as a specific phase in a defined crop-rotation sequence (Davies 1960). Although this concept has been applied in temperate areas, very little application or research has been undertaken in tropical cropping systems.

In this chapter the potential advantages of including a pasture phase in rotations are discussed, and the principles of pasture development and application to traditional 'slash and burn' cropping systems, tropical crops, and plantation crops are reviewed.

The potential advantages of a pasture phase

The term 'potential advantage' is used advisedly in this discussion, as the realization of advantage claimed for ley pastures in tropical cropping systems depends on many factors and most importantly on economic ones. Farmers with limited land resources will only occupy arable land with ley pasture or forage crops if the direct or indirect benefits give equivalent economic returns to crop production. Unless a pasture phase can be demonstrated to give real benefits in terms of higher subsequent crop yields, or in better animal production, then farmers will be reluctant to apply these systems. However, in other situations to be discussed in this chapter, namely pasture sowing in slash-and-burn systems, and pastures in plantation crops, pastures are an alternative form of land use normally more productive than the existing land use.

The temperate concepts of ley farming are based on the improvement in subsequent crop yield following a legume or legume–grass pasture due to improved soil fertility, improved soil structure, and reduction in pests and weeds of arable crops. Other advantages are demonstrated in reduction in soil erosion, and in effects on improved animal production. The application of these principles is examined in relation to tropical cropping systems, although there is less data than for temperate areas.

Effects on soil fertility

In the temperate environment of southern Australia the important role of short term *Trifolium subterraneum* pastures on subsequent wheat yields has been well demonstrated by the work of Watson and Lapins (1969). They showed that the major effect of nitrogen build up on wheat yield is achieved after only one year of pasture, as follows:

Years of sub-clover ley pasture	0	1	2	3	4	5
Subsequent wheat yield (kg ha^{-1})	1080	1820	1950	2080	2220	2350

Also the major effect of the pasture phase was evident in the first year of cropping. After eight years under a sub-clover ley pasture the following wheat yields were achieved with continuing cultivation:

Years of cropping after pasture phase	1	2	3	4
Wheat yield (kg ha^{-1})	1880	1080	810	600

Finally Watson and Lapins (1969) examined the effects of sward management on subsequent wheat yield. They compared swards which over a three-year period had either been grazed with sheep or cut and the hay removed, or cut and the material left on the plots (Table 4.22).

Table 4.22. Wheat yields (kg ha^{-1}) following a three year sub-clover pasture ley subjected to different management (Watson and Lapins 1969)

Year	Grazed	Cut and return	Cut and removed
1	2290	2110	2200
2	1980	1540	1730
3	1680	1870	1510
4	960	920	870
Mean	1730	1610	1580

Here again wheat yields decline rapidly after the first year, but the important point is that wheat yields differ little between management systems. Even cutting and removing the pasture had little effect, and Watson and Lapins concluded that much of the nitrogen returned to the soil surface is lost. Most of the nitrogen increments came from roots, nodules, and stubbles in the ground.

On the basis of these studies, Watson and Lapins (1969) suggested a system of alternate years of crop and pasture, and this was shown also to give the highest economic returns. This concept might be applied in the tropics with a self regenerating annual legume such as *Stylosanthes humilis*, but so far has not been rigorously tested, although the effects of undersowing in sorghum crops on *S. humilis* establishment have been examined (Ive 1970). A *S. humilis* pasture phase has been shown to markedly increase the yield of subsequent crops in the Northern Territory, Australia (Norman and Wetselaar 1960) (Table 4.23).

This study demonstrates the ability of the pasture phase to improve the nitrogen status for subsequent crops. Much of this nitrogen had been leached through this sandy surface soil and accumulated in deeper soil horizons. This nitrogen was exploited more effectively by the deeper rooting bullrush millet, as demonstrated by its much higher nitrogen yield, and was not tapped by the *Sorghum* species.

Where perennial pasture species are used the concept of alternative years of crops and pasture could not apply, as a longer pasture establishment phase is required, and tropical pasture seed is expensive. A number of studies have shown that crop yields may be improved following a

Table 4.23. Effects of a *S. humilis* pasture phase on the dry matter and nitrogen yield of forage crops at Katherine, Northern Territory, Australia (Norman and Wetselaar 1960)

Treatment	Yield (kg ha^{-1})					
	Sorghum[1]		Bullrush millet[2]		Sudan grass[3]	
	dry matter	nitrogen	dry matter	nitrogen	dry matter	nitrogen
Native grass (6 years)	2230	120	2620	126	2240	110
Bare fallow (5 years)	11 020	475	12 100	1076	6830	353
S. humilis pasture (7 years)	10 740	370	11 280	650	9800	440

[1] *S. vulgare*;   [2] *Pennisetum typhoides*;   [3] *S. sudanense*.

tropical pasture phase. Jones (1967) showed increased sorghum yields following siratro (*M. atropurpureum*) pastures, but not after *L. bainesii* (Table 4.24). *L. bainesii* appeared to be competitive for soil nitrogen, and soil levels were little increased.

Table 4.24. Effects of four years pasture pretreatments on sorghum yields and total soil nitrogen levels at Samford, Queensland (Jones 1967)

Pretreatment (kg N ha^{-1} year^{-1})	Sorghum dry matter yield (kg ha^{-1})	Total soil nitrogen (%)
Paspalum plicatulum + N0	1260	0.089
Paspalum plicatulum + N100	2560	0.108
Paspalum plicatulum + N200	6260	0.114
Paspalum plicatulum + Siratro	5530	0.125
Paspalum plicatulum + *L. bainesii*	1411	0.103

In a similar study, Balanchandran and Whiteman (1975) showed that a pure Desmodium sward gave higher subsequent crop yields than nitrogen fertilized Setaria, or Setaria–Desmodium pastures (Table 4.25). This study also demonstrated that nitrogen losses from applied fertilizers may be high, as there was little increase in crop yield comparing the N0 and N300 treatments. Similar effects of legume pasture phase have been demonstrated on maize yields in Ghana (Kannegieter 1969) and sorghum in Kenya (Clarke 1962).

Table 4.25. Effects of three years pasture pretreatments on the total dry matter
          yield of a subsequent Sorghum and oat crop at Mt. Cotton, Queens-
          land (Balachandran and Whiteman 1975)

Pretreatment (kg N ha^{-1} a^{-1})	Total (sorghum and oat crop) yield kg ha^{-1}
Setaria anceps N0	3390
Setaria anceps N300	3560
Setaria–Desmodium N0	4140
Setaria–Desmodium N300	3190
Desmodium N0	4930
Desmodium N300	5040

Thus tropical pastures are demonstrated to improve soil fertility
through legume nitrogen fixation, and this improvement in fertility is
not reduced when the pasture is grazed. In fact Stobbs (1969) showed
in Uganda that crop yields following a *Chloris gayana-Hyparrhenia
rufa-S. guyanensis* pasture were higher in the grazed than in the ungrazed
swards.

Effects on soil structure
Under continuous cultivation there is often a breakdown in soil structure which is demonstrated by the following effects: a loss of soil
aggregate cohesion when wet, leading to increased bulk density and
compaction which may reduce water infiltration rates and also reduced
root penetration; rapid dispersion of finely broken aggregates at the
soil surface leading to surface sealing and crusting, which causes increased
run-off and may also affect seedling emergence; and soils which are
compacted become more difficult to cultivate to form a satisfactory
seed bed and require greater energy for cultivation.

Although these conditions are widely recognized it is often difficult
to demonstrate beneficial effects of a pasture phase on soil structural
properties. Firstly there is the difficulty of measuring and defining soil
structure, and secondly it is difficult to demonstrate direct effects of
different soil structures on plant growth as these are usually confounded
with other effects on soil fertility. Nevertheless, effects of ley pastures
have been demonstrated on structural properties such as bulk density,
aggregate cohesion, and water stable aggregates. These properties are
important in maintaining the long-term stability and productivity of

soils, and especially in the tropics, where many soils have inherently poor structural properties.

In a temperate environment, Low, Piper, and Roberts (1963) have demonstrated the effects of pasture leys on soil physical properties. They showed that the percentage of water-stable aggregates increased with time under a ley pasture, but that the water-stable aggregates declined markedly after two years cropping as follows:

Time after ley phase	% Water-stable aggregates (> 0.5 mm)			
	after 0	1	2	3 years ley
1. Before ploughing at end of ley	1.3	6.1	14.3	19.0
2. After first crop (kale) Year 1	1.7	4.3	8.4	10.7
3. After second crop (wheat) Year 2	1.3	1.5	2.8	3.6

Low *et al.* (1963) also showed that the pasture phase improved soil cohesion properties as follows:

Soil property	Years of ley			
	0	1	2	3
1. No. of water drops required to cause disintegration of aggregates	11.6	13.0	15.0	20.3
2. Draw-bar pull required for cultivation (kg cm^{-2})	46.5	47.6	45.5	42.7
3. Soil crushing strength (g cm^{-2})	58	52	50	42

These data demonstrate that after three years pasture, the soils were more resistant to disintegration by raindrop impact, and were also easier to cultivate to a seed-bed tilth. Also they showed that water infiltration rate was doubled after a one year pasture phase.

Andrew (1965) showed that perennial grasses had a much greater effect in increasing the proportion of water-stable aggregates than an annual legume. After four years under pasture, he recorded the following percentage water-stable aggregates — control 44 per cent, *Trifolium subterraneum* 44 per cent, *L. perenne* 55 per cent, *Phalaris tuberosa* 55 per cent, *L. rigidum* 57 per cent. The effect of the grasses may be related to their more extensive fibrous root systems which break down less rapidly in the soil.

In tropical soils Monier (1965) suggests that the maximum effect in increasing water-stable aggregates is achieved within the first year of incorporating organic materials, but emphasizes that structure can be maintained by maintaining the level of stable humus. These views are supported by Maher (1951) and Webster (1954) who showed in East Africa that although ley pasture had beneficial effects on soil structure, these effects were lost after one or two years cropping. As in the case of improvement in soil fertility these studies suggest that short cycles of pastures and cropping may be more beneficial than long cycles.

Effect on weed and pest control

In many cropping areas weed problems tend to increase with time, leading to reduced crop yields and increasing weed-control costs. A crop rotation involving a pasture phase can be advantageous in reducing arable weed problems as the pasture phase provides a break in the life-cycle of the weed component. The success of a ley pasture in reducing a weed problem will depend on the longevity of the weed seeds in the soil, on the competitive ability of the pasture species and on whether the weed is eaten by the grazing animals.

There are some good examples of weed control in temperate crop-pasture rotations, but little work has been done in tropical cropping systems. Skeleton weed (*Chondrilla juncea*) is a notorious weed in wheat crops in southern Australia. It competes with the wheat crop for nitrogen and water, reduces yield, and causes problems in harvesting. Moore and Robertson (1964) found that a sub-clover (*T. subterraneum*) pasture was more effective in reducing *C. juncea* density than a wimmera rye grass (*Lolium rigidum*) or a mixed rye grass–sub-clover pasture. After four years, sub-clover reduced skeleton weed density by 80 per cent, whereas wimmera rye reduced numbers by only 3 per cent and a rye–sub-clover pasture by 43 per cent compared with unsown control plots. The stronger effect of sub-clover alone was due to a greater shading of weed seedlings and selective grazing by the sheep. A further advantage of the legume ley pasture was due to the increased soil nitrogen level (0.047 per cent cf. control 0.039 per cent) which increased the yield of wheat and its ability to compete with the weed.

Another important weed in wheat areas is cotton thistle (*Onopordum acanthium*). Michael (1968a) has compared a range of grass ley pastures for their effectiveness in reducing the weed, with the following results after three years under pasture in grass ley.

Grass ley	control	cocksfoot	fescue	phalaris	ryegrass	brome
Weed yield (kg ha^{-1})	1932	2307	526	928	1806	1857

Fescue (*Festuca arundinacea*) was superior to other grass species in reducing the weed. In another experiment Michael (1968*b*) found that lucerne (*M. sativa*) and *Phalaris tuberosa* ley pastures were equally effective in reducing variegated thistle (*Silybum marianum*). The final example is the effect of a lucerne ley in controlling wild oats (*Avena fatua*) in wheat. Littler (1967), showed that after four years of lucerne the numbers of wild oats were markedly reduced compared with continuous wheat plots, but that wild oats increased with time after returning to cultivation as follows:

	Continuous wheat	Years after pasture					
		1	2	3	4	5	6
No. *A. fatua* per plot	646	10	44	109	281	522	360

In the first year after the pasture phase *A. fatua* plant numbers are so reduced that chemical control could be very effectively applied to almost eliminate the weed.

In a similar experiment Lloyd (1975) found that a four-year lucerne ley also reduced the disease common root rot (*Cochliobolus sativus*) in wheat. In north Queensland Rhodes grass (*C. gayana*) pastures are employed in rotation with tobacco to reduce root knot nematode (*Heterodera marione*) levels. Mukhopadhyaya and Prasad (1969) have also shown that nematode populations were less where wheat, cotton, and maize are grown in rotation with berseem clover (*T. alexandrianum*) than on plots where each crop was grown in monoculture.

Few studies of pasture crop rotations in the tropics have discussed weed and pest control, although Kannegieter (1969) in Ghana reported that a *Pueraria phaseoloides* ley, after spray killing, provided effective soil protection, weed control, and moisture conservation in the following maize crop. However, Newton and Jamieson (1968) suggest that even with cover-crop rotations, including *P. phaseoloides*, original levels of food-crop production could not be maintained in lowland New Guinea owing to a decline in the soil fertility and a serious build-up of weed, disease, and pest problems.

The temperate ley-pasture–crop rotations discussed above demonstrate the role ley pastures can play in breaking weed and pest cycles. The data show that pasture species differ in their effectiveness in controlling particular weeds, and that a range of pasture species and mixtures must be evaluated in transferring these principles to tropical cropping systems.

Effects in soil conservation

In many cropping systems bare fallowing and continuous monoculture are widely employed. While the fallow period may provide for weed control, preparation of a seed bed and mineralization of soil-stored organic nitrogen, serious penalties in soil erosion may be incurred, particularly in tropical regions where high intensity rainfall events are frequent. In many tropical regions calamitous erosion has become the enduring monument in man's attempts at arable agriculture.

Well-grown pasture swards resist soil erosion by providing a surface cover, preventing raindrop impact, reducing volume of runoff and velocity of flow, improving surface-soil structure and hence rate of infiltration, and in trapping the moving soil particles. These effects are shown by the data of Bennett (1955) (Table 4.26).

Table 4.26. Comparison of annual soil loss and time required to erode the top 17.5 cm of top-soil for a range of land uses and soil types (after Bennet 1955)

Treatment	Soil loss (t. ha^{-1} a^{-1})	Years to erode 17.5 cm
(i)　Fayette Silt Loam (16% slope)		
Grass pasture	0.25	10 000
Rotation (maize, barley, clover)	70.3	36
Maize alone	281	9
Bare fallow	479	5
(ii)　Kirvin Fine Sandy Loam		
Grass pasture (16.5% slope)	0.01	200 000
Forest (12.5% slope)	0.1	20 000
Rotation (cotton maize, clover)		
(8.75% slope)	43	59
Cotton (8.75% slope)	58	43
Cotton on subsoil (8.75% slope)	165	15
(iii) Cecil Sandy Clay Loam (10% slope)		
Virgin forest	0.005	500 000
Grass pasture	0.75	3225
Rotation (cotton, maize, Lespedeza)	35	70
Cotton	78	32
Bare fallow	165	15

Soil loss and rate of erosion was highest in bare fallow, followed by monoculture row crops such as maize and cotton. Although the use of arable rotations considerably reduced the rate of erosion they were greatly inferior to grass pastures. The rate of soil loss under grass pastures was almost negligible.

Recent studies have suggested that bare fallowing is an inefficient method of increasing soil-moisture storage. For example, Bennison and Evans (1968) comparing various crop rotations in Kenya, reported that although rain was the main determinant of yield within each season, the use of fallowing to store water from one season to the next was inefficient. Thus in many situations there may be a good case for replacing bare fallows with a ley pasture phase. Replacing the long fallow in maize rotations with grass or lucerne pastures greatly reduced erosion and run-off in southern Africa (Scott 1967). Stephens (1967) found in Uganda that grass fallows in arable crop rotations increased soil nitrate and potassium supply, and also improved infiltration and moisture status in the top-soil.

The use of legume-based tropical pastures should have wider application in place of bare fallow in crop rotations as an important aid in erosion control. The value of pastures is widely recognized in contour strip cropping on sloping land, where strips of pasture alternate with strips of arable crop to prevent excessive downward water flow. This practice employs the principles of a ley pasture as the pasture strips become the arable crop areas in subsequent years. Finally on land too steep for safe arable cropping, perennial pastures may provide for productive permanent land use while preventing erosion and excessive run-off.

Application in tropical cropping systems
Following the discussion of the potential role of pasture leys in tropical cropping practices, this section examines the principles and practices of undersowing annual crops, and plantation crops with pasture species.

(i) Undersowing annual crops
Undersowing temperate crops, mainly the winter cereals with pasture species, has been widely applied as a technique of pasture establishment (Charles 1958). Now that adapted tropical species are more widely available, these methods deserve closer attention in tropical cropping systems. Undersowing of pasture species has advantages over conventional establishment methods in many situations. The cost of land development for pasture is absorbed in the cropping venture and the normal weed-control methods applied to a cash crop will benefit the establishment of the undersown pasture species.

An undersown legume will have particular value in many smallholder cropping systems where the legume forage developing after the crop is harvested will considerably improve the grazing value of the crop stubbles. The development of a pasture ground cover after harvest, where the ground might normally be left bare, may have important advantages in erosion control. Shelton and Humphreys (1975) also consider the advantages where animals graze dry season stubbles under-sown with legumes and then transfer seed of the adapted pasture plants to other communal grazing areas; the undersown crop becomes a source of seed for wider dissemination at low cost.

The undersown pasture may be used as a short-term ley and ploughed up for the next cropping season, or may become a longer term pasture. It is in the latter context that undersowing has particular application in the 'slash and burn' or *ladang* systems of cultivation. Thomas and Humphreys (1970) and Shelton and Humphreys (1972) have suggested that in the forest cultivation areas of South-East Asia undersowing of the final upland rice crop with legume-based pastures would provide an alternative to the forest regeneration or 'bush fallow' phase. The pastures would not only provide much needed forage, but fertility build-up may be enhanced through the legume nitrogen fixation. Also in many areas of the wet tropics exposed to shifting agriculture and to uncontrolled grazing and burning, previously cultivated areas revert to an *Imperata cylindrica* grassland (Walker 1966). This grassland has a low animal production potential and soil fertility steadily declines. Establishment of a sown grass–legume sward in the final stages of cropping would be a far better alternative.

The objective in pasture development through undersowing must be to obtain adequate establishment of the pasture species with the minimum reduction in yield of the main cash crop. Where undersown pastures are planted in cereal crops the balance of competition is usually strongly in favour of the faster growing, leafier cereal crop (Donald 1963). A range of management options are available to ensure that the competitive advantage is maintained in favour of the cash crop. A number of these options have been elucidated by Shelton and Humphreys (1975) with upland rice undersown with *Stylosanthes guyanensis* under field conditions in central Laos and north east Thailand.

The primary factor in the competitive relationship is the influence of the crop on the light environment of the undersown pasture and it is this factor which is most readily adjusted by – (a) relative timing of establishment of crop and pasture; (b) crop and pasture sowing density; (c) crop row direction; (d) the vigour and maturity type of the crop variety; (e) application of nutrients, particularly nitrogen. This latter factor will also affect the competitive relationship for nutrients and soil moisture.

(a) *Relative timing of crop and pasture establishment*   Delaying the sowing of the undersown species confers a great advantage to the crop species, as shown by the data of Shelton (1972) with *S. guyanensis* sowing in upland rice in a 'slash and burn' system in Laos, as follows:

Time of sowing stylo	0	31	60 days after rice sowing
Mean rice grain yield (g m^{-2})	113	130	134
Mean stylo yield (g m^{-2})	164	22	3
Mean stylo density (No. m^{-2})	24	15	6

In another experiment, Shelton and Humphreys (1975) found that simultaneous sowing reduced grain yield by 19 per cent, but it was not significantly reduced if stylo undersowing was delayed by ten days. Stylo growth was reduced by 50 per cent by delayed sowing and was negatively related to the dry matter yield of the rice variety. These data show that delayed sowing markedly alters the balance in favour of the crop species.

(b) *Sowing densities*   The degree of shading of the undersown pasture is influenced by the initial density or row spacing of the crop, but sowing effects tend to even out as compensatory tillering takes place to fill the canopy space. Thus Shelton and Humphreys (1975) recorded no significant effect of rice density on rice yield, but rice at a density of 20 plants m^{-2} reduced stylo yield by 56 per cent compared with monoculture stylo, and only by a further 10 per cent (66 per cent) at a rice density of 120 plants m^{-2}. On the other hand increasing sowing density of stylo, sown at the same time as the rice significantly reduced rice yield as follows:

Stylo density (no. plants m^{-2})	0	3	9	27	81
Rice grain yield (g m^{-2})	98	85	82	72	63

Although in most crops individual plant size tends to increase through tillering or branching as plant density decreases, increased row spacing may allow increased light penetration with consequent increase in growth of undersown species and a greater depression in crop yield (Schaller and Larson 1955).

(c) *Row direction*   North–south orientation of rows has been shown to allow greater and more uniform light penetration into the interrow space than east–west row direction. The effects of row spacing and row direction on grain yield of *Sorghum vulgare*, and on density of undersown *Stylosanthes humilis* are shown in Table 4.27 from the work of Ive (1970) in northern Australia.

Table 4.27. Effects of row spacing and row direction on (a) the grain yield of *Sorghum vulgare* (kg ha^{-1}) and (b) on density of undersown *Stylosanthes humilis* ($\times$ 10^6 plants ha^{-1}) (Ive 1970)

	Row spacing (cm)			
Row direction	18	54	90	Mean
(a) North–south	1960	2250	2200	2140
East–west	2260	2610	2440	2440
(b)				(no sorghum)
	0.5	2.5	4.2	13.6

These data show that sorghum yield was reduced in the north–south rows, and this was related to a greater growth of the undersown species, and that the undersown *S. humilis* density was greater in the wider row spacings. However, at all row spacings when undersown in Sorghum, *S. humilis* density was much reduced compared with a monoculture situation.

(d) Vigour and maturity type of the crop   It is difficult to predict the effects of different crop varieties on the undersown species. It might be expected that taller, leafier, later-maturing varieties would have a greater depressing effect on the undersown species. However, Shelton and Humphreys (1975) found that although stylo yield under an early, a mid-season and a late rice variety varied by a factor of two, these differences were not related to leaf area indices or light transmission. They suggest that crop variety effects on stylo growth were due more to competition for moisture or nutrients than to competition for light. Differences in seedling growth rate, vigour, and tolerance of shading between pasture species might also be expected to affect competition with the crop species.

(e) Application of nutrients   Usually the competitive advantage for fertilizer uptake lies with the companion crop, leading to increased above and below ground crop growth further increasing the crop competitive advantage. Where a legume is undersown in a tropical cereal crop uptake of applied nitrogen will be strongly in favour of the crop (Vallis, Haydock, Ross, and Henzell 1967; Henzell, Martin, Ross, and Haydock 1968). In their studies of the rice–stylo association, Shelton and Humphreys (1975) found that the growth of stylo was negatively related to the level of urea applied, while rice grain yield increased with increasing nitrogen level (Table 4.28).

Table 4.28. Effect of nitrogen application on rice dry matter and grain yield, light transmission, and *S. guianensis* dry matter yield (Shelton and Humphreys 1975)

	Nitrogen rate (kg ha^{-1})			
	0	20	40	80
Rice dry matter yield (kg ha^{-1})	1700	2010	2280	3150
Rice grain yield (kg ha^{-1})				
with *S. guianensis*	1820	2010	2220	2530
without *S. guianensis*	1820	2710	2540	2630
% light transmission	68	51	47	33
S. guanensis yield (kg ha^{-1})	3380	2920	2160	1240

The reduction in stylo yield was related to increased level of shading with increasing nitrogen levels. In contrast to these data, Thomas and Bennet (1975) in studies of undersowing maize in Malawi with *Chloris gayana* and *Desmodium uncinatum* mixtures found no reduction in legume yield in the establishment year with nitrogen applications up to 112 kg ha^{-1}, even though there was a significant increase in maize dry-matter yield. There was no reduction in maize yield when pastures were undersown at 23 days (first weeding) or 40 days (second weeding) after the maize planting. Although grass and legume yields were considerably lower in the establishment year with the later planting, by the second year there were no significant differences in grass or legume yields. Grass and legume establishment was far better and yields in the second year were twice those recorded from pastures directly sown without maize. This was related to the better weed control in the undersown maize. Thomas and Benett (1975) and Thomas (1975) suggest that the better legume establishment when undersown was due to the shading effect which alters the competitive balance in favour of the *D. uncinatum* relative to the *C. gayana*. Once the maize canopy died back at maturity, the grass growth improved with the increase in light intensity. These effects are supported by the data of Bauer and Okech (1975) where *C. gayana* and *D. uncinatum* were undersown with a *Lupinus albus* crop at Kitale, Kenya. Here again legume content was considerably higher in the undersown swards than in direct-sown plots.

This research supports the view that tropical pastures can be readily integrated into annual cropping systems. With proper adjustment of sowing time of the pasture species, crop density and variety, and application of nutrients, little or no penalty in main crop yield may be incurred. Considerable advantages in the time to utilization of the pasture, in legume establishment and subsequent legume content, and in the grazing value of crop stubbles have been achieved. The incorporation of forage legumes or legume-based pastures into the cropping cycle does not require sophisticated agronomic practices, and will give marked benefits in livestock production and long-term maintenance of crop yields.

(ii) Undersowing plantation crops

Plantation agriculture with permanent tree crops is an extremely important and widespread land use in the tropics. Multiple land use involving either undersown crops or natural grazing is a common feature of many plantation areas, but the development of undersown grass and legume pastures has been a more recent innovation. While pasture development has been mainly undertaken in coconut (*Cocos nucifera*) plantations, increasing interest is being shown in the possibilities of grazing in other plantation crops, oil palm (*Elaeis guineensis*), rubber (*Hevea braziliensis*), kapok (*Ceiba pentandra*), clove (*Eugenia caryophyllus*), cashew (*Anacardium occidentale*), and in tropical reforestation projects (Kirby 1976).

The advantages in terms of more efficient land use and income from animal production are obvious provided the pastures and grazing animals are compatible with the plantation management systems. In the case of coconut plantations livestock are compatible with the management system, and cattle have been widely employed as 'brushers' to control weeds (Ohler 1969). A sown legume-based pasture in this situation may not only provide a greater forage supply, but also a more effective control of weeds. Furthermore many plantation crops have traditionally been undersown with cover crops such as *Centrosema*, *Pueraria*, and *Calopogonium*, to provide a ground cover, to control weeds, and to improve soil nitrogen status. Studies in rubber have shown that a legume cover crop maintained over ten years gave a 37 per cent higher root mass and 20 per cent higher rubber yields (Anon. 1968). Nitrogen input from legume pastures will become available to the tree crop, and as shown in the work of Watson and Lapins (1969), total nitrogen accretion may be little affected by cycling the nutrients through the grazing animal. In any case in cover-crop areas not grazed by ruminants, much of the above-ground material passes through insect and other arthropod grazers.

The cycling of nutrients through the grazing animal may, in fact, improve nutrient availability. In an ungrazed pasture of *Brachiaria brizantha* under coconuts in Ceylon fertilized with 25 kg N ha^{-1} Ferdinandez (1970) obtained an average nut yield over four years of 5780 nuts ha^{-1}, while in the grazed pasture nut yield was 10 180 ha^{-1}. In pastures fertilized with 50 kg N ha^{-1} at a low stocking rate (5 head ha^{-1}) nut yield was 10 360 nuts ha^{-1}, while at a high stocking rate (11 head ha^{-1}) nut yield was 11 580 nuts ha^{-1}. While part of the effect of grazing may be due to reduced competition from the pasture sward, the pasture yields in the ungrazed 25 kg N ha^{-1} treatments were similar. Increased nutrient availability through grazing appears to be important.

Another advantage of sown pastures under plantations may be achieved through improved soil structure. In areas having a distinct dry season coconut yield reductions due to moisture competition have been documented (Krishnamarar 1961). In West Africa it was observed that moisture competition was most severe with young palms, but that older palms were little affected by undersown *Centrosema* (Fremond and Brunin 1966). Improved infiltration due to improved soil structure with undersown pastures led to better water relations for the tree crop after a few years (de Silva 1961).

In coconut plantations introduction of cattle and pastures in the Solomon Islands gave a 10 per cent better recovery of fallen nuts previously lost in rank weeds and grass. Another minor advantage in grazing under plantations may be in the lower temperatures for the grazing animal. Sajise (1973) recorded air temperatures 4 °C lower within a coconut plantation compared with outside under sunny conditions, but little difference under cloudy conditions. Relative humidities were higher under the plantation only in the afternoon period during sunny conditions. Thus animal comfort factor may be improved under plantations (McDowell 1972).

One of the main disadvantages quoted for cattle grazing in plantations is damage to young trees. This is of little consequence in coconut plantations once the terminal growing point is above grazing height and where adequate forage is available. Young rubber trees may be more susceptible to mechanical damage, and livestock in producing plantations may dislodge collecting cups. Cattle may eat oil-palm fruits, although observations in young oil palm in Malaysia suggest that cattle cause little damage to the leaves (C. Samuel, personal communication).

Soil compaction by trampling may be a problem on heavier soils during wet conditions, especially at high grazing pressure where pasture cover is reduced (Plucknett 1972). Cattle may have to be removed to higher areas or lighter soils, or hand-fed during wet periods, but maintenance of adequate pasture cover should be an important management objective.

Competition for light and nutrients are major considerations in development of pastures in plantations. These factors are discussed below.

Light   In the undersown annual crop situation crop dominance of the light environment is a temporary phase. After crop maturity or harvest full radiation reaches the undersown pasture allowing rapid pasture growth. In the plantation situation the radiation levels reaching the undersown pasture in most cases declines as the crop ages and the canopy closes, and the plantation crop permanently dominates the light environment.

In plantation crops such as cocoa (*Theobroma cacao*) radiation levels are so low that ground flora is mostly excluded. Light transmission in mature oil palm and rubber is also low and only very shade-tolerant species (such as *Ottochloa nodosum*) survive. However, light transmission in most coconut plantations is adequate to support reasonable levels of growth of adapted pasture species. This is the major reason why pasture development in the plantation cropping system has mainly occurred in coconut plantations.

The light transmission through the coconut canopy depends upon the canopy characteristics of the variety, the density of planting and the age of the stand. Shading is at a maximum in closely spaced palms of intermediate age (10–20 years), but light transmission tends to improve in older plantations as the canopy is carried higher and losses of palms occur. Some effects of age, density, and time of day on light transmission are demonstrated by the limited data available (Table 4.29).

Sajise's (1973) data give the percentage of the ground area receiving light transmissions above certain values; the values given in Table 4.29 are estimates based on these data. Light transmission in the morning and afternoon is reduced. Sajise's (1973) data show that in early morning and late afternoon 90 per cent of the ground area under a coconut canopy receives less than 50 per cent light, but at noon 60 per cent

Table 4.29. Mean values of percentage light transmission in coconut stands as affected by density, variety, age, height, and time of day

Location	Density (no. palms ha^{-1})	Age of stand (years)	Approx height (m)	Time	Trans-mission (%)	Reference
1. Solomon Islands						
Liapari	180	7	4.5	0900	38	Gutteridge,
				1200	51	Whiteman,
				1500	47	and Watson
						(1976)
Lever point	175	70	15	0900	31	
				1200	97	
				1500	74	
2. Bali	120	30	10	0900	77	Steel and
				1200	80	Humphreys
				1500	61	(1974)
3. Jamaica						
(i)Malayan	560	5	–	–	14	Smith
Dwarf	440	5	–	–	29	(1972)
	348	5	–	–	–	
	281	5	–	–	59	
	230	5	–	–	60	
	187	5	–	–	82	
(ii) dwarf × tall	560	5	–	–	5	Smith
	440	5	–	–	6	(1972)
	348	5	–	–	9	
	281	5	–	–	19	
	230	5	–	–	43	
	187	5	–	–	–	
4. Philippines						
Laguna	218	–	8–9	0800	17	Sajise
				1200	50	(1973)

of the ground area was receiving more than 61 per cent light. The data of Gutteridge *et al.* (1976) clearly shows that in older, less dense stands, light transmission is little reduced, especially at noon, and would not be a major limitation to pasture growth, but in younger denser stands light reduction is highly significant (Smith 1972).

In the relationship between palm density and copra yield per hectare, there is the compensation of increasing yield per tree as density is reduced, over a limited range of densities (Whitehead and Smith 1968; Smith 1972). Where the aim is to maximize pasture growth, the optimum density of palms will be the minimum density required to give maximum copra yield per hectare. Depending on variety, the data from Jamaica (Smith 1972) suggest that for tall varieties optimum density is about 200 palms ha^{-1} and for Malayan Dwarfs about 320 palms ha^{-1}. Density trials in the Solomon Islands (Friend 1975) suggest that densities in the range of 287–340 palms ha^{-1} for dwarf, 200–35 palms ha^{-1} for tall, and 235–87 palms ha^{-1} for hybrids (dwarf X tall) had little effect on early growth and yield up to six years from planting so that lower densities might be used with little reduction in total yield. However, the philosophy should be to plant the coconuts at the appointed optimum spacing for maximum copra yield, and use species adapted to shaded conditions with proper control of stocking rate.

Nutrients   Unless a soil is extremely fertile and all nutrients are in excess of the demands of the tree crop, higher yields are usually obtained with clean weeding (Childs and Groom 1964; Santhirasegaram 1966*a*). In most cases any understorey vegetation whether weeds, naturalized pasture, sown pasture, or crop will compete with the palms for nutrients and water. Even though grazing animals will recycle nutrients in a grazed pasture, a certain proportion of the nutrients will be immobilized in the standing pasture biomass, and some will be removed in animal products. Therefore sufficient fertilizer must be applied to meet the needs of a productive pasture if nut yield is to be maintained.

Nutrient requirements will depend on soil fertility and in most coconut-producing areas fertilizer recommendations are made on the basis of soil types (e.g. Nathaneal 1967). Estimates are available for the amount of nutrients immobilized in a years growth of the trunk, plus spathes and leaves, and removed in nuts for a stand of 173 palms ha^{-1} (70 per acre) (Pillar and Davis 1963) as follows: N 54 kg, P 12 kg, K 68 kg, Ca 32 kg, Mg 12 kg. Where soil supply is limiting these nutrient losses must be restored and sufficient nutrients supplied to meet current growth requirements. In young palms, requirement for nitrogen and phosphorus is very important, while in older bearing palms potassium often becomes limiting.

With undersown pure grass swards competition for nitrogen may become acute, particularly with vigorous high yielding species. In Sri Lanka, fertilized *Panicum maximum* swards reduced nut yield by up to 500 nuts ha^{-1} a^{-1}, whereas with fertilized *Brachiaria brizantha* and *B. milliformis* sward nut yields were increased by 235 and 545 nuts ha^{-1} a^{-1} respectively. However, dry-matter yields of *P. maximum* were twice those of *Brachiaria* (Rajaratnam and Santhirasegaram 1963). But given adequate fertilization yield reductions due to undersown pasture can be reduced as shown by Rodrigo (1945). Without fertilizer undersown *Pennisetum purpureum* reduced copra yield by 39 per cent, but with adequate nitrogen, phosphorus, and potassium fertilizer copra yield was increased by 59 per cent and fodder yield increased eight-fold.

In plantations undersown with grass pastures higher copra yields were obtained when the fertilizer (NPK) was broadcast in circles around the palms than when broadcast in strips or buried in circles (Santhirase-garam 1959). However, where economic pasture responses to fertilizer are obtained then sufficient nutrients should be applied to the inter-sown pasture area to stimulate pasture growth. Where grass and legume pastures are undersown and nitrogen application is required for palm growth, then the nitrogen should be applied only around the palms. Nitrogen applied generally to the area may be rapidly utilized by the grass component leading to grass dominance and a reduction in legume content.

Pasture species for undersowing

Choice of pasture species for undersowing will depend firstly on general suitability to climatic and soil conditions of the area. In most coconut areas rainfall and temperature will not be important limitations, although ability to withstand prolonged dry seasons may be important in some areas. In these conditions the commonly sown grasses such as *Brachiaria brizantha*, *Panicum maximum* (green panic cv. Petrie) have good toler-ance. Although *Centrosema pubescens* has the ability to withstand dry seasons, siratro (*M. atropurpureum*) may be more productive. The annual *Stylosanthes humilis* or the perennial stylos (*S. guianensis* or *S. hamata*) may also have a place under seasonally dry conditions.

Adaptation to soil conditions is important where coconuts are grown on coastal coralline sands. Gutteridge *et al.* (1976) evaluated a range of grasses and legumes undersown in coconuts on a coralline sand, pH 8.0, in the Solomon Islands. On this soil most of the species suffered from iron chlorosis and grew poorly. Of eight legumes tested only siratro (*M. atropurpureum*) performed well and was little affected by iron chlorosis, although *Vigna luteola* yielded well in the first year, but declined thereafter. In pot trials *Desmodium heterophyllum* also appeared tolerant of low iron conditions. Out of five grasses tested Koronivia (*Brachiaria humidicola* = *dictyoneura*) was the only species to perform well and formed a vigorous weed-free stand. These data demonstrate the importance of evaluating species suitability to particu-lar soil conditions, as well as adaptation to conditions in plantations.

The other important adaption is tolerance of reduced light. The tropical grasses will be most affected by low-light conditions because of their ability to respond in photosynthesis up to full sunlight intensities. The legumes will be relatively less disadvantaged as their photosynthesis is saturated at lower light intensities. Within the grasses *Brachiaria milliformis* appears to be more shade tolerant than *B. brizantha* (*decumbens*) (Sajise 1973). In the Phillipines, Guinea grass (*P. maximum*) Alabang X (*Dicanthium aristatum*) and para grass (*B. mutica*) are widely recommended for undersowing (McEvoy 1974). In the Pacific area, koronivia (*B. humidicola*) and batiki blue (*Ischaemum indicum*) persist well in plantations (Ranacou 1972). In many plantations throughout South-East Asia and the Pacific, the naturalized grasses *Paspalum conjugatum*, and *Axonopus affinus* are widely present and persistent under shade and heavy grazing.

The plantation cover crop legumes *Centrosema pubescens*, *Pueraria phasioloides*, and *Calopogonium mucunoides* appear to have been selected, in part, for their shade tolerance and form the basis of many pasture mixtures. *Desmodium heterophyllum* also appears tolerant of shading, while siratro has an important place under drier conditions.

Another consideration in choice of grass species is the competitive effects on palm yield. Reduction in palm yield appears to be related to the vigour and yield of the sown grasses as noted earlier (Rajarantnam and Santhirasegaram 1963; Santhirasegaram 1966*b*). Thus it may be better to accept lower grass yields than major reductions in copra yield. In this regard also plantation managers suggest that tall tussock-forming species such as *P. maximum* should be avoided in favour of sward forming species to facilitate nut collection.

Management of undersown pastures In many plantation areas the main purpose of the grazing animals has been to control grass and weed growth to allow easier collection of fallen nuts. In the Solomon Islands stocking rates of approximately 2 head ha^{-1} can be maintained on naturalized pastures of *Axonopus*, *Paspalum conjugatum*, *Mimosa pudica* with some *Centrosema pubescens*, with adequate rates of live-weight gain.

With sown grass–legume pastures stocking rates will depend on soil fertility, degree of shading, fertilizer inputs, and the species sown. The stocking rate should be adjusted to meet the year-round carrying capacity of the pastures, while maintaining adequate ground cover and a balance between grass and legume. Overstocking in plantations usually leads to increased woody weed invasion, and on heavier soils to surface compaction and pugging during wet conditions. On these soils cattle may have to be moved during wet periods to higher ground or lighter soils so that reserve grazing areas may be required.

Grazing management may be more important in undersown pastures than in the full sunlight conditions outside the plantation. Under shading, because of the relatively greater reduction in photosynthesis, the grasses lose their competitive growth advantage over the legumes. In a recent grazing trial under coconuts in the Solomon Islands, where a sown mixture of *Brachiaria mutica*, *B. decumbens*, and *B. humidicola* with the legumes *C. pubescens*, *P. phaseoloides*, and *S. guianensis*, was compared with the naturalized pasture which contains *A. compressus*, *P. conjugatum*, *M. putica*, and *C. pubescens*, the sown grass component was eliminated after one year's grazing at all stocking rates (1.5, 2.5, and 3.5 head ha^{-1}) (Watson and Whiteman, unpublished data). In this trial all treatments were set-stocked and continuously grazed. There were no significant differences in liveweight gain per head or per hectare between the sown and naturalized pasture at any stocking rate. In contrast to this there was one paddock which was rotationally grazed (4 weeks on/4 weeks off) at the highest stocking rate. In this paddock there was a good balance between sown grass and legume. These results seem to demonstrate that under the shaded conditions with continuous grazing, and where the animals were selectively grazing the grasses, the grasses could not compete and the swards became legume-dominant. However, where a rest period was allowed after each grazing the grasses had sufficient time to build up their leaf area and better compete with the legumes. Thus we would recommend that a rotational grazing system be employed in undersown tropical pastures. Other authors, Plucknett (1972), McEvoy (1974), Javier (1974) also recommend rotational grazing systems in plantations, but these recommendations were not based on comparative data. However, it must be recognized that the sown grasses and legumes will decline with overgrazing under any grazing system.

REFERENCES

Albrecht, S., Okon, Y., Lonnquist, J., and Burrie, R. (1976). Corn roots fix nitrogen. *Crops Soils Mag.* (Aug.), p. 16.
Andrew, C. S. (1966). A kinetic study of phosphate absorption by excised roots of *Stylosanthes humilis*, *Phaseolus lathyroides*, *Desmodium uncinatum*, *Medicago sativa* and *Hordeum vulgare*. *Aust. J. agric. Res.* **17**, 611–24.

Andrew, C. S. and Norris, D. O. (1961). Comparative responses to calcium of five tropical and four temperate pasture legume species. *Aust. J. agric. Res.* **12**, 40.

— and Robins, M. F. (1969*a*). The effect of phosphorus on the growth and chemical composition of some tropical pasture legumes. I. Growth and critical percentages of phosphorus. *Aust. J. agric. Res.* **20**, 665.

— — (1969*b*). The effect of potassium on the growth and chemical composition of tropical and temperate pasture legumes. *Aust. J. agric. Res.* **20**, 1009.

— — (1971). The effect of phosphorus on the growth and chemical composition and critical phosphorus percentages of some tropical pasture grasses. *Aust. J. agric. Res.* **22**, 693.

Andrew, W. D. (1965). The effects of some pasture species on soil structure. *Aust. J. exp. Agric. Anim. Husb.* **5**, 133.

Anon. (1966). *International Rules for Seed Testing*. Proc. Int. Seed Test. Assoc. Vol. 31, pp. 1–152.

Anon. (1968). Proc. Natural Rubber Conf., Kuala Lumpur, August, 1968.

Anon. (1970). *Crop and pasture planting guide*. Advisory leaflet No. 1058. Div. Plant Ind., Qld Dept. Primary Industries.

Austin, J. D. A. (1970). Townsville stylo — looking for companion grasses. *Turnoff* (NTA Animal Industry and Agriculture Branch) **2**, 28.

Bailey, D. R. (1969). Weedkillers for tropical pastures. *Qd agric. J.* **95**, 625.

Balachandran, N. and Whiteman, P. C. (1975). Nitrogen yield and recovery by setaria and desmodium swards during the pasture phase and a subsequent crop phase. *Proc. Aust. Conf. Tropical Pastures*, Townsville, Vol. 2, Section 5, p. 5.

Bauer, P. and Okech, A. (1975). Annual Report, Min. of Ag. Pasture Research Project, Kenya, p. 28.

Beckett, P. H. T. and Webster, R. (1971). Soil variability: a review. *Soils Fertil.* **34**, 1.

Bennett, H. H. (1955). *Elements of soil conservation*. McGraw Hill, New York.

Bennison, R. H. and Evans, D. D. (1968). Some effects of crop rotations on the productivity of crops on a red earth in a semi-arid tropical climate. *J. agric. Sci., Camb.* **71**, 365.

Bergersen, F. J. (1961). Haemoglobin content of legume root nodules. *Biochim. biophys. Acta* **50**, 576.

Bisset, W. J. and Marlowe, G. W. C. (1974). Productivity and dynamics of two siratro based pastures in the Burnett coastal foothills of south-east Queensland. *Trop. Grassl.* **8**, 17.

Bland, B. F. (1968). White clover. *Wld Crops* **20**, 3.

Blaxter, K. L. and Wilson, R. S. (1963). Assessment of crop husbandry technique in terms of animal production. *Anim. Prod.* **5**, 27.

Blunt, C. G. and Humphreys, L. R. (1970). Phosphate response of mixed swards at Mt Cotton, South Eastern Queensland. *Aust. J. exp. Agric. Anim. Husb.* **10**, 431.

Bowen, G. D. (1959). Field studies on the nodulation and growth of *Centrosema pubescens* Benth. *Qd J. agric. Sci.* **16**, 253.

Bruce, R. C. (1965). The effects of *Centrosema pubescens* Benth. on soil fertility in the humid tropics. *Qd J. agric. Sci.* **16**, 221.

Butler, G. W. and Bathurst, N. O. (1956). The underground transference of nitrogen from clover to associated grass. *Proc. 7th Int. Grassld. Cong. N.Z.*, pp. 168–78.

—— Greenwood, R. M., and Soper, K. (1959). Effects of shading and defoliation on the turnover of root and nodule tissue of plants. *N.Z. Jl agric. Res.* **2**, 415.

Campbell, M. H. and Swain, F. G. (1973). Factors causing losses during the establishment of surface sown seed. *J. Range Mgmt* **26**, 355.

Charles, A. H. (1958). The effect of undersowing on the cereal companion crop. *Fld Crops Abstr.* **11**, 233.

Childs, A. H. B. and Groom, G. G. (1964). Balanced farming with coconuts and cattle. *E. Afr. agric. For. J.* **29**, 206.

Clarke, R. T. (1962). The effect of some resting treatments on a tropical soil. *Emp. J. exp. Agric.* **30**, 57.

Coaldrake, J. E. (1970). The Brigalow. In *Australian grasslands* (ed. R. M. Moore) pp. 123–40. ANU Press, Canberra.

Collis-George, N. and Hector, J. B. (1966). Germination of seeds as influenced by matric potential and by area of contact between seed and soil water. *Aust. J. Soil Res.* **4**, 145.

Crofts, F. C. (1969). Nitrogen fertilizers for balancing pasture production with animal needs. In *Intensive utilization of pastures* (ed. B. J. F. James) pp. 76–89. Sydney University Press.

Crozier, J. F. (1976). Use of fertilizers, considering especially nitrogen. In *Energy use and British agriculture*. Proc. 10th Ann. Conf. Reading Agric. Club. (eds. D. M. Butler and H. Day) pp. 11–14.

Davies, W. (1960). *The grass crop, its development, use and maintenance*. Ind. Rev. Ed. London, Spon.

Day, J. M. and Dobereiner, J. (1976). Physiological aspects of N_2-fixation by a *Spirillum* from Digitaria roots. *Soil Biol. Biochem.* **8**, 45.

—— Neves, M. C. P., and Dobereiner, J. (1975). Nitrogenase activity on the roots of tropical forage grasses. *Soil Biol. Biochem.* **7**, 107.

Diatloff, A. (1971). Pelleting of tropical legume seeds. *Qd agric. J.* **97**, 363.

—— (1974). Factors involved in the amelioration of retarded symbiosis in Tinaroo glucine. *Aust. J. agric. Res.* **25**, 577.

Dilz, K. and Mulder, E. G. (1962). Effect of associated growth on yield and nitrogen content of legume and grass plants. *Pl. Soil* **16**, 229.

Dixon, R. O. D. (1969). Rhizobia (with particular reference to relationships with host plants. *A. Rev. Microbiol.* **23**, 137.

Dobereiner, J. (1966). Evaluation of nitrogen fixation in legumes by the regression of total plant nitrogen with nodule weight. *Nature, Lond.* **210**, 850.

— and Day, J. M. (1974). In *Soil management in tropical America.* Proc. Seminar, CIAT, Colombia, pp. 197–210.

Donald, C. M. (1960). The impact of cheap nitrogen. *J. Aust. Inst. agric. Sci.* **26**, 319.

— (1963). Competition among crop and pasture plants. *Adv. Agron.* **15**, 1.

Epstein, E. (1972). *Mineral nutrition of plants: principles and perspectives,* pp. 257–83. J. Wiley, New York.

Evans, T. R. (1970). Some factors affecting beef production from subtropical pastures in the coastal lowlands of south-east Queensland. *Proc. 11th Int. Grassl. Cong.,* Australia, p. 803.

Ferdinandez, D. E. F. (1970). Report of the agrostologist – 1969. *Ceylon Cocon. Q.* **21**, 49.

Firth, J. A., Evans, T. R., and Bryan, W. W. (1975). Effects of soils, fertilizer and stocking rates on pastures and beef production on the Wallum in south-eastern Queensland. 4. Budgetary appraisals of fertilizer and stocking rates. *Aust. J. exp. Agric. Anim. Husb.* **15**, 531.

Fremond, Y. and Brunin, C. (1966). Cocotier et couverture der sol. *Oléagineaux* **21**, 361.

Friend, D. (1975). The experimental programme, 1971–1974, of the Joint Coconut Research Scheme. Ministry Agriculture & Lands, Solomon Is. Mimeo Rep. 1975.

Gates, C. T. (1970). Physiological aspects of the rhizobial symbiosis in *Stylosanthes humilis, Leucaena leucocephala,* and *Phaseolus atropurpureus. Proc. 11th Int. Grassld. Cong.,* p. 442.

— (1974). Nodule and plant development in *Stylosanthes humilis* H.B.K.: symbiotic responses to phosphorus and sulphur. *Aust. J. Bot.* **22**, 45.

Gibson, A. H. (1971). Factors in the physical and biological environment affecting nodulation and nitrogen fixation by legumes. *Pl. Soil* (special) 139–52.

Gillard, P. (1970). Pasture development in the dry tropics of north Queensland. *Proc. 11th Int. Grasslds Cong.,* p. 807.

Gray, S. G. (1968). A review of research on *Leucaena leucocephala. Trop. Grassl.* **2**, 19.

Grof, B. (1968). Viability of seed of *Brachiaria decumbens. Qd J. Agric. Anim. Sci.* **25**, 149.

Gutteridge, R. C. and Whiteman, P. C. (1977). Pasture research on the grassland foothills of the Kongga land system. In *Proc. Regional Seminar on Pasture Research and Development in the Solomon Islands and Pacific Region.* University of Queensland, Department of Agriculture.

— — and Watson, S. E. (1976). *Final report on the regional programme of pasture species evaluation and soil fertility assessment, 1973–1976.* University of Queensland, Mimeo Report.

Haggar, R. J., Leeuw, P. N. de, and Agishi, E. (1971). The production and management of *Stylosanthes gracilis* at Shika, Nigeria. II. In savanna grassland. *J. agric. Sci., Camb.* 77, 437.

Hagstrom, G. R. and Berger, K. C. (1963). Molybdenum status of three Wisconsin soils and its effect on four legume crops. *Agron. J.* 55, 399–401.

Hall, R. L. (1971). Some implications of the use of potassium chloride in nutrient studies. *J. Aust. Inst. agric. Sci.* 37, 249.

— (1974). Analysis of the nature of interference between plants of different species. II. Nutrient relations of Nadi *Setaria* and Greenleaf *Desmodium* association with particular reference to potassium. *Aust. J. agric. Res.* 25, 749.

Harty, R. L. (1972). Germination requirements and dormancy effects in seed of *Urochloa mosambicensis. Trop. Grassl.* 6, 17.

Henzell, E. F. (1963). Nitrogen fertilizer responses of pasture grasses in south-eastern Queensland. *Aust. J. exp. Agric. Anim. Husb.* 3, 290.

— (1968). Sources of nitrogen for Queensland pastures. *Trop. Grassl.* 2, 1.

— Fergus, I. F., and Martin, A. E. (1966). Accumulation of soil nitrogen and carbon under a *Desmodium uncinatum* pasture. *Aust. J. exp. Agric. Anim. Husb.* 6, 157.

— Martin, A. E., Ross, P. J., and Haydock, K. P. (1968). Isotopic studies on the uptake of nitrogen by pasture plants. *Aust. J. agric. Res.* 19, 65.

Holmes, J. H. G., Franklin, H. C., and Lambourne, L. J. (1966). The effect of season supplementation and pelleting on intake and utilisation of some sub-tropical pastures. *Proc. Aust. Soc. Anim. Prod.* 6, 354.

Hubble, G. E. and Martin, A. E. (1962). The nitrogen status of Queensland soils. In *A review of nitrogen in the tropics with particular reference to pastures.* Comm. Agric. Bur. Bull. No. 46.

Humphreys, L. R. (1974). A guide to better pastures for the tropics and sub-tropics, 3rd edn. Wright Stephenson, Australia.

Hutchinson, K. J. and King, K. L. (1970). Sheep numbers and soil arthropods. *Search* 1, 41.

Hutton, E. M. (1968). Australia's pasture legumes. *J. Aust. Inst. Agric. Sci.* 34, 203.

— and Bonner, I. A. (1960). Dry matter and protein yields in four strains of *Leucaena glauca* Benth. *J. Aust. Inst. Agric. Sci.* 26, 276.

Imrie, B. C. (1972). Effect of seed size on germination and seedling yield of *Desmodium. SABRAO Newsl.* 4, 85.

Ive, J. R. (1970). Undersowing Townsville stylo. *Turnoff* 2, 13.

Janssen, K. A. and Vitosh, M. L. (1974). Effect of lime, sulphur, and molybdenum on N_2 fixation and yield of dark red kidney beans. *Agron. J.* 66, 736.

Javier, E. Q. (1974). Improved varieties for pastures under coconuts. ASPAC, Food & Fert. Tech. Centre. Extension Bull. No. 37.

Jensen, H. L. and Betty, R. C. (1943). N fixation in leguminous plants.

III. Importance of Mo in symbiotic N fixation. *Proc. Linn. Soc. N.S.W.* **68**, Pts 1–2, 1–8.

Jones, M. B. and Quagliato, J. L. (1973). Response of four tropical legumes and alfalfa to varying levels of sulphur. *Sulphur Inst.* **9**, 6.

— Ruckman, J. E., and Lauder, P. W. (1972). Critical levels of sulphur in Bur clover. *Agron. J.* **64**, 55.

Jones, R. J. (1965). The use of cyclodiene insecticides as liquid seed dressings to control bean fly (*Melanagromyza phaseoli*) in species of *Phaseolus* and *Vigna marina* in south-east Queensland. *Aust. J. exp. Agric. Anim. Husb.* **5**, 458.

— (1967*a*). The effects of some grazed tropical grass–legume mixtures and nitrogen fertilised grass on total soil nitrogen, organic carbon and subsequent yields of *Sorghum vulgare*. *Aust. J. exp. Agric. Anim. Husb.* **7**, 66.

— (1967*b*). Effects of close cutting and nitrogen fertilizer on growth of siratro (*Phaseolus atropurpureus*) pasture at Samford, south-eastern Queensland. *Aust. J. exp. Agric. Anim. Husb.* **7**, 157.

— (1969). Germination studies. Ann. Rep. CSIRO. Div. Trop. Past. (1968–69), p. 53.

— (1970). The effect of nitrogen fertilizer applied in spring and autumn on the production and botanical composition of two sub-tropical grass–legume mixtures. *Trop. Grassl.* **4**, 97.

— (1972). The place of legumes in tropical pastures. ASPAC Food & Fert. Tech. Centre Tech. Bull. No. 9.

— (1973). The effect of cutting management on the yield, chemical composition and *in vitro* digestibility of *Trifolium semipilosum* growth with *Paspalum dilatatum* in a sub-tropical environment. *Trop. Grassl.* **7**, 277.

Jones, R. K. (1968). Initial and residual effects of superphosphate on Townsville lucerne pastures in north-eastern Queensland. *Aust. J. exp. Agric. Anim. Husb.* **8**, 521.

Jones, R. M. (1973). Seedling death of *Desmodium intortum* in the Queensland wallum with special reference to potassium chloride fertilizer. *Trop. Grassl.* **7**, 269.

— (1975*a*). Effect of soil fertility, weed composition, defoliation and legume seeding rate on establishment of tropical pasture species in south-east Queensland. *Aust. J. exp. Agric. Anim. Husb.* **15**, 54.

— (1975*b*). Pasture establishment. In Aust. Inst. Agric. Sci., Qld Branch *Refresher course in management of improved tropical pastures*, pp. 68–80.

— and Date, R. A. (1975). Studies in the nodulation of Kenya white clover (*Trifolium semipilosum*) under field conditions in south-east Queensland. *Aust. J. exp. Agric. Anim. Husb.* **15**, 519.

Kannegieter, A. (1969). The combination of a short term *Pueraria* fallow, zero cultivation and fertilizer application. Its effect on a following maize crop. *Trop. Agric., Colombo* **125**, 77.

Keya, N. C. O., Olsen, F. J., and Holliday, R. (1971). The role of superphosphate in the establishment of oversown tropical legumes in natural grasslands of western Kenya. *Trop. Grassl.* **5**, 109.

Keya, N. C. O., Olsen, F. H., and Holliday, R. (1972). Comparison of seed beds for oversowing of *Chloris gayana* (Kunth) *Desmodium uncinatum* (Jacq) mixture in *Hyparrhenia* grassland. *E. Afr. agric. For. J.* **37**, 286.

Kirby, J. M. (1976). Forest grazing, a technique for the tropics. *Wld Crops* **28**, 248.

Koller, D. (1972). Environmental control of seed germination. In *Seed biology* (ed. T. T. Kogloniski) Vol. II, pp. 1–101. Academic Press, London.

Krigel, S. (1967). The early requirements for plant nutrients by subterranean clover seedlings. *Aust. J. agric. Res.* **18**, 879.

Krishnanarar, N. M. (1961). Trial of intercultivation practices in coconut gardens. *Indian Cocon. J.* **14**, 87.

Levin, A. P., Funk, H. B., and Tendler, M. D. (1954). Vitamin B$_{12}$, rhizobia and leguminous plants. *Science, N.Y.* **120**, 784.

Linehan, P. A. and Lowe, J. (1960). Yielding capacity and grass/clover ratio of herbage swards as influenced by fertilizer treatments. *Proc. 8th Int. Grassl. Cong., Reading, UK*, p. 133.

Littler, J. W. (1967). Pasture–wheat rotation trial, 1958–1967. Mimeo Publ. Qld D.P.I. Oct. 1967.

Lloyd, D. L. (1975). Pasture rotation research. Ann. Rep. Qld Wheat Research Inst. 1974–75. p. 33.

Low, A. J., Piper. F. J., and Roberts, P. (1963). Soil changes in ley-arable experiments. *J. Agric. Sci., Camb.* **60**, 229.

McDowell, R. E. (1972). *Improvements of livestock production in warm climates.* Freeman, San Francisco.

McEvoy, M. G. (1974). Establishment and management of pastures under coconuts. ASPAC Food & Fert. Tech. Centre, Extension Bull. No. 38.

McLean, D. and Grof, B. (1968). Effect of seed treatments on *Brachiaria mutica* and *B. ruziziensis. Qd J. agric. Anim. Sci.* **25**, 81.

McWilliam, J. R., Clements, R. J., and Dowling, P. M. (1970). Some factors influencing the germination and early seedling development of pasture plants. *Aust. J. agric. Res.* **21**, 19.

Magadan, P. B., Janier, E. Q., and Madamba, J. C. (1974). Beef production on native (*Imperata cylindrica* (L.) Beauv.) and para grass (*Brachiaria mutica* (Forsk.) Stapf.) pastures in the Philippines. Proc. 12th Intl. Grassld. Congr. Moscow, pp. 370–8.

Maher, C. (1951). Soil conservation in Kenya colony. Part I. Factors affecting erosion, soil characteristics and methods of conservation. *Emp. J. exp. Agric.* **18**, 137.

't Mannetje, L., Shaw, N. H., and Elich, T. W. (1963). The residual effect of molybdenum fertilizer on improved pastures on a prairie-like soil in sub-tropical Queensland. *Aust. J. exp. Agric. Anim. Husb.* **3**, 20.

Mar, R. K. and Nielson, R. G. H. (1970). Urea reduces emergence of lucerne. *Qd agric. J.* **96**, 454.

Marshall, B. and Long, M. I. E. (1971). Calcium intake and excretion of
    Zebu cattle used for digestibility trials. *E. Afr. agric. For. J.* **37**, 46.
Martin, T. W. (1960). The role of white clover in grasslands. *Herb.
    Abstr.* **30**, 159.
Michael, P. W. (1968*a*). Control of the biennial thistle, *Onopordum*, by
    amitrole and five perennial grasses. *Aust. J. exp. Agric. Anim. Husb.*
    8, 332.
— (1968*b*). Perennial and annual pasture species in the control of
    Silybum marianum. Aust. J. exp. Agric. Anim. Husb. **8**, 101.
Middleton, C. H. (1970). Some effects of grass–legume sowing rates
    on tropical species establishment and production. *Proc. 11th Int.
    Grassl. Cong.*, p. 119.
— (1973). Oversowing legumes into grass swards. *Qd agric. J.* **99**, 217.
Milford, R. (1967). Nutritive values and chemical composition of seven
    tropical legumes and lucerne grown in sub-tropical south-eastern
    Queensland. *Aust. J. exp. Agric. Anim. Husb.* **7**, 540.
— and Haydock, K. P. (1965). The nutritive value of protein in sub-
    tropical pasture species in south-east Queensland. *Aust. J. exp.
    Agric. Anim. Husb.* **5**, 13.
— and Minson, D. J. (1965). The relation between the crude protein
    content and the digestible crude protein content of tropical pasture
    plants. *J. Br. Grassl. Soc.* **20**, 177.
— — (1968*a*). The digestibility and intake of six varieties of Rhodes
    grass (*Chloris gayana*). *Aust. J. exp. Agric. Anim. Husb.* **8**, 413.
— — (1968*b*). The effect of age and method of haymaking on the
    digestibility and voluntary intake of the forage legumes *Dolichos
    lablab* and *Vigna sinensis. Aust. J. exp. Agric. Anim. Husb.* **8**, 409.
Miller, H. P. and Perry, R. A. (1968). Preliminary studies on the estab-
    lishment of Townsville lucerne (*Stylosanthes humilis*) on uncleared
    native pasture at Katherine, N.T. *Aust. J. exp. Agric. Anim. Husb.*
    8, 26.
Minson, D. J. and Milford, R. (1967). The voluntary intake and digesti-
    bility of diets containing different proportions of legume and mature
    Pangola grass (*Digitaria decumbens*). *Aust. J. exp. Agric. Anim. Husb.*
    7, 546.
— — (1968). The nutritional value of four tropical grasses when fed
    as chaff and pellets to sheep. *Aust. J. exp. Agric. Anim. Husb.* **8**, 270.
Monier, G. (1965). Effect of organic matter on soil structural stability.
    Sols Afr. **10**, 29.
Moore, R. M. and Robertson, J. A. (1964). Studies on skeleton weed –
    competition from pasture plants. *CSIRO Field Station Rec.* **3**, 69.
Motooka, P. S., Plucknett, D. L., Saiki, D. F., and Younge, O. R.
    (1967). Pasture establishment on tropical bushlands by aerial herbi-
    cide and seeding treatments in Kuai. Hawaii. Agric. Expl. Sta. Tech.
    Prog. Rep. No. 165.
Mukhopadhyaya, M. C. and Prasad, S. K. (1969). Nematodes as affected
    by rotations and their relation with yields of crops. *Indian J. agric.
    Sci.* **39**, 366.

Munns, D. N. (1968). Nodulation of *Medicago sativa* in solution culture. IV. Effects of indole-3-acetate in relation to acidity and nitrate. *Pl. Soil* **29**, 257.

Nathaneal, W. R. N. (1967). The application of fertilizers to adult coconut palms in relation to theoretical concepts. *Ceylon Cocon. Q.* **18**, 5.

Nelson, M. (1973). *The development of tropical lands. Policy issues in Latin America*. John Hopkins University Press, Baltimore.

Newton, K. and Jamieson, G. I. (1968). Cropping and soil fertility studies at Keravat, New Britain, 1954–1962. *Papua New Guin. agric. J.* **20**, 25.

Nitis, I. M., Rika, K., Supardjata, M., Nurbudhi, K. D., and Humphreys, L. R. (1976). Productivity of improved pastures grazed by Bali cattle under coconuts. Publication Dept. Anim. Husb., Province of Bali, Indonesia. May 1976.

Norman, M. J. T. (1961). The establishment of pasture species with minimum cultivation at Katherine, N.T. CSIRO Div. Land Res. Reg. Surv. Tech. Paper No. 14.

— (1965). The response of birdwood grass–Townsville lucerne pasture to phosphate fertilizers at Katherine, N.T. *Aust. J. exp. Agric. Anim. Husb.* **5**, 120.

— (1966). Katherine Research Station 1956–64: a review of published work. CSIRO Div. Land Res. Tech. Paper No. 28.

— (1970). Relationship between liveweight gain of grazing beef steers and availability of Townsville lucerne. *Proc. 11th Int. Grassl. Cong.*, p. 829.

— and Wetselaar, R. (1960). Performance of annual fodder crops at Katherine, N.T. CSIRO Div. Land Res. Reg. Surv. Tech. Paper No. 9.

Norris, D. O. (1956). Legumes and the *Rhizobium* symbiosis. *Emp. J. exp. Agric.* **24**, 247.

— (1964). Techniques used in work with *Rhizobium*. In *Some concepts and methods in sub-tropical pasture research*. Common. Agric. Bur. Bull. No. 47.

— (1967*a*). The intelligent use of inoculants and lime pelleting for tropical legumes. *Trop. Grassl.* **1**, 107.

— (1967*b*). In *Glycine javanica* – its strength and weaknesses. CSIRO Rural Research No. 59, p. 11.

— (1970). Nodulation of pasture legumes. In *Australian grasslands* (ed. R. M. Moore) pp. 339–48. ANU Press, Canberra.

— (1973). Seed pelleting to improve nodulation of tropical and subtropical legumes. 5. The contrasting response to lime pelleting of two *Rhizobium* strains on *Leucaena leucocephala*. *Aust. J. exp. Agric. Anim. Husb.* **13**, 98.

Nutman, P. S. (1956). The influence of the legume in root–nodule symbiosis. A comparative study of host determinants and functions. *Biol. Rev.* **31**, 109.

Ohler, J. G. (1969). Review article – cattle under coconuts. *Trop. Abstr.* **24**, 639.

Oke, O. L. (1969). Sulphur nutrition of legumes. *Expl. Agric.* **5**, 111.

Olsen, F. J. and Moe, P. G. (1971). The effect of phosphate and lime on the establishment productivity, nodulation and persistence of *Desmodium intortum*, *Medicago sativa* and *Stylosanthes gracilis*. *E. Afr. agric. For. J.* **37**, 29.

Ozanne, P. G., Greenwood, E. A. M., and Shay, T. C. (1963). The cobalt requirement of subterranean clover in the field. *Aust. J. agric. Res.* **14**, 39.

Partridge, I. J. (1975). The improvement of mission grass (*Pennisetum polystachyon*) by top dressing superphosphate and oversowing a legume (*Macroptilium atropurpureum*). *Trop. Grassl.* **9**, 45.

Pate, J. S. (1958). Nodulation studies in legumes. II. The influence of various environmental factors on symbiotic expression in the vetch (*Vicia sativa* L.) and other legumes. *Aust. J. biol. Sci.* **11**, 496.

Pillar, N. G. and Davis, T. A. (1963). Exhaust of micronutrients by the coconut palm – a preliminary study. *Indian Cocon. J.* **16**, 81.

Playne, M. J. (1969a). The nutritional value of intact seed pods of Townsville lucerne (*Stylosanthes humilis*). *Aust. J. exp. Agric. Anim. Husb.* **9**, 502.

— (1969b). The effect of dicalcium phosphate supplements on the intake and digestibility of Townsville lucerne and spear grass by sheep. *Aust. J. exp. Agric. Anim. Husb.* **9**, 192.

— (1972). Nutritional value of Townsville stylo (*Stylosanthes humilis*) and of spear grass (*Heteropogon contortus*)-dominant pastures fed to sheep. 2. The effect of superphosphate fertilizers. *Aust. J. exp. Agric. Anim. Husb.* **12**, 373.

Plucknett, D. L. (1972). Management of natural and established pastures for cattle production under coconuts. *Proc. Regional Seminar on Pastures and Cattle under Coconuts*. South Pacific Commis., Noumea, New Caledonia, p. 109.

Powrie, J. K. (1960). A field response by subterranean clover to cobalt fertilization. *Aust. J. Sci.* **23**, 148.

Prodonoff, E. T. (1965). *Seed testing in Queensland*. Standards Branch, Qld Dept. Primary Ind., p. 58.

Purcell, D. L. (1964). Gidyea to grass in the Central West. *Qd agric. J.* **90**, 304.

Rajaratnam, D. T. and Santhirasegaram, K. (1963). Comparison of grasses. *Ceylon Cocon. Q.* **14**, 35.

Ranacou, E. (1972). Pasture species under coconuts. *Proc. Regional Seminar on Pastures and Cattle under Coconuts*. South Pacific Commis., Noumea, New Caledonia, p. 95.

Rees, M. C., Minson, D. J., and Smith, F. W. (1974). The effect of supplementary and fertilizer sulphur on voluntary intake, digestibility, retention time in the rumen and site of digestion of Pangola grass in sheep. *J. Agric. Sci., Camb.* **82**, 419.

Reisenauer, H. M. (1960). Cobalt in nitrogen fixation by a legume. *Nature, Lond.* **186**, 375.

Robertson, J. A. and Young, N. D. (1976). Thinning *Eucalyptus micro-carpa* woodlands. *Trop. Grassl.* **10**, 129.

Rodrigo, E. (1945). Fodder grass experiment (Lunwila). *Ann. Rep. Coconut Res. Scheme*, 1945. p. 11.

Roe, R. (1970). Ann. Rep. CSIRO Div. Trop. Past. (1969–70), p. 60.

Rovira, A. D. (1961). *Rhizobium* numbers in the rhizospheres of red clover and paspalum in relation to soil treatment and the numbers of bacteria and fungi. *Aust. J. agric. Res.* **12**, 77.

Sahise, P. E. (1973). Competition within the pasture coconut plant community. ASPAC Training Course on Pasture Production under Coconut Palms. Davao City, Philippines. Oct. 1973.

Santhirasegaram, K. (1959). Report of the agrostologist – 1958. Ann. Rep. Ceylon Coconut Rest. Inst., p. 56.

— (1966*a*). Utilisation of spacing among coconuts for intercropping. *Ceylon Cocon. Planters' Rev.* **4**, 43.

— (1966*b*). The effect of monospecific grass swards on the yield of coconuts in the north-west province of Ceylon. *Ceylon Cocon. Q.* **17**, 73.

Schaller, F. W. and Larson, W. E. (1955). Effect of wide spaced corn rows on corn yield and forage establishment. *Agron. J.* **47**, 271.

Schank, S. C., Day, J. M., and de Lucas, E. D. (1977). Nitrogenase activity, nitrogen content, *in vitro* digestibility and yield of 30 tropical forage grasses in Brazil. *Trop. Agric., Trin.* **54**, 119.

Scott, J. D. (1967). Advances in pasture work in South Africa. 2. Cultivated Pastures. *Herb. Abstr.* **37**, 159.

Shaw, N. H. (1961). Increased beef production from Townsville lucerne (*Stylosanthes sundiaca* Taub.) in the spear grass pastures of central coastal Queensland. *Aust. J. exp. Agric. Anim. Husb.* **1**, 73.

— Gates, C. T., and Wilson, J. R. (1966). Growth and chemical composition of Townsville lucerne (*Stylosanthes humilis*). I. Dry matter yield and nitrogen content in response to superphosphate. *Aust. J. exp. Agric. Anim. Husb.* **6**, 150.

Shelton, H. M. (1972). Factors influencing competition between upland rice (*Oryza sativa* Linn.) and undersown stylo (*Stylosanthes guyanensis* Auld.). Ph.D Thesis, University of Queensland, Australia.

— and Humphreys, L. R. (1972). Pasture establishment in upland rice crops at Na Pheng, C. Laos. *Trop. Grassl.* **6**, 223.

— — (1975). Undersowing rice (*Oryza sativa*) with *Stylosanthes guyanensis*. *Expl Agric.* **11**, 89–111.

Silva, M. A. T. de (1961). Cover crops under coconuts. *Ceylon Cocon. Planters' Rev.* **11**, 17.

Simpson, J. R. (1965). The transference of nitrogen from pasture legumes to an associated grass under several systems of management in pot culture. *Aust. J. agric. Res.* **16**, 915.

Slatyer, R. O. (1960). Agricultural climatology of the Katherine area, N.T. CSIRO Div. Land Res. Reg. Surv. Tech. Paper No. 13.

Smith, C. J. (1967). Sowing dryland pastures. *Rhodesia agric. J.* **64**, 69.

Smith, R. W. (1972). The optimum spacing for coconuts. *Proc. Conf. Cocoa & Coconuts in Malaysia*. Inc. Soc. Planters, Malaysia, p. 429.

Souto, S. M. and Dobereiner, J. (1968). Effects of phosphorus, soil temperature, and soil humidity on nodulation and development of two varieties of perennial soybean (*Glycine javanica* L.). *Pesquisas* **3**, 215.

— — (1970). Soil temperature effects on nitrogen fixation and growth of *Stylosanthes gracilis* and *Pueraria javanica*. *Pesquisas* **5**, 365.

Steel, R. J. H. and Humphreys, L. R. (1974). Growth and phosphorus response of some pasture legumes sown under coconuts in Bali. *Trop. Grassl.* **8**, 171.

Stephens, C. G. and Donald, C. M. (1958). Australian soils and their response to fertilizers. *Adv. Agron.* **10**, 167.

Stephens, D. (1967). Effects of grass fallow treatments in restoring fertility of Buganda clay loam in south Uganda. *J. agric. Sci., Camb.* **68**, 391.

Stobbs, T. H. (1969). The effect of grazing resting land upon subsequent arable crop yields. *E. Afr. agric. For. J.* **35**, 28.

— (1975). Palatability of siratro. CSIRO Div. Trop. Crops & Past. Ann. Rep. 1974–75, p. 109.

Stocker, G. C. and Sturtz, J. D. (1966). The use of fire to establish Townsville lucerne in the Northern Territory. *Aust. J. exp. Agric. Anim. Husb.* **6**, 277.

Stonard, P. (1969). Effect of sowing depth on seedling emergence of three species of *Stylosanthes*. *Qd J. agric. Anim. Sci.* **26**, 65.

Swain, F. G. (1959). Responses to molybdenum three years after previous application on red basaltic soils on the far north coast of N.S.W. *J. Aust. Inst. agric. Sci.* **25**, 51–4.

Swanson, A. F. and Hunter, R. (1936). Effect of germination and seed size on sorghum stands. *J. Am. Soc. Agron.* **28**, 997.

Thairu, D. M. (1972). The contribution of *Desmodium uncinatum* to the yield of *Setaria sphacelata*. *E. Afr. agric. For. J.* **37**, 215.

Thomas, D. (1975). Grass–legume establishment under maize. *Span* **18**, 63.

— (1976). Effects of close grazing or cutting on the productivity of tropical legumes in pure stands in Malawi. *Trop. Agric., Trin.* **53**, 329.

— and Bennet, A. J. (1975). Establishing a mixed pasture under maize in Malawi. *Expl. Agric.* **11**, 257–76.

— and Whiteman, P. C. (1971). The effect of soil type on the establishment, early growth and nodulation of *Glycine wightii*. *Aust. J. exp. Agric. Anim. Husb.* **11**, 513.

Thomas, R. and Humphreys, L. R. (1970). Pasture improvement at Na Pheng, Central Laos. *Trop. Grassl.* **4**, 229.

Tietzel, J. K. (1969). Responses to phosphorus, copper and potassium on granite loam of the wet tropical coast of Queensland. *Trop. Grassl.* **3**, 43.

Tudsri, S. and Whiteman, P. C. (1977*a*). Effects of initial and maintenance of phosphorus levels on the establishment of four legumes oversown into *Setaria anceps* swards. *Aust. J. exp. Agric. Anim. Husb.* **17**, 629.

— — (1977*b*). The effect of cultural treatments on establishment of *Macroptilium atropurpureum* cv. Siratro oversown in a *Setaria anceps* sward. *Trop. Grassl.* **11**, 49.

Vallis, I. (1972). Soil nitrogen changes under continuously grazed legume–grass pastures in sub-tropical coastal Queensland. *Aust. J. exp. Agric. Anim. Husb.* **12**, 495.

— (1973). Sampling for soil nitrogen changes in large areas of grazed pastures. *Commun. Soil. Sci. Pl. Anal.* **4**, 163.

— (1978). Nitrogen relationships in grass/legume mixtures. In *Plant relations in pastures* (ed. J. R. Wilson) p. 190. CSIRO, Melbourne.

— and Jones, R. J. (1973). Net mineralisation of nitrogen in leaves and leaf litter of *Desmodium intortum* and *Phaseolus atropurpureus* mixed with soil. *Soil Biol. Biochem.* **5**, 391.

— Haydock, K. P., Ross, P. J., and Henzell, E. F. (1967). Isotopic studies on the uptake of nitrogen by pasture plants. III. The uptake of small additions of ^{15}N-labelled fertilizer by Rhodes grass and Townsville lucerne. *Aust. J. agric. Res.* **18**, 865.

Vicente-Chandler, J., Silva, C., and Figarella, J. (1961). Effects of nitrogen fertilization and frequency of cutting on the yield and composition of a range of tropical grasses. *J. Agric. Univ. Puerto Rico* **45**, 1.

Vincent, J. M. (1970). *A manual for the practical study of the root-nodule bacteria.* I.B.P. Handbook No. 15. Blackwell, Oxford.

Walker, B. (1975). Stocking rates – effects on pasture quantity and quality. Aust. Inst. Agric. Sci. Refresher Course, Management of Improved Tropical Pastures, p. 104.

— and Potere, J. K. (1974). Effects of stocking rate on plant populations in tropical legume grass pasture. *Proc. 12th Int. Grassl. Cong.*, Moscow, p. 388.

Walker, D. (1966). Vegetation of the Lake Ipea region, New Guinea highlands. *J. Ecol.* **54**, 503.

Walker, J., Moore, R. M., and Robertson, J. A. (1972). Herbage response to tree and shrub thinning in *Eucalyptus populnea* shrub woodlands. *Aust. J. agric. Res.* **23**, 405.

Wallace, M. M. H. (1970). Insects of grasslands. In *Australian grasslands* (ed. R. M. Moore) pp. 361–70. ANU Press, Canberra.

Watson, E. R. and Lapins, P. (1969). The mini-ley concept. In *Rural research*, CSIRO No. 67 (Sept. 1969), p. 24.

Webster, C. C. (1954). The ley and soil fertility in Britain and Kenya. *E. Afr. agric. For. J.* **20**, 71.

Whitehead, D. C. (1970). *The role of nitrogen in grassland productivity.* Comm. Agric. Bur. Bull. 48.

Whitehead, R. A. and Smith, R. W. (1968). Results of a coconut spacing trial in Jamaica. *Trop. Agric. Trin.* **45**, 127.

Whiteman, P. C. (1968). Effects of temperature on the vegetative growth of six tropical legume species. *Aust. J. exp. Agric. Anim. Husb.* **8**, 528.

— — (1969). The effects of close grazing and cutting on the yield, persistence and nitrogen content of four tropical legumes with Rhodes grass at Samford, south-eastern Queensland. *Aust. J. exp. Agric. Anim. Husb.* **9**, 287.

— (1970*a*). Seasonal changes in the growth and nodulation of perennial pasture legumes in the field. III. Effects of flowering on nodulation of three *Desmodium* species. *Aust. J. agric. Res.* **21**, 215.

— (1970*b*). Distribution and weight of nodules in tropical pasture legumes in the field. *Expl Agric.* **7**, 75.

— (1970*c*). Seasonal changes in growth and nodulation of perennial tropical pasture legumes in the field. II. Effects of controlled defoliation levels on nodulation of *Desmodium intortum* and *Phaseolus atropurpureus*. *Aust. J. agric. Res.* **21**, 207.

— and Lulham, A. (1970). Seasonal changes in growth and nodulation of perennial tropical pasture legumes in the field. I. The influence of planting date and grazing and cutting on *Desmodium uncinatum* and *Phaseolus atropurpureus*. *Aust. J. agric. Res.* **21**, 195.

Whitney, A. S. and Kanehiro, Y. (1967). Pathways of nitrogen transfer in some tropical legume–grass associations. *Agron. J.* **59**, 585.

Wildin, J. H. and Hall, R. L. (1975). The influence of phosphorus and potassium on grass–legume relations. Proc. Aust. Conf. on Tropical Pastures, Townsville, Qld, Section 4–20.

Williams, W. T. and Gillard, P. (1971). Pattern analysis of a grazing experiment. *Aust. J. agric. Res.* **22**, 245.

Winkworth, R. E. (1969). Germination of Townsville lucerne (*Stylosanthes humilis* H.B.K.) in relation to weather at Katherine, N.T. *J. Aust. Inst. agric. Sci.* **35**, 201.

Wooding, F. J., Poulsen, G. M., and Murphy, L. S. (1970). Response of nodulated and non-nodulated soybean seedlings to sulphur nutrition. *Agron. J.* **62**, 277.

5

Animal production from tropical pastures

The value of a pasture must be determined by the output of animal products. While total yield is important as a determinant of the number of animals which may be supported per unit area, equally the efficiency with which it is utilized and the amount of animal product produced per unit of pasture consumed determine the yield of animal product per unit area of pasture. Thus animal production is a function of pasture yield, stocking rate, efficiency of pasture utilization, and pasture quality, plus a range of extrinsic factors such as breed and adaptation of the animal, animal health, internal and external parasites, and the level of husbandry.

Where the pasture is the sole source of food, then the pasture must provide all the requirements of energy, protein, minerals, and vitamins for animal production. If protein, mineral, and vitamin levels are adequate, the rate of animal production is very largely a function of quantity of energy digested, which depends on the amount eaten or ingested and the proportion of each unit of feed that is digested. Thus the two most important pasture parameters related to pasture quality are digestibility and voluntary intake.

Digestibility

The digestibility is a measure of the proportion of the feed consumed which is digested and metabolized by the animal. Minson (1976) defines apparent digestibility of a feed (A) as:

$$A = P - (E + M)$$

where P = potential digestibility of the feed, E = fraction of the feed which escaped digestion, and M = metabolic excretions which appear in faeces. Minson (1976) suggests that the potential digestibility of all chemical components in plants, except lignin, cutin, and silica, is 100 per cent. Complete digestion never occurs owing to lignin encrustations protecting cellulose and hemicellulose from degradation by rumen micro-organisms. Thus digestibility declines as lignin content increases, but the relationship between lignin content and digestibility varies between species (McLeod and Minson 1974).

Estimation of digestibility

The standard technique against which the accuracy of other methods is compared, is based on measuring the intake and digestion of cut pasture samples fed to animals in metabolism pens (Minson 1971). This is termed the *in vivo* estimation.

Sheep are most commonly used because of economics of size. Cattle tend to give a higher estimate of digestibility on feeds of lower digestibility than sheep, which tend to give higher estimates on high quality feeds, but both rank a range of feeds in the same order (Playne 1970). Forages to be compared must be grown under the same conditions and cut and fed at the same time. The standard methods described by Minson (1971) are as follows:

(i)   Forages are cut and dried in a forced draught drier at 100 °C. This does not alter digestibility compared with fresh cut feed or stored frozen feed (Minson 1966).

(ii) Forage is coarse chaffed to 2–5 cm lengths.

(iii) Each forage being compared is fed to a set of 8–10 sheep, each in a separate cage. There is a preliminary feeding period of 7–14 days to accustom the animals to the feed. Animals are fed *ad libitum*, usually 10–20 per cent more feed being supplied than eaten in the previous day. The *measurement* period is ten days, during which period feed eaten is measured, and rejected feed removed daily and weighed. Faeces are collected daily, in harnessed bags attached to the animal, and their oven dry weight is determined.

(iv) Dry matter yield digestibility (DMD) in % units can be calculated:

$$\text{DMD (\% units)} = \frac{\text{DM eaten} - \text{DM in faeces}}{\text{DM eaten}} \times \frac{100}{1}$$

Laboratory techniques

Because of the time and expense of *in vivo* feeding trials many chemical laboratory methods have been developed to attempt to predict digestibility of forages. Most of these methods lack precision compared with the *in vivo* method, as shown in Table 5.1 (Minson 1971).

The *in vitro* method has given the lowest errors in prediction, and has become the most widely used method for the rapid screening of

Table 5.1. Errors in predicting digestibility from some laboratory analysis
methods (Minson 1971)

Analysis	Residual standard deviation
Nitrogen	± 6.2
Crude fibre	± 4.3–5.2
Normal acid fibre	± 4.3
Acid detergent fibre	± 3.9–5.6
Lignin (Van Soest method)	± 5.0
In vitro digestibility	± 2.0–2.3

large numbers of samples since its development by Tilley and Terry (1963). The basic steps in the method used by Minson and McLeod (1972) are:

(i) A 0.5 g sample of oven-dry ground forage (milled to pass a 1 mm screen) is fermented anaerobically with a rumen fluid: artificial saliva mixture (1 : 4 v/v) for 48 hours at 39 °C.
(ii) The dry matter residue is then hydrolysed with acid pepsin for a further 48 hours.
(iii) Dry matter digestibility is then calculated after correcting for any indigestible material that may have been added in the rumen fluid by subtracting the weight of dry matter in blank containing only the reagents:

$$\text{DMD (\% units)} = \frac{\text{DM sample} - (\text{DM residue} - \text{DM blank})}{\text{DM sample}} \times \frac{100}{1}$$

The rumen fluid is obtained from rumen fistulated sheep fed a standard diet of lucerne (*M. sativa*) and grass hay. With each batch of samples are included standards of known *in vivo* digestibility. Results can then be corrected by adding or subtracting the average difference between the *in vitro* and *in vivo* results.

It is most important that all successive analyses are made under strictly standardized conditions, and appropriate standards of known *in vivo* digestibility must be included in every batch.

A variant of the *in vitro* test-tube method is the *nylon bag technique*. In this, 6g samples of ground material are placed in small nylon bags (mesh size 18–32 μm) attached to an iron ring, and placed in the rumen through a rumen fistula. After 48 hours the bags are removed, washed for 30 minutes by a standardized method, and then subjected to 48 hours acid pepsin digestion. The weight of dry matter residue is determined. Problems with this method are the accretion of suspended particles from the rumen fluid into the bag, and losses of suspended solids from the bag.

Cellulase digestion technique

A new technique has been developed which promises to be more rapid, convenient, and precise than the *in vitro* method (Jones and Hayward 1975). In this technique 0.2 g of ground herbage sample is first incubated with acid pepsin (0.2 per cent pepsin in 0.1N hydrochloric acid) for 24 hours at 40 °C, The supernatant is then filtered off and the sample then digested with a cellulase enzyme solution prepared from the fungus *Trichoderma viride*, for 48 hours at 40 °C. The residue is then isolated by filtration and the amount digested calculated as a percentage of the original dry matter.

For a range of grass samples varying in *in vivo* DMD from 55–76% units, correlation of ($r = 0.93$) and a residual standard deviation (r.s.d.) of 2.99 was obtained. In a number of rye grass cuts an r.s.d. of 2.15 was obtained, and for a number of cocksfoot cuts r.s.d. was 1.88. In all these comparisons r.s.d. for the cellulase method was lower than for the standard *in vitro* method.

This method does not require collection of rumen fluid which often leads to problems of variability in fluid activity. However further testing of this technique is required, particularly with tropical feeds.

Voluntary intake

The daily intake of dry matter is the most important single factor controlling the nutritional value of tropical pastures (Milford and Minson 1966). Although digestibility and intake are generally correlated (r values 0.66–0.76, Minson, Stobbs, Hegarty, and Playne 1976), the relationship is far from perfect, and of little predictive value. There is no satisfactory laboratory method for predicting voluntary intake.

The standard method for estimating voluntary intake is by feeding cut forages to individual animals in pens indoors as described for estimation of *in vivo* digestibility. Because of variation in intake between individuals of similar body weight (coefficient of variation of 10 per cent), each feed is fed to 8–10 sheep.

Intake is usually expressed on the basis of metabolic body weight (liveweight in $kg^{0.75}$) in the units, g per day per (kg $W^{0.75}$). Even though intakes are expressed in terms of metabolic body weight, large differences have been found in the intake of feed by sheep and cattle (Playne 1970) (Table 5.2). However, the ranking order of feeds is generally the same.

Table 5.2. Comparison of digestibility and intake values of three cuts of *Cenchrus ciliaris* with cattle and sheep (Playne 1970)

	Dry matter intake (g per day (per kg $W^{0.75}$))		Dry matter digestibility (% units) *in vivo*		*in vitro*
Cuts	Cattle	Sheep	Cattle	Sheep	
1	102	43	64.5	62.0	62.5
2	81	34	59.9	52.8	51.6
3	82	32	57.4	52.8	51.3

Intake values measured in pen feeding trials demonstrate the relative differences between cultivars and species, but may bear little relationship to intake in the field where factors such as sward structure, availability of leaf, density of leaf, and botanical composition of a sward affect actual amounts eaten. Various methods have been developed to estimate intake under grazing conditions.

Herbage intake is a function of the time spent grazing, the rate of biting and the intake per bite,

$$I = T \times R \times S$$

where I = intake, T = grazing time, R = rate of biting, S = bite size. For measuring grazing time the vibracorder grazing clock attached to the animal has proved accurate and reliable (Stobbs 1970).

Recently an accurate method for measuring the number of grazing bites has been developed by Stobbs and Cowper (1972). A micro-switch attached to the jaw operates through a two-way switch which separates jaw movements in the head down (grazing) position from the head up position (ruminating and mastication bites). Biting movements are recorded on an electronic counter. Thus if the total number of grazing bites are measured:

$$I = N \times S$$

where N = total number of grazing bites per day.

Bite size (S) can be determined using oesophageal-fistulated animals, fitted with a foam rubber plug in the oesophagus and a collecting bag under the fistula. The amount of material collected divided by the number of bites during the sampling periods allows average bite size to be calculated.

Measurement of the components of eating behaviour are providing valuable comparisons between different pasture conditions. Stobbs (1970) has shown that grazing time on tropical pastures are longer than for temperate pastures. On good quality pastures, grazing times of less than 7 hours per day may be recorded, while on low quality pastures or low yielding pastures, grazing times may extend to 10–12 hours per day. Thus grazing bites can vary between 12 000 and 36 000 per day.

Number of grazing bites per day depends upon the availability of feed and the amount of feed which can be prehended per bite. Stobbs (1973) found bite size to vary from 0.05–0.80 g organic matter (OM) per bite. He concluded that daily intake of adult cattle is below maximum if bite size is less than 0.3 g OM because grazing cattle rarely exceed 36 000 grazing bites per day.

Assessment of nutritive value by grazing evaluation

Estimation of digestibility and voluntary intake of forage samples by the methods discussed above are only an indirect measure of potential of that forage for animal production. Animal production in the field is a function of two components — production per head and stocking rate (or number of animals per unit area per unit time). Production per head is closely related to the daily intake of digestible energy and hence is very much a function of pasture quality, while the stocking rate is more related to the yield of the pasture, but of course both parameters are interrelated.

Animal production levels estimated from pen feeding trials are modified in the field by many factors such as — pasture yield; the sward structure and relative proportions of components such as dead matter, green leaf and stem which affects ease of harvesting, and selection by the animal (Stobbs 1973); effects of the grazing animal itself through defoliation, trampling and fouling of the pasture; energy used in grazing depends on distance walked and time spent grazing; and differences due to the outdoor environment, particularly temperature.

Thus in the final analysis, although estimation of digestibility and voluntary intake are important in screening and selection of species, animal production from chosen pasture species must be measured in the field under grazing. Usually species or mixtures will be evaluated under a range of stocking rates, using the type of animal appropriate to the form of animal production for which the pasture is designed.

The designs and methods used in grazing trials are discussed in Section 5.3.

Pasture factors affecting nutritive value

The nutritive value of a pasture is basically a function of the species in the pasture and the stage of growth, but may be modified by climatic factors during growth, soil factors which affect nitrogen and other mineral status, and management factors which affect pasture regrowth rate, sward structure and botanical composition.

Species

(a) Tropical grasses   The nutritive value of pasture species, even at similar stages of growth, varies widely, both in DMD and voluntary intake. Reviewing comparisons of over 1000 tropical and temperate grass samples, Minson and McLeod (1970) showed that tropical grasses were, on average, 13% units lower in DMD (Fig. 5.1). Most samples of temperate grasses had digestibilities above 65 per cent but few tropical grass samples were in that category.

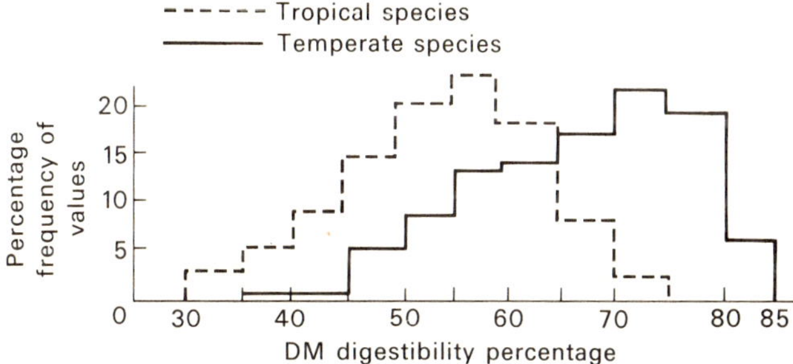

Fig. 5.1. The frequency distribution of tropical and temperate grass digestibilities (Minson and McCleod 1970).

Minson and McLeod (1970) suggested that lower DMD% values of tropical grasses may, in part, be due to higher growing temperatures, but the data of Reid, Post, Olsen, and Nugerwa (1973) supports the view that for selected or improved species of tropical grasses such as *Brachiaria*, *Chloris*, *Setaria*, and *Panicum*, DMD values are comparable to those of similarly managed temperate grasses. Dry matter digestibilities for unimproved natural tropical grasses such as *Cymbopogon*,

Hyparrhenia, and *Themada* were lower as shown by the regression equations taken from a study of *in vitro* digestibility of 42 grasses grown in Uganda:

'Unimproved' tropical grasses $Y = 66.02 - 0.16x$
'Improved' tropical grasses $Y = 80.70 - 0.31x$
Temperate grasses $Y = 75.92 - 0.22x$

where Y = dry matter digestibility (%) and x = days after initiation of primary growth.

The equations show that the rate of decline in DMD of the 'improved' tropical grasses is higher than for temperate or 'unimproved' species. After 16 weeks growth most of the tropical grasses had declined to very low DMD values in the range of 30–50% units. In many tropical environments and particularly those with a well-defined dry season, animals may be presented with a large mass of dry matter of low mean digestibility for much of the year and in this regard differ markedly from most temperate pasture situations.

Within the tropical grasses quite large differences in digestibility between genera and species have been found (Marshall, Long, and Thonston 1969; Minson 1971; Reid *et al.* 1973). Reid *et al.* (1973) found that in the improved grasses *Brachiaria* had a consistently higher digestibility at all stages of growth than the genera *Chloris*, *Setaria*, and *Panicum*. *Chloris gayana* had a markedly higher rate of decline in DMD with advancing maturity, which may have been associated with the early emergence of flowers.

Minson (1971) has shown that while quite significant differences in *in vivo* DMD are found between species, perhaps more importantly, larger differences may be found in voluntary intake (Table 5.3). In many cases differences in intake were completely unrelated to differences in digestibility. The data in Table 5.3 suggest that there is considerable potential for selection or breeding of tropical grasses for improved nutritive value, but this will be difficult because of the lack of rapid economical methods for predicting voluntary intake for screening large numbers of genotypes.

One factor which does appear to be related to intake is the percentage of leaf in the pasture. This relationship is shown in data reported by Minson (1971) in a comparison of three cultivars in the genus *Panicum* (Table 5.4). While DMD values were similar, voluntary intake increased with increasing leaf percentage.

Table 5.3. Maximum differences between genera, species, and cultivars of tropical grasses in *in vivo* dry matter digestibility and voluntary intake compared at the same age of regrowth (after Minson 1971)

Species compared	Differences in DMD (% units)	Differences in voluntary intake (%)	Differences in intake of digestible dry matter (%)
(a) Comparisons between genera			
(i) *Cynodon plectostachyum* 2 *Brachiaria* spp and 2 *Digitaria* spp	4.7	34	30
(ii) *C. dactylon* and *S. anceps*	2.5	49	39
(iii) *C. gayana, S. splendida, D. decumbens, P. maximum,* var. *trichoglume* and *P. purpureum*	6.0	20	24
(b) Comparison of species within genera			
(i) *Digitaria* (2 spp)	0.5	12	15
(ii) *Panicum* (6 spp)	3.0	37	36
(iii) *Setaria* (6 spp)	4.0	7	17
(c) Comparison of cultivars within species			
(i) *C. ciliaris* (7 cvs)	1.8	24	28
(ii) *C. gayana* (6 cvs)	2.6	4	6
(iii) *Paspalum notatum* (4 cvs)	5.0	39	43
(iv) *Paspalum plicatulum* (2 cvs)	4.0	45	56

Table 5.4. Comparison of mean leaf percentage, dry matter digestibility, and voluntary intake in *Panicum coloratum* cv. Kabulabula and cv. Burnett and *P. maximum* cv. Hamil (after Minson 1971)

	Kabulabula	Burnett	Hamil
Leaf percentage	28	38	53
Dry matter digestibility (% *in vivo*)	55.2	57.0	55.5
Voluntary intake (g per day per $W^{0.75}$)	49.1	58.1	67.3

More recently Laredo and Minson (1973) have shown that when the leaf and stem of five tropical grasses were fed separately, the voluntary intake of the leaf fraction was 46 per cent higher than the stem fraction, even though there was little difference in DMD or the level of chemically determined fibre in the two fractions. The difference in intake was caused by the much longer retention time of stem fraction in the rumen. This longer passage time may be related to the smaller surface area per unit volume of stem ingested thus reducing the rate of breakdown by rumen micro-organisms compared with leaf material.

Field grazing studies also point to the importance of green leaf content of a pasture. Cattle selectively graze the leaf fraction in preference to the stem, and the intake per bite and time spent grazing can be related to the yield and density of leaf in tropical pastures (Stobbs 1973). These observations are supported by a number of animal production studies where levels of production were closely related to the yield of green leaf in the pasture. Willoughby (1958) demonstrated an asymptotic relationship between sheep liveweight gain and the amount of green matter on offer, with a maximum gain at 1120–570 kg of green dry matter per hectare. Similarly Sharkey and Hedding (1964) found that wool growth tended to increase proportionally with amount of green pasture available up to 2000 kg DM per hectare.

Studies with beef cattle have also shown that the relationship between liveweight gain and the quantity of green leaf in the pasture is curvilinear ('t Mannetje 1974). Yates, Edye, Davies, and Haycock (1964) found that 80 per cent of the variation in cattle performance when grazing *S. almum* in late summer was related to changes in greenleaf yield. In another study (Halim and Whiteman, unpublished) comparing three tropical grass pastures, liveweight gain per head in winter was highly correlated with percentage of green leaf and was unrelated to total dry-matter yield of the pastures.

These studies indicate that the ability of a pasture species to produce a high density of green leaf readily accessible in the sward to the grazing animal is a most important component in the nutritive value of tropical grasses. Also small differences between species in percentage of green leaf produced during the dry or cool season can have large effects on animal production.

(b) Tropical legumes

By comparison with the tropical grasses the data of Reid *et al.* (1973) suggest that the young growth of tropical legumes may in fact have lower DMD values. However, DMD declines at a slower rate as shown by the regression of *in vitro* DMD over 20 weeks for nine tropical legumes grown in Uganda:

$$Y = 68.9 - 0.11x$$

DMD values were lower than for the temperate legumes *Medicago sativa* and *Trifolium semipilosum* included in the study. *T. semipilosum* had the highest DMD values of any species ($Y = 84.2 - 0.09x$).

As the pasture sward matures, tropical legumes generally have higher DMD and VI values than the associated grass as shown in Table 5.5 (Playne and Haydock 1972). Here *S. humilis* was grown with spear grass (*H. contortus*) and *in vivo* DMD and VI measured at the same time. Values for some other tropical legumes at a range of regrowth ages are also shown in Table 5.5, from Milford (1967) and Milford and Minson (1968).

Table 5.5. Comparison of *in vivo* dry matter digestibility and voluntary intake values of *S. humilis* and *H. contortus* (Playne and Haydock 1972) and values for other tropical legumes at different ages of regrowth (Milford 1967; Milford and Minson 1968).

Species	Age (days)	DMD (%)		VI g per kg $W^{0.75}$ per day	
		S. humilis	*H. contortus*	*S. humilis*	*H. contortus*
(a) *S. humilis* and					
H. contortus	44	57	55	58	53
	110	58	43	67	31
	145	55	40	69	34
(b) Other legumes					
Macroptilium					
lathyroides	120	62		44	
	164	57		28	
	209	42		24	
Macroptilium					
atropurpureum	160	50		37	
Desmodium					
uncinatum	120	54		56	
	164	51		57	
	187	53		53	
Stylosanthes					
guianensis	164	48		34	
Lotononis bainesii	204	60		59	
Lablab purpureus	70	59		64	
	111	57		66	
Vigna sinensis	70	64		76	
	111	59		68	

Comparisons of nutritive value of tropical legumes must be interpreted with some caution. Jones (1969) determined *in vitro* organic-matter digestibility and Stobbs (1971) dry-matter digestibility of *D. intortum* and *M. atropurpureum*. Both found that *D. intortum* had 10–14 units lower digestibility, which was also found by 't Mannetje (1975). Vallis and Jones (1973) found that the rate of decomposition and nitrogen mineralization from leaves and litter of *D. intortum* was many times slower than of *M. atropurpureum* when incubated in soil at 24 °C for up to 32 weeks. They postulated that leaf protein in *D. intortum* was protected against microbial decomposition and that this may also apply to decomposition by rumen bacteria. Protection from rapid decomposition may be related to higher tannin contents in *D. intortum*.

The low VI value for *M. atropurpureum* cv. Siratro shown in Table 5.6 must also be qualified. Stobbs (1977) has shown that acceptability and intake of siratro is very much higher for autumn-grown than summer-grown material. In a field grazing study at 2.96 heifers ha^{-1} when the percentage of dry green matter of siratro in the swards were: spring 2 per cent, summer 18 per cent, autumn 41 per cent, and winter 28 per cent, corresponding percentages of siratro intake from oesophageal fistulae samples were: spring 2 per cent, summer 9 per cent, autumn 62 per cent, and winter 7 per cent. Thus in autumn cattle selected a higher proportion of siratro than was in the swards. In pen feeding trials with summer and autumn grown siratro stored in a deep freeze and fed at the same time, autumn grown material was clearly preferred as shown in Table 5.6.

Table 5.6. Herbage consumed by cattle fed six-week regrowth of siratro leaf harvested in summer and autumn (Stobbs 1977).

Experiment	Dry matter consumed (g cow^{-1} 30 min^{-1})	
	Summer-grown	Autumn-grown
1	495	651
2	730	1980
3	837	1380

The reasons for differing acceptability are not well understood. However, intake of legumes is commonly higher than for grasses at similar levels of digestibility. Thornton and Minson (1973) have shown that this is due to the shorter time legume material is retained in the rumen, and the higher density to which ingested legume can be packed in the rumen.

Effect of age and maturity

The general effect of increasing age or stage of growth of pastures on nutritive value has been referred to and demonstrated in Table 5.5 and the regression equations of Reid *et al.* (1973). With increasing age the proportion of potentially digestible components comprising soluble carbohydrates, proteins, and other cell contents tends to decline, while the proportion of lignin, protected cellulose and hemicellulose and other indigestible fractions such as cuticle and silica, increase. Minson (1971) has shown a linear decline in the digestibility of cellulose as the percentage of lignin in the cellulose increased.

Another important change as pastures mature is that the proportion of stem compared with leaf increases. The stem fraction at a given stage of growth is often of lower digestibility than the leaf as shown by the data of Jones (1969):

		56	84	112
Age of regrowth (days)		56	84	112
In vitro organic matter digestibility (%)	Leaf	64.8	66.4	68.5
	Stem	59.5	63.1	59.8

However, even at similar levels of digestibility intake of leaf is up to 46 per cent higher than for stem, as previously discussed (Laredo and Minson 1973). Thus increasing proportion of stem in the sward reduces the overall nutritive value of the dry matter on offer.

The effects of increasing plant age are further reinforced by the differences in the physiological characteristics of leaves produced later in the life of the plant. Wilson (1976) studied changes in cell-wall content, nitrogen concentration, and dry-matter digestibility of leaves at the same stage of development, produced successively at increasing levels of insertion in green panic (*P. maximum* var. *trichoglume*) grown in a constant environment. In leaf blades with increasing level of insertion from base to top, cell-wall content increased from 33–64 per cent, lignin content tended to increase, nitrogen concentration decreased from 4.7–2.5 per cent, and *in vitro* DMD decreased from 77–65 per cent. DMD of leaf sheaths was lower than for leaf blades and declined more rapidly with time. Thus upper leaves produced by the physiologically older plant appear to be of lower nutritive value than earlier produced leaves.

One important selection criterion for improved pasture species is selection for a slow rate of change in the factors associated with declining nutritive value, often associated with a long vegetative period and delayed onset of flowering. These changes are modified by environmental factors and by pasture management. The effects of age of regrowth on animal production are widely recognized and grazing management aims to maintain pstures in a young leafy growth stage for as long as possible.

Environmental factors

Minson and McLeod (1970) have reviewed the data of over 1000 determinations of digestibility of tropical and temperate grasses and reported that the tropical grasses are on average about 13 units lower in *in vivo* DMD, as shown in Fig. 5.1. They suggested that the lower values in tropical grasses were related to the higher growth temperatures, and when a number of temperate and tropical grasses were grown together under the same conditions in southern Queensland there was little difference in DMD. Digestibility was correlated with mean growing temperature (r = –0.76) and with evaporation (r = –0.64) and the correlation was higher when temperature and evaporation were combined (r = –0.83).

Wilson and Ford (1973) grew 13 tropical and 11 temperate grasses at day/night temperatures of 21/13, 27/19, and 32/24 °C. Each plant was harvested at the same physiological growth stage, two days after the fifth leaf reached maximum length. Most of the temperate grasses declined an average of five units in *in vitro* DMD with increase in temperature from 21/13 to 32/24 °C. This decline was associated with a fall in the percentage soluble carbohydrates. In contrast temperature had little effect on DMD of the tropical grasses. Digestibility of the temperate grasses was generally higher than that of the tropical grasses at 21/13 °C, but there was little difference between the two groups at 32/24 °C.

't Mannetje (1975) compared eight tropical grasses and twelve legumes at temperatures of 32/24, 26/15, and 20/6 °C, at two day-lengths of 14 and 11 hours. In this experiment, where plants were harvested at the same age of regrowth (28 days), in contrast to the same physiological stage of development, *in vitro* DMD generally declined with increasing temperature. At lower temperatures plants had a higher leaf percentage and higher nitrogen concentrations.

Day and night temperatures have differential effects on digestibility of leaf and stem. Ivory, Stobbs, McLeod, and Whiteman (1974) found in *Cenchrus ciliaris* and *Pennisetum clandestinum* that low day temperatures (15 °C) gave significantly higher leaf and stem digestibilities than higher temperatures (up to 30 °C). In buffel grass, night temperatures had no effect on leaf digestibility and in kikuyu high night temperatures increased leaf digestibility, while in both species stem digestibility was decreased with increasing night temperature. These differential temperature responses caused stem digestibility to be higher than leaf at low night temperature, and leaf digestibility higher than stem at high night temperatures. Thus the relative digestibility of leaf and stem fraction may be determined by the night temperature experienced during growth.

Increasing temperature appears to have rather direct effects on the digestibility of temperate grasses through declining soluble carbohydrate levels. In the tropical grasses temperature effects are indirect, whereby increasing temperature increases the rate of growth and hastens the progress towards maturity. Thus at a given age at a higher growth temperature digestibility values are lower (Ivory *et al.* 1974). Temperature effects on the tropical legumes were variable, but appear to be similar to the tropical grasses ('t Mannetje 1975).

Tropical grasses, in general, accumulate less soluble carbohydrate than the temperate grasses under all temperature regimes (Wilson and Ford 1973). When compared at their optimum temperatures for growth, the differences are more apparent. Environmental factors of temperature and water stress may also affect cell-wall content. Wilson and Ford (1971) measured consistently higher digestibility coupled with lower cell-wall content in the temperate *Lolium perenne* compared with *Panicum maximum* and *Setaria anceps*. Within the tropical grasses Duble, Lancaster, and Holt (1971) reported that digestibility of five species was inversely related to cell wall content.

Under dry winter conditions in Brazil, Schank, Day, and de Lucas (1977) found in 30 cultivars of *Digitaria decumbens*, other *Digitaria* spp, and *Cynodon dactylon*, *in vitro* digestibility of forage cut every 28 days was about ten units lower, as shown in Fig. 5.2. At this time of the year, growth rates were very much reduced. Although cell-wall contents were not measured it would appear that these may have been much higher than under the more favourable growing conditions in summer.

Sufficiency of proteins, minerals, and vitamins

In the assessment of nutritive value of pastures voluntary intake and digestibility are the most important determinants of pasture quality, provided protein, mineral, and vitamin levels are adequate. While vitamin deficiencies are unusual in grazing ruminants, protein and mineral deficiencies are common, particularly in tropical pasture situations. If

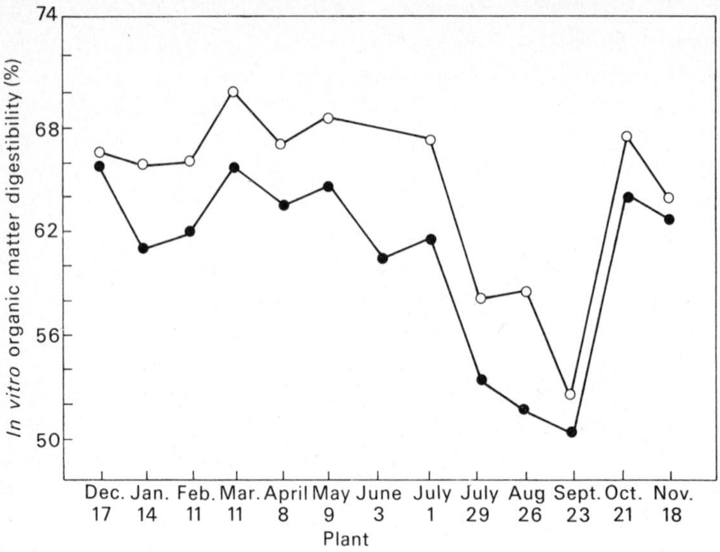

Fig. 5.2. *In vitro* digestibility of hybrid digitgrasses compared with commercial cultivars throughout the year. ○ 'Best' four hybrids; ● four commercial cultivars (Schank *et al.* 1977).

the diet is deficient in protein or mineral elements, then appetite of the animal will be depressed and functions of digestion may be impaired thus reducing voluntary intake.

Protein

Dietary protein requirements depend upon the type and level of animal production. High-yielding dairy cows and rapidly growing fat lambs for example, have higher protein requirements than wool-producing wethers or finishing beef steers. For finishing beef steers of 300–500 kg liveweight, 11 per cent crude protein (CP) is adequate (NRC 1970), whereas for a 550 kg dairy cow producing 20 kg milk per day, 15–16 per cent CP is required.

A striking difference between temperate and tropical pastures is that at similar stages of growth tropical grasses have considerably lower CP percentages than temperate species. This is particularly evident in many tropical natural pasture swards. With well managed temperate pastures, animal production is rarely limited by the lack of dietary protein, whereas in the tropics, deficiencies of protein may limit animal production for much of the year.

Protein deficiency operates at two levels. The pasture on offer may contain less than the requirement for maximum growth rate or production, e.g. 11 per cent CP (1.8 per cent nitrogen) for growing steers. As the pasture matures and protein content declines still further, a critical level of protein content is reached below which voluntary intake of dry matter is depressed. For temperate feeds the critical CP percentage is about 8.5 (Blaxter and Wilson 1963), and for tropical grasses 7 per cent CP (Milford and Minson 1966).

The depressing effect of low crude protein on dry-matter intake acts in two ways. It is suggested that when nitrogen intake is below a critical level activity of rumen micro-organisms is reduced, so that efficiency and rate of digestion and hence intake is inhibited (Minson 1971). However, it has been shown (Egan 1977) that with sheep on low nitrogen diets (1.0–1.5g N per 100 g DOM) dry-matter intake was increased by 15–20 per cent when readily soluble protein was perfused directly into the duodenum. Thus low protein intakes also appear to act indirectly through depressing appetite, probably as a symptom of general ill health in the animal. The feeding of protected protein, such as formaldehyde-treated casein, which escapes digestion in the rumen, but is hydrolysed in the hind gut, may also act in a similar way.

Crude protein deficiencies in the diet may be overcome by supplementing animals with high protein concentrates such as soyabean or peanut meals, or meat or fish meals; by non-protein nitrogen sources such as urea; or by incorporating a legume in the pasture. When the CP percentage is below the critical level in mature native pastures supplementation not only increases protein intake but leads to increased intake of dry matter and hence net digestible energy. As previously dicussed (Section 4.2 and Fig. 4.1) legumes maintain higher crude protein contents with increasing age than the associated tropical grasses, and are of major importance in tropical pastures in maintaining dietary levels of protein intake.

Nitrogen content of grasses may be directly increased by addition of fertilizer nitrogen. Where nitrogen content is below the critical level in tropical grasses increased nitrogen content may markedly increase voluntary intake. For example, with pangola grass (*D. decumbens*) fertilized with 113 kg N ha^{-1} to raise protein content from 3.7 to 7.2 per cent CP, Minson (1967) obtained a 54 per cent increase in voluntary intake.

In temperate grasses increasing levels of nitrogen fertilization leads to a marked decline in soluble carbohydrates, but as total soluble content (when the increased protein levels are added to the digestible fraction) is little changed, dry matter digestibility is also little affected (Minson, Raymond, and Harris 1960). Effects in tropical grasses appear to be variable, although soluble-carbohydrate content appears less sensitive to levels of nitrogen. Wilson (1973) obtained an increase in 3–5 *in vitro* DMD units in *Panicum maximum* at moderate compared with low nitrogen fertilizer levels, and this was related to a reduction in cell-wall content. However Minson (1973) found that the average effect of increased nitrogen content on *in vivo* DMD of *C. gayana*, *D. decumbens*, and *P. clandestinum* was negligible.

Thus we may conclude that direct effects of nitrogen fertilization on digestibility of temperate and tropical grasses are small and variable, but that increasing the nitrogen content, which is more likely to be below the critical level in tropical grasses, can increase voluntary intake by up to 78 per cent (Minson 1971).

Minerals

Where particular mineral elements are deficient in pastures, lowered animal production or ill health may result both from direct effects of the element deficiency on body biochemistry and indirectly through depressed appetite and reduced pasture intake. Recommended levels of essential mineral elements are shown in Table 5.7 (from Minson *et al.* 1976).

Table 5.7. Recommended contents of mineral elements in diets of beef cattle weighing 300–500 kg (from Minson *et al.* 1976)

Element	Concentration in dry matter (%)	Element	Concentration in dry matter (p.p.m.)
Nitrogen	1.8	Iron	30
Phosphorus	0.22	Manganese	40
Potassium	0.31–0.44	Zinc	20–40
Calcium	0.22–0.35	Copper	4–10
Magnesium	0.12	Cobalt	0.1
Sodium	0.05	Iodine	0.12
		Selenium	0.05

Increases in pasture intake have been reported by feeding supplementary nitrogen, phosphorus, sodium, sulphur, selenium, and cobalt (Minson 1976). Mineral nutrition of livestock is a complex study as animals are not only affected by the content of elements in the pasture, but also by the balance or ratios of different elements, as in the problem of grass tetany in dairy cattle.

Direct supplementation of animals has been compared with application of the deficient element to the pasture. Rees, Minson, and Smith (1974) measured voluntary intake and *in vivo* DMD of pangola grass (*D. decumbens*) grown with and without sulphur fertilizer, fed to sheep either with or without sulphur supplementation. Sulphur fertilizer increased sulphur content of the pasture from 0.09 to 0.15 per cent, increased dry-matter yield from 2100 to 4200 kg ha^{-1}, and increased leaf percentage from 13 to 23 per cent. When fed to sheep in pens, sulphur fertilization gave a larger increase in voluntary intake than supplementation, while both increased DMD similarly (Table 5.8).

Table 5.8. Effects of sulphur supplementation and fertilization on voluntary intake and digestion of *Digitaria decumbens* (Rees *et al.* 1974)

	−S Fertilizer		+S Fertilizer	
	−S supplement	+S supplement	−S supplement	+S supplement
Total sulphur in diet (g day^{-1})	0.6	1.4	1.4	2.0
Voluntary intake (g per kg W$^{0.75}$ day^{-1})	44	57	64	65
Dry matter digestibility (%)	55	61	60	59
Retention time in rumen (hours)	24	−	20	−

Sulphur deficiency apparently depresses microbial activity in the rumen so that increasing the sulphur level increased DMD, similarly. However, the increased intake due to sulphur fertilization could be related to the higher leaf content in the pasture and a shorter retention time in the rumen.

In a similar study Rees and Minson (1976) examined the effect of calcium fertilizer on *D. decumbens* (Table 5.9).

In this case supplementation had no effect on intake or digestibility, whereas fertilization increased both. Increased digestibility resulted

Table 5.9. Effects of calcium supplementation and fertilization on voluntary intake and digestion of *Digitaria decumbens* (Rees and Minson 1976)

| | −Ca Fertilizer | | +Ca Fertilizer | |
	−Ca supplement	+Ca supplement	−Ca supplement	+Ca supplement
Voluntary intake (g per kg $W^{0.75}$ day^{-1})	39	38	43	43
Dry matter digestibility (%)	45.8	45.2	48.0	47.2
Retention time in rumen (hours)	33.3	−	27.4	−

mainly from increased digestibility of the hemicellulose fraction, while increased intake was again related to the shorter retention time in the rumen. Rees and Minson (1976) suggest that calcium fertilizer effects may be related to changes in the cell wall components allowing an increased rate of microbial breakdown in the rumen.

These two examples demonstrate the important interactions between mineral nutrients, pasture components, and the nutritive value of pastures. Correction of nutrient deficiencies in pastures may not only increase pasture growth and yield, but also improve the nutritive value of the pasture through effects on intake and digestibility.

The nutritive value of pastures has been discussed mainly in relation to digestibility and intake, and these factors are seen to be modified by interactions between species, stage of growth, environmental and soil nutrient effects. Nutritive value may also be modified by management practices. These effects are the subject of the next Section in which interactions of stocking rate and grazing methods on animal production and pasture parameters are explored.

5.2 STOCKING RATE AND GRAZING MANAGEMENT

In any area being grazed, whether improved pasture, natural pasture, or rangeland, the output of animal products per unit area is a function of the production per animal and the number of animals per unit area, i.e.

Animal production ha^{-1} = production head^{-1} × number animals ha^{-1}

Production per head reflects the genetic potential of the particular animal, the standards of animal management and husbandry, and more importantly the quality of the pasture. The number of animals per ha which can be supported by the pasture for a given period of time (the stocking rate) is basically a function of pasture yield. Of course there is an interaction between the factors of quality and yield, and changes in these parameters are strongly affected by stocking rate but may be modified to some extent by grazing management. However, adjustment of stocking rate is the most important management factor, and is the major determinant of animal production and pasture composition. This chapter examines the relationship between stocking rate and production, describes some of the effects of stocking rate on pasture characteristics, and makes some comparisons between the major grazing management systems.

1. *Stocking rate*

A number of terms are used to describe the relationship between the number of animals supported on an area of land and the land resources providing that support. Very often the definition of this relationship is imprecise and it is therefore important that the terms are clearly defined.

Stocking rate

In its simplest form stocking rate is defined as the number of animals grazing a unit of area (usually 1 hectare) at a particular time. This definition is obviously too general and the type of animal should be defined (sheep, cattle, etc.), and more precisely the class of animal must be defined (e.g. mature wether sheep, or yearling steers, etc.). In order that stocking rate comparisons can be standardized, animal units (a.u.) or livestock units (l.u.) have been defined for the two main groups of domestic animals, sheep and cattle. Different breeds or classes or age groups within each category can be related to a standard animal unit, as shown in Table 5.10.

It is usual to distinguish between the *instantaneous* stocking rate (a.u. ha^{-1}) and the *sustained* stocking rate (a.u. ha^{-1} time^{-1}, usually 1 year).

Carrying capacity

This aims to define the long-term stocking rate and integrates short-term fluctuations in stocking rate. The defined carrying capacity of an area usually involves the idea of an *optimum* stocking rate which can be *safely sustained* and will be related to the forage resources available in the deficit season of the year, or indeed to long-term deficits, as in years of drought.

Table 5.10. Relationship between classes of livestock and defined animal units in cattle and sheep (after Moorhouse 1975)

Cattle		Sheep	
1 animal unit	= 400 kg steer	1 dry sheep equiv. (d.s.e.)	= 40 kg merino wether
a 400 kg steer	= 1 a.u.		
calf (1–8 months)	= 0.35 a.u.	wether	= 1 d.s.e.
weaner (8–12 months)	= 0.40 a.u.	maiden ewe	= 1 d.s.e.
steer (1–2 years)	= 0.87 a.u.	lambs (up to 2 tooth)	= 0.5 d.s.e.
breeding cow	= 2.0 a.u.	breeding ewe	= 1.7 d.s.e.
bull	= 2.0 a.u.	ram	= 1.7 d.s.e.
	1.0 a.u. (steer) = 8 d.s.e. (wethers)		

Grazing pressure

The above definitions relate animal numbers to area of land, whereas the most important aspect is not the area but the amount of forage produced. The term grazing pressure is used to relate the number of animal units to the amount of forage available, e.g. kg of dry matter on offer per animal unit, or some component of total forage such as kg of dry green leaf on offer per animal unit ('t Mannetje 1974; Willoughby 1958). Grazing pressure has also been defined (Mott 1960) as the ratio of feed demand to feed supply:

$$GP = \frac{\text{Dry matter demand animal}^{-1}\ \text{day}^{-1} \times \text{no. animals ha}^{-1}}{\text{Dry matter available day}^{-1}\ \text{ha}^{-1}}$$

Where many animal types graze a pasture, as in the case of wild animal populations on rangeland, zoologists define stocking rate, carrying capacity, or grazing pressure in terms of *biomass* per unit area, or unit of forage. In grazing studies with domestic animals, since animal liveweight changes with time on a trial, Bryan and Evans (1973) have suggested that stocking rate may be more precisely defined in terms of *biomass* ha^{-1}. However, the definition of standardized animal units does relate to differences in biomass per unit area.

The expression of stocking rate

Where stocking rates are low or less than one animal per hectare, as in rangeland or extensive grazing situations, stocking rate is often expressed as hectares per animal (ha an^{-1}). Under more intensive stocking, and where stocking rate is greater than one animal per hectare, it is more usually expressed as animals per hectare (an ha^{-1}). Shaw (1970) has shown that the relations between stocking rate and either animal production per unit area or grazing pressure will depend on which expression is used.

Shaw (1970) analysed the data from Owen and Ridgman (1968) to show the relationship between animal production per hectare, and stocking rate expressed as an ha^{-1} or ha an^{-1} (Fig. 5.3). The an ha^{-1} function gives a quadratic type curve, whereas the ha an^{-1} curve is strongly asymmetrical. As will be seen later in the analysis of Jones and Sandland (1974), the an ha^{-1} curve can be described in a simple mathematical quadratic function, whereas the ha an^{-1} function will be mathematically more complex. Furthermore, Shaw (1970) points out that in the design of a grazing trial, treatments equally spaced in terms of ha an^{-1} will not cover the animal production range as well as a set of an ha^{-1} treatments. Shaw (1970) concludes that there are very real advantages in choosing treatments on the basis of the an ha^{-1} expression because grazing pressure and production per hectare are more directly related to an ha^{-1} than to ha an^{-1}

The relationship between stocking rate and animal production
The value of new pasture species, or recommendation of improved management practices should be tested under field grazing and evaluated in terms of animal output. Early attempts to compare effects of different pasture treatments in terms of animal production proved very difficult to interpret (Wheeler 1962). In hindsight it can be seen that problems arose often because of an arbitrary choice of a common single stocking rate for all treatments. Meaningful conclusions were not obvious because of the lack of a unifying model of animal production relationship in grazed pastures and the lack of appreciation of the major interaction between pasture treatments and stocking rate (Owen and Ridgman 1968). Stocking rate models seek to define the relationship between the effects of increasing stocking rate and the production per animal and also the production per hectare (product animal^{-1} × no. animals ha^{-1}).

Mott (1960) proposed a general model to describe the relationship between animal output and stocking rate, based on available data of stocking rate trials and theoretical considerations. This model was developed from the consideration that under relatively uniform pasture conditions intake of animals will remain constant over a wide range of low stocking rates (Owen and Ridgman 1968). As stocking rate is

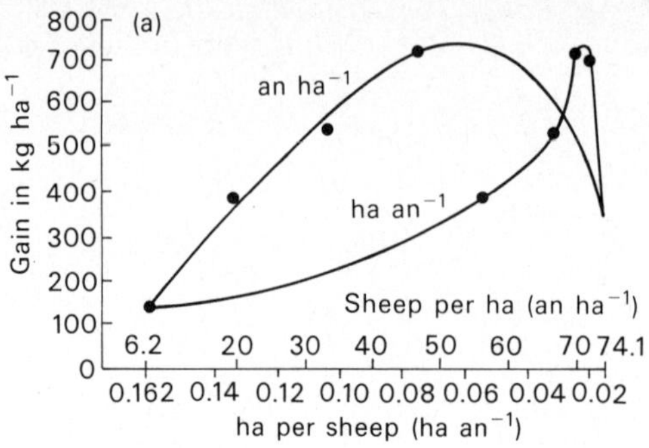
(a)
Gain in kg ha^{-1}
800 700 600 500 400 300 200 100 0
an ha^{-1}
ha an^{-1}
Sheep per ha (an ha^{-1})
6.2   20   30   40   50   60   70 74.1
0.162 0.14 0.12 0.10 0.08 0.06 0.04 0.02
ha per sheep (ha an^{-1})

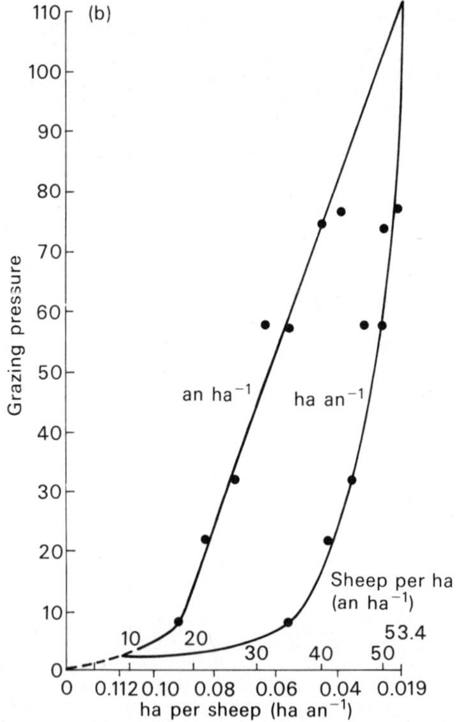
(b)
Grazing pressure
110 100 90 80 70 60 50 40 30 20 10 0
an ha^{-1}
ha an^{-1}
Sheep per ha (an ha^{-1})
10   20
30   40   50
53.4
0   0.112 0.10   0.08   0.06   0.04   0.019
ha per sheep (ha an^{-1})

increased a point is reached where the amount of forage available equals the requirement of the animals, and at stocking rates above this point intake per animal is limited and production per animal declines.

Mott's (1960) model unified the results of different trials by expressing the results in terms of ratios to the values at presumed optimum stocking rates, as shown in Fig. 5.4.

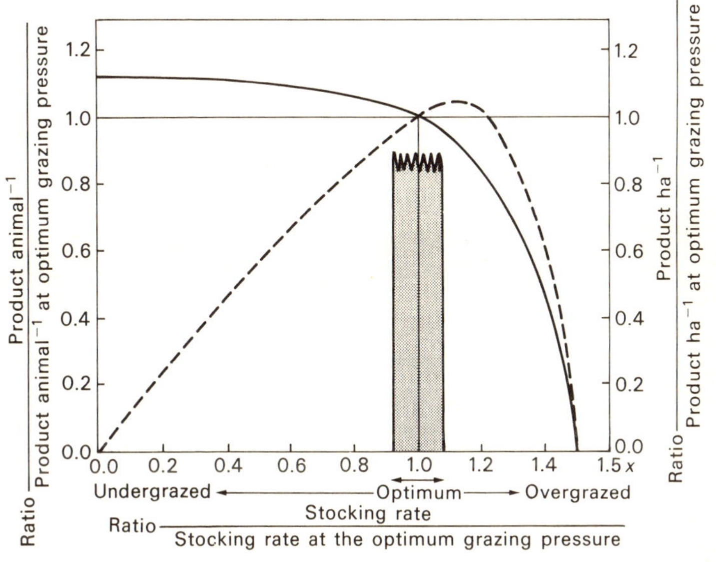

Fig. 5.4. The relation between stocking rate and gain per animal (——) and per hectare (– –) after Mott 1960). (From Jones and Sandland 1974.)

This model, which could be termed an *exponential* model (of the form $y = k - ab^x$, where k, a, and b are constants) predicts an optimum stocking rate range for production per head, and a maximum production per hectare at a slightly higher stocking rate. At stocking rates below the optimum the model predicts little change in production per head

Fig. 5.3. (a) The relation between weight gain per hectare by sheep and stocking rate expressed both as sheep per hectare and hectare per sheep (adapted from data by Hodgson in Owen and Ridgman 1968). (From Shaw 1970.) (b) The relation between grazing pressure and stocking rate expressed both as sheep per hectare and hectare per sheep (adapted from data by Campbell 1966). (From Shaw 1970.)

with changes in stocking rate, while above the optimum, production declines rapidly with increasing stocking rate.

This model was further developed by Petersen, Lucas, and Mott (1965) which predicted a clearly defined optimum stocking rate (Fig. 5.5(a)). This model might be termed a *critical* stocking rate model. A similar model was developed by Conniffe, Browne, and Walshe (1970) (Fig. 5.5(b)).

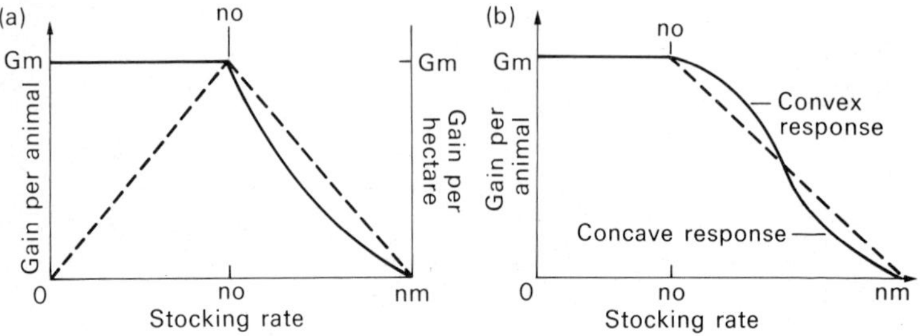

Fig. 5.5. Theoretical relation between stocking rate and animal performance (after Peterson, Lucas, and Mott 1965). – – gain per hectare; — gain per animal; no, optimum stocking rate; nm stocking rate at which animals only maintain weight; Gm, maximum liveweight gain per animal or per hectare. (b) Theoretical relation between stocking rate and animal gain (after Conniffe, Brown, and Walshe 1970). no, optimum stocking rate; nm, stocking rate at which animals only maintain weight; Gm, maximum liveweight gain per animal. (From Jones and Sandland 1974.)

Recently Jones and Sandland (1974) analysed the data from a grazing trial comparing tropical grass–legume, and grass plus nitrogen pastures conducted by Jones (1974a). They found the relationship between stocking rate and liveweight gain per head was linear, as expressed by the linear regression equation:

$$Y_a = a - bx$$

where Y_a is gain per animal, stocking rate is x, and a and b are constants. The linear relationships were highly significant with r values of –0.999, –0.980, and –0.973 for each of the pastures (Fig. 5.6).

If the relationship between stocking rate and gain per animal is truly linear then the relation between stocking rate (x) and gain per hectare (Yh) is defined by the quadratic equation:

$$Y_n = ax - bx^2.$$

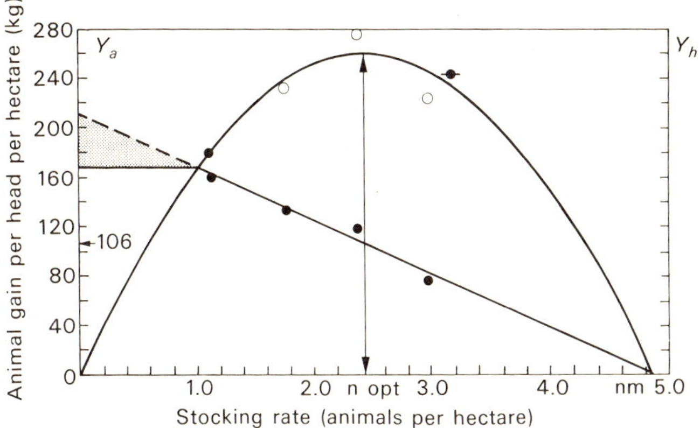

Fig. 5.6. The relation between gain per animal and gain per hectare in response to increasing stocking rate for a Setaria–Siratro pasture (the effect of holding animal gain constant below 1 animal ha^{-1} is indicated). $Y_a = 212 - 43.7x$ ($r = -0.98$, $P < 0.001$); $Y_h = 212x - 43.7x^2$. (Jones 1974.)

The most important feature of this model is that the stocking rate giving the maximum gain per hectare, or optimum stocking rate, can be accurately defined by the equation:

$$X = a/2b,$$

or is half the X-value of which Y_a or $Y_h = 0$. Armed with a method for accurately assessing optimum stocking rate, rather than presuming an optimum rate as in earlier models, Jones and Sandland (1974) tested the general application of the model by reassessing the earlier data used by Mott (1960) and others. In all, they used data from 33 different pastures, again unifying the data by expressing the data as ratios at the *calculated* optimum stocking rate for each set of trial results, as shown in Fig. 5.7.

All data, from both temperate and tropical pastures, with both sheep and cattle data, fitted the linear model with almost no departure from linearity ($r = -0.992$). However, the authors do suggest that at very low stocking rates, further reduction in stocking rate may have little effect in reducing gain per animal. This is more likely to occur in temperate high quality pastures than in tropical swards where even at very low stocking rates ability for increased selections may lead to improved nutrition. Nevertheless, a small departure from linearity at this level does not affect the linear relationship over most of the range.

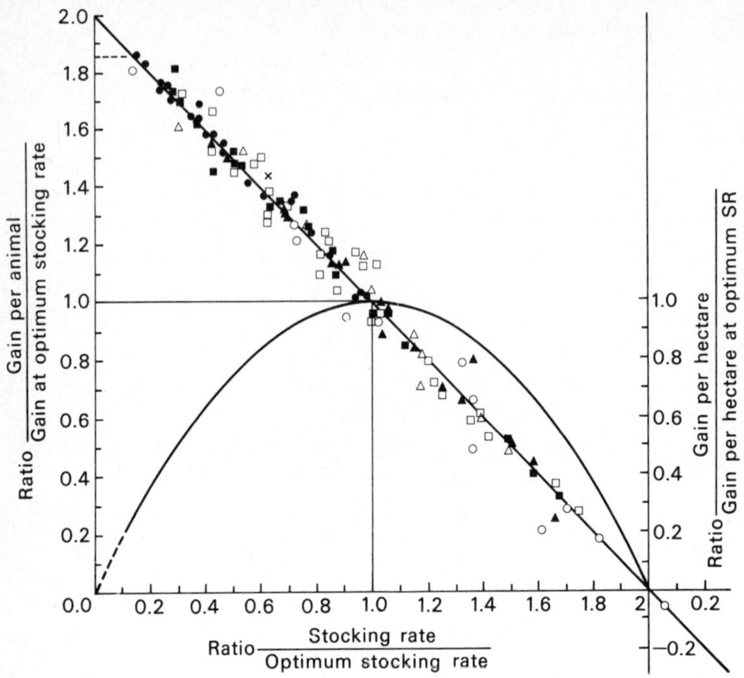

Fig. 5.7. The relation between stocking rate and both gain per head and gain per hectare from grazing experiments conducted with a variety of pasture species in a wide range of environments. Y_a = 1.999 − 0.999x (r = −0.0992, P < 0.001); Y_h = 1.999x − 0.999x². (From Jones and Sandland 1974.)

The general applicability of the linear model represents a most important advance in our understanding of the stocking rate relationship and the analysis of grazing trials. First, while allowing an accurate assessment of optimum stocking rate to give maximum gain per hectare, the model also shows that zero animal gain is reached when the stocking rate is double that required for maximum gain per hectare. This was earlier suggested by Riewe (1961).

In a linear model there is no 'critical' stocking rate, as suggested by earlier models. Therefore production per hectare only changes gradually on either side of the optimum stocking rate, and does not decline sharply at stocking rates above the optimum. Fortunately for the pasture manager, this suggests there is some flexibility in the assessment of optimum stocking rate without incurring too large a penalty for a slightly optimistic assessment.

Finally the linear model can be most usefully applied in the comparison of different pastures or treatments under grazing. Examining the linear regression equation:

$$Y_a = a - bx$$

the value 'a' represents the potential animal production at infinitely low stocking rate, and thus reflects the nutritive value of the pasture and the genetic potential of the animals to produce. The value '$-b$', which expresses the slope of the regression, indicates the rate of decline in gain per animal as stocking rate increases, and thus reflects the ability of the pasture to yield and withstand the other effects of increasing grazing pressure. In comparing different pastures, the pasture with the highest 'a' value and smallest '$-b$' value will be superior. However there may be compensation between these two factors; that is the pasture with the highest 'a' value may not have the lowest '$-b$' value, in which case animal production is compared at the optimum stocking rate.

Jones and Sandland (1974) suggest that the linear model has its best application under longer term grazing studies, and that there may be some differences in predicting optimum stocking rate with short grazing periods (around 100 days) on high quality pastures. Here, for short periods, animal gain at the optimum stocking rate, and at lower stocking rates may be at the genetic potential of the animals. This is unlikely to occur in tropical pasture swards.

However, there are some important exceptions to the linear model which were not considered by Jones and Sandland (1974). Where there is an interaction between increasing stocking rates causing marked changes in botanical composition of the pasture, animal production may markedly increase or decrease depending on the direction of change. The main examples are where stocking rate affects the legume content.

In the data of Austin (1970; previously discussed, on page 241) increasing stocking rate caused a reduction in competition from associated grasses, so that the proportion and yield of the oversown legume *Stylosanthes humilis* increased. In this case, gain per animal and per hectare increased as stocking rate increased, although if stocking rate had been further increased above the range used in the trial, then at some point production would decline.

Walker (1974) studied the effects of six stocking rates on a *Setaria*

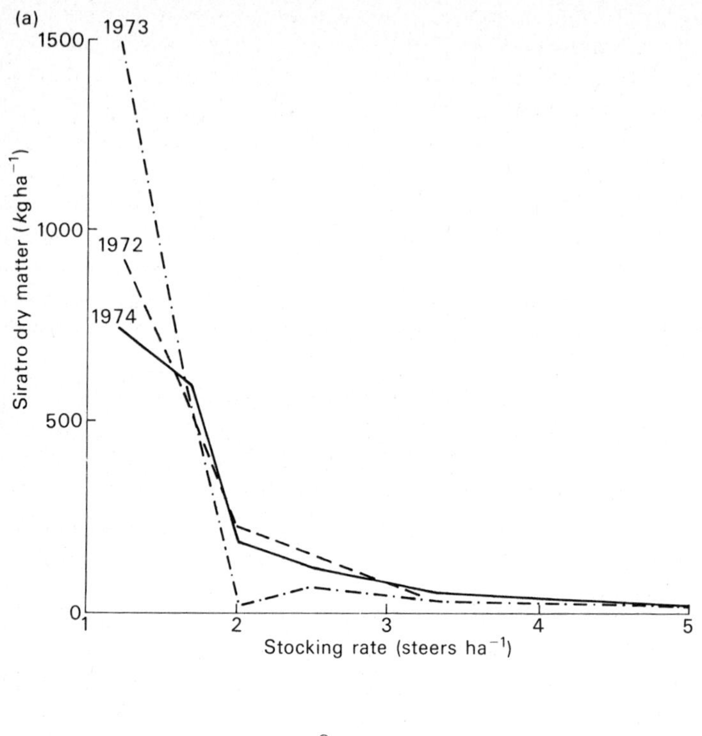

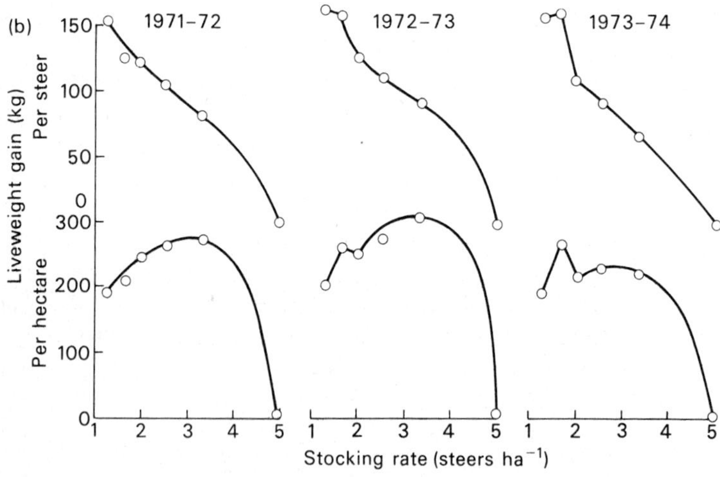

Fig. 5.8. (a) Effect of stocking rate on legume yields in July 1972, 1973, and

anceps–Macroptilium atropurpureum (cv. Siratro) pasture. During the first year of grazing, at stocking rates above 1.7 steers ha^{-1} siratro content and yield declined dramatically to negligible levels. In this year liveweight gain per animal followed the linear model, but by the third year of grazing, where effects of low legume content at higher stocking rates had accumulated, maximum liveweight gain per animal and per hectare were obtained at the same stocking rate of 1.7 steers ha^{-1}, as shown in Fig. 5.8.

These departures from linearity do not detract in any way from the general application of the linear model. They do reinforce the requirement to use an appropriate range of stocking rates in grazing studies. Analysis of reasons for departure from linearity will provide important data for better grazing management of different species in pastures.

2. Stocking rate effects on sward characteristics

The linear relationship between animal production and increasing stocking rate demonstrates that under longer term grazing conditions as animal numbers increase per unit area of pasture, the net intake of metabolizable energy per animal declines. The rate of decline with increasing stocking rate is a function of pasture growth rate and the availability of high nutritive value components in the sward.

Animals graze selectively in pastures, generally seeking species and plant parts of higher nutritive value (Theurer 1970). Sheep are more selective than cattle, but even cattle demonstrate a high degree of discrimination during grazing. The grazing process causes marked changes in the structure of the sward canopy, in the proportions of components (leaf, stem, inflorescence, and dead material), and the botanical composition of the pasture. These effects become increasingly evident with increasing stocking rate, or with increasing time at given stocking rate when pasture growth rate is less than the total intake of the grazing animals.

Stobbs (1975a) has shown that when cattle intensively graze a pasture the uppermost leaves are selected first, followed by leaf-bearing stem, and finally, if forced to it, the almost leafless stem. Cattle rarely graze a small area down to leafless stem whilst more leafy herbage is still available, except patches fouled by excreta.

1974 on Siratro-Kazungula setaria pastures (Walker 1974). (b) Effect of stocking rate on liveweight gain per steers and per hectare over three years on Siratro-Kazungula setaria pastures at Tedlands, Central Queensland (Walker 1974).

Pen feeding studies have shown that intake of leaf is up to 46 per cent higher than for stem, even when stem had a slightly higher digestibility (Laredo and Minson 1973). Many field studies have shown that grazing animals prefer leaf to stem, and green material (or young herbage) in preference to dry material (or old herbage). The material selected is usually higher in nitrogen and other minerals, and lower in fibre than the mean value for the total pasture on offer (Chacon 1976). It is not surprising that animal production levels are more closely correlated with leaf or green material than with total pasture dry matter, as previously discussed (p. 286). Stobbs (1973) found that the bite size of grazing cattle was influenced by the leaf yield on offer and suggested that the distribution and accessibility of leaf within the sward profile is an important factor controlling intake of grazing cattle. Thus effects of stocking rate and differences between pasture species which affect sward structure are important determinants of pasture nutritive value.

Sward structure parameters

Pasture swards are heterogenous as a result of differences in anatomical, morphological, and nutritive value of the plant parts and species in the sward. Pasture species may be tall growing or prostrate, but sward structure at any time will be modified by the stage of growth, grazing management particularly stocking rate, environmental and edaphic conditions including fertilizer usage.

Total dry-matter yield has been the major selection criteria for pasture plants. The ability of grazing animals to harvest this yield is affected by the way the yield is distributed in the height of the pasture and by the distribution of components (stem, leaf, and dead material) making up the yield (Allden and Whittaker 1970; Stobbs 1975b). The distribution of yield in a sward may be determined by taking stratified cuts in layers from the top to the bottom of the sward, and by sorting the material into species and components (green leaf, green stem, dead matter). Pastures may be compared in terms of: mean sward bulk density — total yield per unit area/mean sward height, or layer bulk density — yield in each layer per unit area/height of the stratified cutting interval, expressed as kg ha^{-1} cm^{-1}. Yield of individual components (e.g. dry matter in green leaf) can be similarly determined.

Recent studies (Stobbs 1973, 1975b) show that tropical swards, in spite of high dry matter yields, have lower bulk densities (range 14–200 kg ha^{-1} cm^{-1}) than temperate swards (160–410 kg ha^{-1} cm^{-1}). Also density of leaf material is lower, particularly in the uppermost layers of the sward. Stobbs (1973) compared the leaf area index (LAI) per centimetre of plant height and showed that temperate pastures had a mean of 0.33 LAI cm^{-1} compared with 0.11 LAI cm^{-1} for tropical swards. Within tropical pasture swards there is a marked gradient in the distribution of dry matter and chemical components, as shown in Table 5.11 (Stobbs 1973). Temperate swards tend to be more uniform.

Table 5.11. Gradient in nutrients within a *Setaria anceps* cv. Kazungula sward (Stobbs 1973)

Sward layer	*In vitro* DMD%		Nitrogen (% DM)		Phosphorus (% DM)		Calcium (% DM)	
	Leaf	Stem	Leaf	Stem	Leaf	Stem	Leaf	Stem
Top	59.1	58.5	1.60	0.64	0.24	0.25	0.27	0.12
Middle	57.1	54.4	1.10	0.42	0.21	0.29	0.29	0.12
Bottom	55.6	49.2	0.67	0.21	0.21	0.22	0.41	0.12

Since animals prefer to graze the uppermost layers of the sward, the amount and nutritive value of this material is important. Also, as the pasture is grazed down, animals are forced to eat forage of lower DMD and nitrogen content. While low sward bulk densities may limit the ability of the grazing animal to harvest an adequate amount of material with each bite (Stobbs 1975b), high sward bulk densities in tropical pastures may only reflect a high stem yield (Laredo and Minson 1973). Where stem yield is high, leaf may be inaccessible to the grazing animal.

Because of the easier prehension of leaf, and higher intake of leaf than stem (Laredo and Minson 1973), the yield, distribution, and density of leaf and leaf:stem ratio in the sward are important determinants of intake. The ability of grazing animals to satisfy their intake requirements is therefore markedly affected by type of pasture and stage of growth as shown in Table 5.12 (Stobbs 1975a).

Cattle graze tropical pastures for a longer time each day than temperate pastures, even when large quantities of herbage are available (Stobbs 1975a). However, grazing times seldom exceed about 720 minutes per day, so that cattle on mature pastures may have difficulty in satisfying their nutrient requirements when the amount of feed harvested at each bite is restricted by accessibility of green leaf. Under these conditions, cows compensate for small bite size by increasing the number of bites or rate of biting (Stobbs and Cowper 1972; Stobbs 1975c).

Chacon (1976) has examined the changes in sward structure with

Table 5.12. Mean grazing time, rate of biting, and size of bite prehended by Jersey cows grazing pastures of varying age and quality (Stobbs 1975a)

Pasture	Grazing time (min/24 h)		Rate of biting (bites/24 h)		Bite size (g O.M./bite)‡	
	Total	Range	Total	Range	Mean	Range
Immature* temperate	464	410–521	28 800	25 020–36 420	0.43	0.31–0.71
Immature tropical	561	418–595	44 530	32 500–54 200	0.34	0.17–0.50
Mature† tropical grass	677	483–734	61 700	48 700–75 220	0.17	0.05–0.31
Mature tropical legume	719	588–788	–	–	–	–

* <3 week-old regrowths.
† >3 week-old regrowths.
‡ Calculated from 3–4 animals each grazing 3–6 pasture replicates.

time under grazing and with increasing stocking rate. The diets selected by cows when first introduced into the pastures contained more than 90 per cent leaf, and by the last days of grazing were 55–60 per cent leaf. Thus leaf yield declined rapidly, and leaf in the uppermost layer of the sward was the first to be removed (Fig. 5.9).

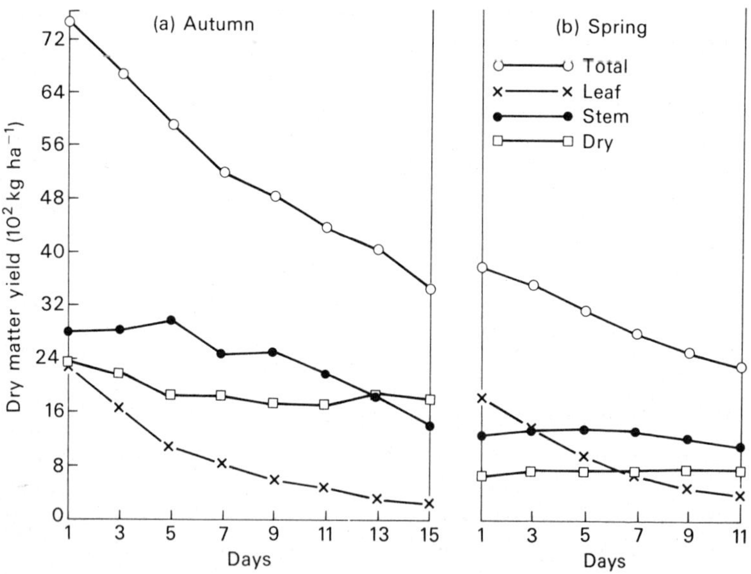

Fig. 5.9. Mean dry-matter yield of leaf, stem, and dead material of Kazungula setaria pastures in (a) autumn and (b) spring at different stages of defoliation (Chacon 1976).

With increasing defoliation with time, bite size, intake, and rate of intake declined (Fig. 5.10). In an apparent attempt to compensate, grazing time, total number of eating bites and rate of biting increased to maximum values after 6-9 days of grazing, but with further defoliation then declined (Fig. 5.10). In another experiment with increasing grazing pressure (lenient, medium, and heavy grazing) Chacon (1976) again found that under heavy grazing and a high level of defoliation, grazing time was reduced, and bite size markedly lower (lenient, 0.165g, heavy, 0.079g, organic matter bite^{-1}).

The major animal factor influencing intake was bite size, which Chacon (1976) found was determined largely by leaf yield, leaf to stem ratio, and bulk density of the swards. On heavily grazed pastures, grazing time was reduced and animals appeared reluctant to spend time selecting out the small quantity of leaf from stem and excreta. This was not caused by increased retention of bulk in the rumen as manual removal of rumen contents did not increase grazing time. In all three experiments maximum grazing times were found at about 1000 kg dry green leaf ha^{-1} , and at 600 kg dry green leaf ha^{-1} grazing time declined markedly. Thus grazing time, bite size, and intake at any point in time were not related to grazing pressure (leaf yield cow^{-1}) but to actual leaf yields per unit area.

Stobbs (1975b) has shown that leaf yield and density of grass swards is increased by nitrogen application (Table 5.13). Bite size increased linearly with increasing nitrogen applications and averaged 0.22, 0.26, and 0.34 g organic matter bite^{-1} at 0, 40, and 60 kg N ha^{-1} , respectively.

From studies of sward structure effects on grazing, Stobbs (1975a) suggests that animal production can be improved by selecting pasture plants for high leaf yield which is easily harvested; by fertilizing pastures to produce dense leafy swards; and by grazing management practices which maintain swards as near as possible to optimum for easy defoliation.

3. *Grazing management systems*

Recommendations for different methods of grazing management of pasture have led to a great deal of controversy and often confusion. Many comparisons of different grazing systems are confounded by effects of stocking rate *per se* and by other treatments imposed differentially, leading to inconclusive or erroneous results. For a given area of pasture, the linear model of Jones and Sandland (1974) demonstrates clearly the major effect of stocking rate in determining the production

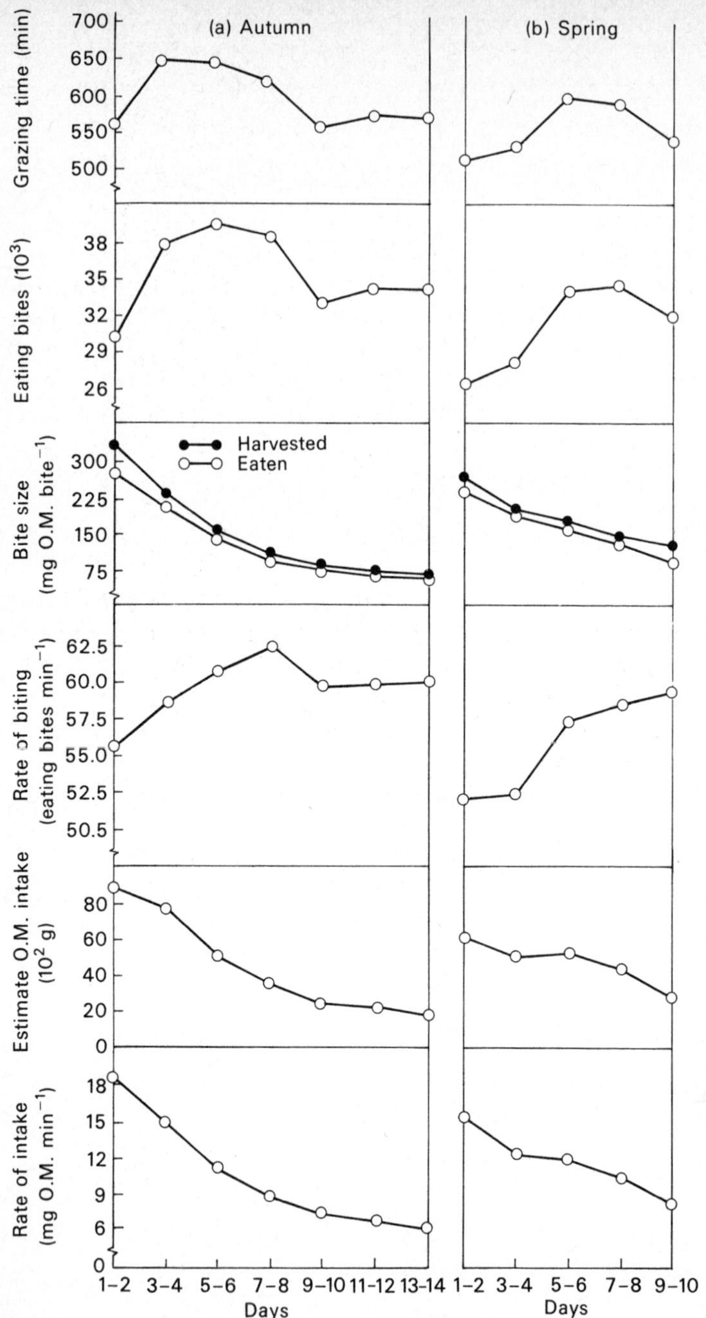

Fig. 5.10. Variation in grazing time, eating bites, bite size, rate of biting, esti-mated intake, and rate of intake of cattle grazing Kazungula setaria pastures in (a) autumn and (b) spring at different stages of defoliation (Chacon 1976).

Table 5.13. Effect of nitrogen upon mean leaf yield and mean bulk density of *Setaria anceps* cv. Kazungula (from Stobbs 1975*b*)

| | No nitrogen | | | | 40 kg nitrogen per ha | | | | 60 kg nitrogen per ha | | | | LSD | |
Vertical layer (cm)	Leaf yield (kg ha⁻¹)	As % total leaf	Leaf (% of sward)	Bulk density (kg ha⁻¹ cm)	Leaf yield (kg ha⁻¹)	As % total leaf	Leaf (% of sward)	Bulk density (kg ha⁻¹ cm)	Leaf yield (kg ha⁻¹)	As % total leaf	Leaf (% of sward)	Bulk density (kg ha⁻¹ cm)	$P = 0.05$ Leaf yield	Bulk density
>60	–	–	–	–	40	1	55	3	154	4	74	9		
45–60	10	1	47	1	108	4	67	7	359	9	73	24		
30–45	41	3	59	3	477	17	65	32	975	24	70	65		
15–30	244	16	61	16	830	30	58	55	1147	28	53	76		
GL–15	1217	80	61	81	1314	48	61	88	1399	35	52	93		
Total	1512	100	61	25	2769	100	61	37	4034	100	60	52	984	12

per head and per hectare of that pasture. Unless a particular system of grazing management, compared with some other system leads to either improved pasture growth and yield, or increased nutritive value of the pasture, or to increased efficiency of utilization of the pasture yield, then the system of management will not affect the relationship between stocking rate and animal production.

The main comparisons of stocking method are between continuous grazing and some form of rotational grazing. It is necessary to define these terms clearly. *Continuous grazing* is defined as that type of management whereby grazing animals are confined within a single enclosed pasture area for the entire grazing season, which may be a full year (Booysen 1975). Within this system the pasture may be *set stocked*, i.e. the number of animals kept constant or *variably stocked*, i.e. numbers varied according to pasture growth, but some animals are present on the pasture at all times. This is also termed *put-and-take* grazing.

Rotational grazing requires that the pasture is subdivided into a number of enclosures with at least one more enclosure than groups of animals (Booysen 1975). Again a rotational grazing system may be *set stocked* or *variably stocked*. Within a rotational grazing system there is another variable, the time interval allowed for grazing each unit before moving to the next. Time allowed in each unit may be short or long, relative to the stocking rate and pasture production. Booysen (1975) defined two contrasting management systems – (a) high utilization grazing (HUG), where animals are held in a pasture for a relatively long time in an attempt to graze the grazeable material as completely as possible before moving the animals to the next enclosure; and (b) high production grazing (HPG) where animals are grazed in an enclosure for a relatively short time and the pasture is leniently grazed before moving the animals to the next enclosure. Of course each enclosure will be grazed more often in a year than in the HUG system.

Rotational grazing systems may operate on a *set rotation* where each unit is successively grazed in sequence, or on a *deferred rotational grazing* system. The latter system recognizes that there are critical periods in the phenology of the desirable plants in the pasture, viz. seed germination and seedling establishment, flowering and seed set, and the period of accumulation of root and crown reserves prior to the winter or dry season. Deferred rotational grazing provides for each enclosure to be rested from grazing during at least one of the critical periods, or even for longer periods, during the annual cycle of grazing (Anderson 1967).

Other modifications of rotational grazing systems are *strip grazing* and *'leader–follower' grazing*. Strip grazing is widely used with dairy herds grazing forage crops or lucerne (*M. sativa*) to ration the amount of forage available to the herd. Electric fences are commonly employed to enclose the allowed area of forage, while another fence may be placed behind the herd to fence-off the previously grazed strip. In this way grazing is carefully controlled and rationed, which may be important when feed is scarce, or to prevent bloat.

Creep grazing is a term given to a rotational grazing system whereby the highest producing animals (e.g. milking cows) are allowed the first grazing in a paddock. This allows for maximum selection of highest quality forage. Once opportunity for selection has declined then less demanding classes of livestock, such as the dry dairy cows or beef steers are moved in to graze the 'aftermath' while the milking cows are moved to fresh grazing. This system of grazing aims to ensure that the pasture is grazed down to provide for more uniform regrowth.

Comparisons of grazing systems

A great deal of controversy and argument is generated on this topic by proponents of particular grazing systems. There is no doubt that greater capital and management inputs are required for fencing, watering facilities, and stock movement in rotational grazing systems compared with continuous grazing systems. However 'more management' is not necessarily 'better management', unless rotational grazing leads to higher animal production.

Comparisons of grazing systems are often confused by the use of variable or 'put-and-take' grazing methods. In this, animals are added to or subtracted from a grazing unit, in order to match the growth rate of the pasture. Many scientists, for example Willoughby (1970), would argue that this is inappropriate under real farming conditions. Property size is fixed, and, in general, farmers are unable to change rapidly the number of animals on the farm. In periods of feed deficit, when many farmers are attempting to reduce stock numbers, stock prices are likely to be low, thus discouraging sales. Therefore, on many farming units, stock numbers tend to be relatively fixed, although breeding times are usually adjusted to allow for turnoff of grown animals at the end of the main pasture growth period.

After reviewing a number of stocking rate and grazing method experiments, Wheeler (1962) concluded that stocking rate was the dominant factor affecting production per hectare. This is now clearly demonstrated by the linear response model of Jones and Sandland (1974). In tropical pasture swards the ability of animals to select forage of higher nutritive value is particularly important, so that even modest increases in stocking rate may cause a marked decline in production per head.

The most important difference between continuous grazing and rotational grazing systems is that, except at very low stocking rates, animals on a rotational system will, after a certain period of time on a new pasture, be more restricted in their potential for selecting higher quality forage, as shown in Fig. 5.10 (Chacon 1976). Thus in rotational grazing, animals incur a penalty by being forced to consume forage of lower quality. This penalty will be larger in periods of feed deficit or low pasture quality as shown by Walker and Scott (1968) in Tanzania. In this trial, weight gains of steers under continuous grazing were similar to, or higher than, rotationally grazed steers during the growing season on natural pastures, but during the dry season weight losses were significantly higher in the rotation grazing group. During the dry season the opportunity for animals to select forage over the whole available pasture area is particularly important in maintaining adequate intake.

't Mannetje, Jones, and Stobbs (1976) have reviewed and summarized the data from 16 grazing trials in the tropics on natural pastures, sown mixed pastures, and nitrogen-fertilized grass pastures. Out of 12 trials where liveweight gain data allow comparison of continuous and rotational grazing systems, in eight trials continuous grazing was superior, in two trials rotational grazing was superior, while in another two trials liveweight gains were equal. Differences were often not large, but if there is an advantage in grazing method it would seem to lie with a continuous grazing system. 't Mannetje *et al.* (1976) suggest that currently available evidence from the tropics does not indicate that any grazing management system will give higher animal production than continuous grazing with set stocking. This is also the easiest and least expensive method to implement.

In advocating continuous set-stocked grazing systems, this does not mean that pasture areas will not be subdivided or that a high degree of pasture management will not be required. Pasture areas will be subdivided and animals moved to allow for segregation of different classes of animals (e.g. weaners, maiden heifers, bulls, will be separated from the main breeding herd); for destocking areas for tick control; for pasture burning or fodder conservation; to allow for weed-control measures, or fertilizing or irrigating and other management purposes. But there is little point in instituting rotational grazing in tropical pastures simply for the sake of having a rotational system unless there are particular management requirements. In fact high utilization rotation grazing, as defined by Booysen (1975), has been shown to seriously reduce vigour and to cause high mortality in the tropical twining legumes (Jones 1974*b*; Whiteman 1969). However, where defoliation levels can be closely controlled as in the case of strip grazing with dairy cows, Edgley and Quinlan (1973) have maintained good legume–grass balance with *Desmodium intortum–Setaria anceps* and *Glycine wightii–Panicum maximum* pastures with 3–6 weeks grazing cycles, grazed in strips at about 100 cows ha^{-1} at each grazing, at overall stocking rates around 2 cows ha^{-1}.

Special management requirements may dictate the use of rotational grazing systems. Lucerne (*Medicago sativa*) will not persist under continuous grazing (Peart 1968), and the best management system for this species is a rapid intensive grazing followed by an adequte period for regrowth. The shrub legume *Leucaena leucocephala* is more productive under a simple two-paddock rotation grazing than under continuous grazing ('t Mannetje *et al.* 1976). Rotational grazing may be appropriate with irrigated pastures to avoid problems of pugging damage. In a grazing trial with irrigated pangola grass (*Digitaria decumbens*) Evans and Pulsford (1970) lightly harrowed to spread dung patches, applied nitrogen, and irrigated each unit at the end of the grazing cycle, and obtained extremely high levels of animal production.

In rangeland management, systems of deferred rotational grazing have been widely recommended to improve botanical composition and ground cover of pastures degraded by previous overgrazing (Anderson 1967). Even though range condition may be improved, animal production may be lower with these rotational systems than with continuous grazing as found by Owensby, Smith, and Anderson (1973) and Prajapati (1970).

In temperate pastures, where pasture quality and the requirement for maximum opportunity for selection may be less of a limitation to animal production, rotational grazing may give higher levels of animal production. This question has been examined in some detail in New Zealand and advantages of rotational grazing were evident only at high stocking rates. McMeekan and Walshe (1963) found no difference in per hectare butterfat production at 1.85 cows ha^{-1}, but at 2.30 cows ha^{-1} there was an 8 per cent advantage for rotational grazing, and at 2.90 cows ha^{-1} a 19 per cent advantage. However, they point out that the main advantage of the rotational system was attributable to the greater amount of fodder conserved in the rotational system.

The stocking rate effect is demonstrated by the data of Kissock (1966) who studied ewe and lamb production on hill-country pastures in New Zealand. The stocking rate had to be reasonably high before the rotational grazing showed to advantage. In this trial the higher production resulted from an increased number of lambs per hectare (due to a 7–15 per cent higher lambing percentage) rather than from a faster lamb growth-rate (Smetham 1973) (Table 5.14).

Table 5.14. Effect of stocking rate and grazing method on lamb production from hill pastures in New Zealand (Kissock 1966)

Stocking rate (ewes ha^{-1})	Lamb meat production (kg ha^{-1})	
	Rotational	Set-stocked
12.4	287	314
14.8	393	400
16.1	444	364
17.3	430	333
18.5	442	341

Increased production from rotational grazing is not due to higher levels of pasture dry-matter production (Smetham 1973). In fact in a number of trials dry matter yields are lower (Campbell 1969; Pigden and Greenshields 1960). It has been suggested (Smetham 1973) that advantages of rotational grazing result from higher quality of herbage on offer, owing to more complete utilization at each grazing, and, therefore, less dilution of the subsequent regrowth with residual decayed dead matter. Campbell (1966) showed that set-stocked pastures had a higher content of dead matter than the rotation grazed pastures. The data of Pigden and Greenshield (1960) suggested that yield of total digestible nutrients was 11–22 per cent higher in rotation grazed swards, even though total dry-matter yield was lower. Thus, higher quality forage and better utilization of pasture appears to give some advantage at high stocking rates in temperate pasture regions where there are no major limitations due to long dry seasons with critical forage deficits.

These conditions apply to only a relatively small proportion of the world's grazing lands. Under most tropical grazing conditions continuous grazing at relatively set-stocking rates will provide the most appropriate grazing management. The critical determinant of animal production is the stocking rate and method of grazing has only a minor influence.

The most important role of grazing management must be to determine the optimum stocking rate for sustained production, while the aim of pasture management is to increase the optimum stocking rate through better pasture management, introduction of improved and better adapted pasture species, use of fertilizers, weed control, and efficient utilization of the forage produced.

5.3 BEEF AND MILK PRODUCTION FROM TROPICAL PASTURES

1. *Legume-based pastures*

Introduction

Legume-based pastures have a long history in temperate agriculture, beginning with the widespread adoption of 'four-course rotations' in Britain and Europe in the 1700s, where the value of clover-based ley pastures in the cropping cycle was recognized. Legume–grass temperate pastures have been widely used in many countries, and have found particular application in the highly efficient pastoral industries of New Zealand. The introduction of legume-based pastures in the early 1900s to the generally infertile soils of Southern Australia has had a marked impact on agricultural production. The combination of superphosphate with subclover (*Trifolium subterraneum*) not only markedly increased animal production, but also greatly increased wheat yields in the following cropping phase.

The development of legume-based tropical pastures has been much more recent, with the major developments probably occurring since 1950, following the introduction, selection, and breeding of new legumes adapted to tropical conditions. Already these developments in tropical pasture science have had a major impact on animal production in the tropics, and again this has been particularly evident in the low fertility soils of Northern Australia.

In this section a range of actual production data on tropical legume-based pastures will be reviewed.

Beef and milk production

Direct comparisons of actual animal production data from grazing trials must be approached cautiously as performance will be affected by many factors, the main ones being: (a) Pasture factors — rainfall, soil fertility, fertilizer usage, species and botanical composition, climatic conditions (heat waves, drought, frosts), age of pasture. (b) Animal factors — breed, age, sex of animals, health, parasite control, adaptation to climatic conditions, and supplementation with minerals or licks.

A major factor affecting actual production figures will be the stocking rates chosen, modified further by the grazing methods (whether set stocked — continuous or rotational, or variable stocked — continuous or rotational). Ideally different pastures and environments should be compared at the optimum stocking rates for production per ha, which could be defined from the equations of Jones and Sandland (1974). Allowing for all these complications it is possible to make fairly realistic comparisons between pastures and environments for beef production, but extremely difficult in the case of dairy production. Comparisons of dairy production are further complicated by having to take into account the stage of lactation and breed differences, and by the fact that dairy cows can milk 'off their backs' in times of nutritional stress (Stobbs 1973*b*).

The lactating dairy cow is far more sensitive to nutritional stress than a beef animal, and levels of milk production change rapidly with level of nutrition (Stobbs and Sandland 1972). Also nutrient requirements of the lactating cow are very much higher than for the beef animal (Stobbs 1974*b*). These effects are very evident on tropical pastures where the lower values of DMD and VI compared with temperate pastures, markedly limit milk production per cow, but still allow reasonable levels of liveweight gain per head for beef production.

(a) Beef production

This section presents actual recorded levels of beef production from legume-based tropical pastures, from a wide range of environments (Table 5.15). Obviously levels of production per hectare will depend on the environmental potential for pasture growth, and rainfall is a major factor affecting this. Therefore the data in Table 5.15 are divided into three categories, depending on mean annual rainfall above 600 mm. There has been little success with pasture improvement based on legume introduction in the tropics in areas receiving less than 600 mm annual rainfall.

In the low rainfall environments (Table 5.15(a)) the major pasture improvement has been based on *Stylosanthes humilis* (TS) oversown into native pastures. Under these conditions of extensive grazing production per head is more important than production per hectare. In all cases inclusion of TS has increased liveweight gain (LWG) head^{-1} day^{-1}, and has also led to increased stocking rate, and hence to marked increases in LWG ha^{-1}. In the wet season, TS pastures have maintained high stocking rates and given high LWG ha^{-1} (Norman and Phillips 1970). The advantage of TS-based pastures is clearly demonstrated in the dry season. On native pastures without TS animals lose weight, while with TS animals gain weight (Norman 1970). The rate of gain was related to the amount of TS on offer (Norman 1970).

At Katherine the average liveweight gain (LWG) of steers grazing TS in the dry season is 68 kg head^{-1} at a stocking rate of 2.5 head ha^{-1}, whereas on native pastures animals lose about 20 per cent of their liveweight even at extremely low stocking rates. At both Katherine and Rodd's Bay steers on TS pastures reach slaughter weight at 1½ to 2 years earlier than steers on native pastures (Shaw and Norman 1970).

Oversowing of TS into native pastures, with addition of superphosphate, has led to striking increases in animal productivity. Through increased production per head, higher stocking rates, and more rapid rates of turnoff, not only is beef production increased, but also conception and calving rates are increased on fertilized native pastures oversown with TS (Edye, Ritson, Haydock, and Davies 1971).

In the medium rainfall environments (Table 5.15(b)), LWG head^{-1} exceeded 0.40 kg day^{-1} only with the crossbred heifers in the Uganda trial (Stobbs 1969d). Over the range of pastures and stocking rates most values fall between 0.25 and 0.35 kg head^{-1} day^{-1}. Stocking rates giving maximum LWG ha^{-1} are similar over all trials, approximately 2.5 animal units per hectare (where 1 a.u. = 400 kg steer) taking into account that the East African Zebus used in Stobbs and Oyenuga and Olubaja's experiments are small, and approximately equivalent to 0.5 a.u. The higher animal production recorded in the Uganda trials may be related to the long growing-season there, whereas pasture production at Samford is limited by a cool, dry winter period, and at Ibadan and Darwin by a marked dry season. The data from Stobbs (1969d) demonstrates that high levels of animal production can be achieved on legume-based pastures, particularly with animals with improved genetic potential.

The data from higher rainfall environments is generally similar to that of medium rainfall in the Beerwah, South Johnstone, and Fiji experiments, but much higher in Hawaii. The data of Partridge and Ranacou (1974) demonstrates the value of including a legume in the pasture system and it is likely that higher levels of LWG ha^{-1} would have been recorded if a range of stocking rates had been used.

The potential of legume-based pastures in high rainfall environments is demonstrated by the Hawaii data (Younge and Plucknett 1965).

Table 5.15. Summary of beef animal performance on legume-based tropical pastures

Location	Mean annual rainfall (mm)	Pastures and treatment	Stocking rate (animals ha^{-1})	Liveweight gain kg head^{-1} day^{-1}	kg ha^{-1} a^{-1}	Comments	References
(a) Low rainfall (600–950 mm)							
Kangaroo Hills North Queensland	630	Native pasture (*Heteropogon, Both-riochloa, Chyrosopogon*) + *Stylo-santhes humilis*				Mean values over 3 years Clearing = removal of open woodland trees,	Gillard (1970)
		i. Cleared – Fertilizer	0.22	0.45	37	*Eucalyptus*	
			0.41	0.36	54	Fertilizer = basal at	
		+ Fertilizer	0.22	0.46	38	36 kg P ha^{-1} + Mo	
			0.41	0.35	52	Maintenance – 12 kg P ha^{-1} a^{-1}	
		ii. Uncleared – Fertilizer	0.22	0.19	16	*S. humilis* content about	
			0.41	0.15	22	500 kg ha^{-1}, seldom	
		+ Fertilizer	0.22	0.25	20	exceeded 10% of total	
			0.41	0.15	22	DM	
Rodd's Bay Central Coastal Queensland	780	Native pasture (*Hetropogon, Dicanthium*) alone	0.27	0.11	11	Mean values over 2 years	Shaw (1961)
		Nature pasture + *S. humilis*	0.47	0.30	51	Hereford steers, 1½–2½	
			0.61	0.38	85	years, set stocked, continuous grazing	
Rodd's Bay Central Coastal Queensland	780	Native pasture (Low SR)	0.27	0.25	27	Means over 7 years trial	Shaw and 'Mannetje (1970)
		Native pasture (High SR)	0.62	0.14	32	Hereford steers, 1½ to	
		Native pasture + fertilizer (high SR)	0.62	0.30	68	2½ years old.	
		NP + TS no fertilizer	0.74	0.37	103	Fertilizer – basal 82 kg P	
		NP + TS + fertilizer	0.94	0.46	163	63 kg K	

Maintenance — 12 kg P,
13 kg K ha^{-1} a^{-1}
TS = *Stylosanthes humilis*

Katherine Northern Territory Australia	925	(a) Wet season (135 days) *S. humilis* (with some native grasses	1.6 2.5 3.3	0.88 0.87 0.87	196 291 384	Grazing Jan. to May av. 135 days. At SR 3.3 = 75% (1630 kg), 2.5 = 6.7% (1920 kg), 1.65 = 58% (1460 kg), TS in pasture	Norman and Phillips (1970)
		(b) Dry season (i) Native pasture alone Native pasture + TS	0.17 3.00	−0.29 +0.44	−5.5 +150	Means of 3 dry season grazings. Annual fertilizer — 5–10 kg P Grazing period = 112 days	Norman (1970)
		(ii) Native pasture with 1000 kg ha^{-1} TS Native pasture with 4000 kg ha^{-1} TS Native pasture with 7000 kg ha^{-1} TS	1.64 1.64 1.64	0.19 0.29 0.44	59 92 136	Means of 2 dry season grazings. Grazing period = 190 days. TS yield measured at start of grazing in April	Norman (1970)

(b) Medium rainfall (950–1500 mm)

| Samford South-East Queensland | 1000 | *Setaria anceps*- *Macroptilium atropurpureum* | 1.1 1.7 2.4 2.9 | 0.44 0.37 0.33 0.21 | 180 235 280 225 | Mean values over 3 years Hereford steers, set stocked continuous grazing Basal and maintenance applications of super-phosphate, and micro-elements | Jones (1974) |

(continued)

Table 5.15. (Cont'd)

Location	Mean annual rainfall (mm)	Pastures and treatment	Stocking rate (animals ha^{-1})	Liveweight gain kg head^{-1} day^{-1}	kg ha^{-1} a^{-1}	Comments	References
Ibadan Nigeria	1220	1. *Cynodon plechtostachyus* (66%), *Centrosema pubescens* (20%)	3.2 4.2	0.19 0.19	228 298	Stocking rate raised to 4.2 ha^{-1} in year when annual rainfall was 1780 mm	Oyanuga and Olubajo (1966)
		2. *C. plectostachyus* (10%) *C. pubescens* (14%) *S. guyanensis* (55%)	3.2 4.2	0.19 0.22	227 339	Rotational grazing, 1 week on, 3 weeks off	
		3. *D. decumbens* (16%) *C. pubescens* (39%) *S. guyanensis* (20%)	3.2 4.2	0.19 0.12	222 190	Zebu steers, 12–18 months, Initial LW = 140 kg	
Serere Uganda	1400	1. *Hyparrhenia rufa* + *S. guyanensis*	1.6 2.5 5.0	0.28 0.27 0.23	165 246 422	Means over 3 years trial East African Zebu steers Set-stocked, continuous grazing	Stobbs (1969*a*)
		2. (i) *H. rufa* + *S. guyanensis* (ii) *H. rufa* + *C. pubescens*	4.8 4.8	0.30 0.29	529 500	Means over 14 month trial East African Zebu heifers Rotational grazing, put-and-take	Stobbs (1969*b*)
		3. (i) *P. maximum, C. gayana, H. rufa* (grass)	2.64	0.24	227	Mean values over 3 years trial. East African Zebu steers	Stobbs (1969*c*)
		(ii) Grass + *C. pubescens* + *S. guyanensis*	3.25	0.29	345	Put-and-take grazing system	
		(iii) Grass + legume + fertilizer (PSK)	3.75	0.27	369		

	1400	4. (i) *P. maximum* + *D. intortum* (Zebu heifers)	4.18	0.39	596	Mean values over 18 month trial. East African	Stobbs (1969*d*)
		(ii) *P. maximum* + *S. guyanensis* (Zebu heifers)	4.24	0.35	542	Zebu heifers and X-bred heifers crossed with either Hereford or Aberdeen Angus bulls	
		(iii) *P. maximum* + *D. intortum* (X-bred heifers)	4.21	0.49	753	Rotational grazing, put-and-take	
		(iv) *P. maximum* + *S. guyanensis* (X-bred heifers)	4.28	0.53	829	All treatments received 250 kg ha^{-1} superphosphate	
Darwin Northern Territory Australia	1550	*Urochloa mosambicensis* + *Stylosanthes humilis*	0.6 / 1.25 / 2.50	0.34 / 0.34 / 0.40	74 / 158 / 368	Second year trial. Steers, continuous grazing set-stocked. Seasonal rainfall, 90% of rainfall in Oct.–Mar.	Austin (1970)

(c) High rainfall (1500–)

Beerwah South-East Queensland	1650	*C. gayana, D. decumbens, P. dilatatum, P. commersonii, D. intortum, D. uncinatum, L. bainesii T. repens*				Mean values over 6 years trial set-stocked, continuous grazing Hereford steers. Maintenance fertilizer rates	Evans and Bryan (1973)
		(i) 1P 1K	1.23	0.46	208	1P 1K: 125 kg ha^{-1} super-phosphate + 125 kg ha^{-1} KCl	
			1.65	0.45	271		
			2.47	0.30	271	2P 1K: 250 kg ha^{-1} super-phosphate + 125 kg ha^{-1} KCl	
		(ii) 2P 1K	1.23	0.64	289		
			1.65	0.54	325		
			2.47	0.36	327		
		(iii) 2P ½K	1.23	0.62	278	2P ½K: 250 kg ha^{-1} superphosphate + 63 kg KCl	Evans and Bryan (1973)
			1.65	0.55	331		
			2.47	0.39	356		

(continued)

Table 5.15. (*Cont'd*)

Location	Mean annual rainfall (mm)	Pastures and treatment	Stocking rate (animals ha^{-1})	Liveweight gain kg head^{-1} day^{-1}	kg ha^{-1} a^{-1}	Comments	References
		(iv) 2P 1½K	1.23	0.62	278	2P 1½K: 250 kg ha^{-1} superphosphate + 188 kg ha^{-1} KCl	
			1.65	0.42	254		
			2.47	0.39	351	Mean LWG ha^{-1} over fertilizer rates: 1.23 beasts ha^{-1} : 263 kg ha^{-1} 1.65 beasts ha^{-1} : 295 kg ha^{-1} 2.47 beasts ha^{-1} : 326 kg ha^{-1}	
South Johnstone North Queensland	3000	(i) *P. maximum* alone	4.2	0.22	337	Mean values over 2 year trial	Grof and Harding (1970)
		(ii) *P. maximum* + *C. pubescens*	4.2	0.30	460	Rotational grazing	
Fiji	2030	(i) *Dicanthium caricosum* alone	1.5	0.215	110	Mean values over 3 years Fresian and Zebu steers	Partridge and Ranacou (1974)
		(ii) *D. caricosum* + 10% of area with *Luecuena leucocephala*	1.5	0.300	170	Continuous set stocked with access to *Leucaena*	

		(iii) *D. caricosum* + 20% of area with *L. leucocephala*	1.5	0.500	270	1–2 days per week for 10%, and 3–4 days for 20%. Basal fertilizer 250 kg superphosphate 125 kg K_2SO4 Maint. 225 kg superphosphate	
Hawaii	2300	*Digitaria decumbens + Desmodium intortum*				Put-and-take, rotational grazing. Yearling Hereford steers	Younge and Plucknett (1965)
		Superphosphate 250 kg ha^{-1}	4.45	0.43	711		
		500 kg ha^{-1}	–	–	997		
		Superphosphate 1000 kg ha^{-1}	–	–	1221	Basal fertilizer – lime 5600 kg, KCl 224 kg, Zn 56 kg, B 11 kg, and Mo 2.8 kg ha^{-1}	Younge and Plucknett (1966)
		1500 kg ha^{-1}	5.88	0.61	1304		

Current trials in the British Solomon Island in 1000 mm rainfall region, suggest that production in excess of 800 kg ha^{-1} a^{-1} LWG may be achieved (Watson and Whiteman, unpublished data). Similar levels of production have now been achieved at South Johnstone, Queensland (Tietzel, personal communication).

In a recent review of animal production on tropical pastures, Stobbs (1974*a*) also finds that average LWG head^{-1} day^{-1} is about 0.35 kg, and rarely exceeds 0.60 kg head^{-1} day^{-1}. Since the same animals have a potential gain of 1.2–1.4 kg head^{-1} day^{-1} when fed grain rations, Stobbs (1974*a*) suggests that the nutritional value of tropical pastures is limiting animal production. While this is true, the data in Table 5.15 clearly demonstrate the improvement in animal production by the inclusion of a legume in the pasture system, and where high pasture growth rates are attained, as in Hawaii and Uganda, high levels of per hectare production are achieved.

(b) Milk production

As previously discussed, the limitations of tropical pastures for milk production are particularly evident, because of the higher nutritional requirements of the lactating dairy cow (Stobbs 1974*b*). These factors have been reviewed recently in an article entitled 'Limitations to dairy production in the tropics', in *Tropical Grasslands*, Vol. 5 (3), pp. 145–302. In this review Stobbs (1971*a*) concludes from records of 11 trials in Australia that longer term average production per cow on tropical pastures seldom exceeds 8–9 kg cow^{-1} day^{-1}, although there are some differences in quality between pasture species. From records of ten trials in other tropical countries, production ranged from 6.5 to 15 kg cow^{-1} day^{-1}, with the higher figure for Fresians. Colman (1971) suggests that the quantity of pasture available is not a problem, though seasonality of pasture growth leads to major problems and periods of critical nutritional stress. Apart from quality limitations, climatic stresses (temperature, humidity, and rainfall) can markedly affect dairy production in tropical regions (Payne, Laing, and Raivoka 1951).

By comparison, milk yields per cow in temperate areas are considerably higher, although it is difficult to find data for annual yields from pasture alone without supplementary or winter stall feeding. As an indication average annual milk yield per dairy cow in England (1973–4) was 4106 kg, and for herds in the Recording Scheme, 4645 kg head^{-1} a^{-1} at 3.8 per cent butterfat (Milk Marketing Board 1974). By comparison, in Queensland, Australia, in which more dairying is carried out than in other comparable tropical or subtropical climates in the world, with 500 000 head of dairy cattle (1970), average milk production per cow is only 1500 kg (Hayman and Radcliffe 1973; Ferguson 1973). Both sets of data include supplementary feeding. A general indication of milk yields per cow from temperate pastures alone is given by the following data from Greenhalgh (1970) who examined the effects of amount of dry matter available in a rye grass pasture on milk production.

Amount of dry matter allowed:	11.4	15.9	15.9	kg DM cow^{-1} day^{-1}
% Utilization of pasture:	92	75	58	%
Mean intake of organic matter:	9.9	11.9	10.8	kg cow^{-1} day^{-1}
Dry matter digestibility:	77.9	78.1	76.4	%
Milk production per cow:	16.2	16.6	16.8	kg cow^{-1} day^{-1}
Milk production per ha:	16 250	14 950	13 260	kg ha^{-1}

Hodgson, Combellas, Wade, and Booth (1975) stated that milk yields of 20–25 kg cow^{-1} day^{-1} were sustained satisfactorily on rye grass alone over the first 6–8 weeks of the grazing season. Also Tayler and Aston (1975) produced lactation yields of 4500 kg cow^{-1} on all grass diets based on grass silage plus dried pellets.

Thus temperate pastures alone can support milk yields up to 25 kg cow^{-1} day^{-1}, and between 15–20 kg in the long term, but the following data demonstrate the limitations of tropical pastures for milk production.

(i) Tropical legumes   Stobbs (1971) assessed the feeding value of Siratro (*M. atropurpureum*) and *Desmodium intortum* in a three period switchback design. In the preliminary period Jersey cows grazed pangola (*Digitaria decumbens*) plus concentrate, during the experimental period, Siratro, Desmodium, Siratro, and a post-grazing period on Pangola. Although legumes had higher nitrogen, phosphorus, calcium, and magnesium than the grass, milk yields declined on the legumes to an average 7.7 kg cow^{-1} day^{-1}. Also liveweight declined on the legumes, BF% increased while solids non-fat (SNF%) declined. Intake of digestible energy was the major limitation. Hamilton, Lambourne, Roe, and Minson (1970) found that the nutritive value of *Lablab purpureus* (formerly *Dolichos lablab*) was higher than that of Siratro and Desmodium. Milk yields on *Lablab* were 12 kg cow^{-1} day^{-1} and declined to 11 kg cow^{-1} day^{-1} over eight weeks.

Also quality of milk may be affected by grazing tropical legumes. In cows grazing *Lablab purpureus* and *Leucaena leucocephala*, milk taints were accentuated by homogenization, but only with *Leucaena* did the taint persist after pasteurization (Stobbs and Fraser 1971). Milk from cows grazing legumes in pure stands was higher in BF%, lower in SNF%, and milk from Siratro produced a very poor quality cheese. Milk quality for cheese was improved by feeding an energy supplement of grain. These data indicate that tropical legumes fed alone may markedly reduce both quantity and quality of milk.

(ii) Tropical legume–grass pastures   Dale and Holder (1968) provide a direct comparison of milk yield potential of tropical pastures by comparing identical twins on transfer from Lucerne (*Medicago sativa*) plus concentrate to a kikuyu grass (*Pennisetum clandestinum*)-*Glycine wightii* pasture. Lucerne plus concentrate gave 20 kg cow^{-1} day^{-1}, kikuyu-glycine — 11 kg cow^{-1} day^{-1}.

Cowan (1975) studied grazing time and milk yield of Fresian cows grazing *Panicum maximum-Glycine wightii* pastures at four stocking rates (1.3, 1.6, 1.9, and 2.5 cows ha^{-1}) in North Queensland. Cows grazed approximately 600 minutes per day, with more time spent grazing at night in summer. These grazing times are similar to those reported by Stobbs (1974*b*). Average milk yields were 8.0 ± 0.5 kg cow^{-1} day^{-1}. With a supplement of 4.5 kg crushed maize per day, milk yields increased to 11.0 ± 0.5 kg cow^{-1} day^{-1}, demonstrating the limitation of intake of digestible energy in pasture alone.

In the same environment Cowan, O'Grady, Moss, and Byford (1974) compared milk yields of 108 Jersey and 129 Fresian cows over seven years grazing *P. maximum-G. wightii* pastures. No supplement was fed.

	Jersey	Fresian
kg milk day^{-1}	9.06	12.54
Fat yield (kg cow^{-1} day^{-1})	0.42	0.42
Lactation (days)	272	331
Liveweight gain (kg a^{-1})	7	43

Even though pasture quality limits milk production, differences in breed performance are still expressed. Over a range of tropical pasture conditions Colman and Holder (1968) suggest that milk yields decline in later lactation to 4.5-7.0 kg cow^{-1} day^{-1} and are maintained until drying off.

(iii) Tropical grasses   In comparisons of tropical grasses, nitrogen is usually applied; this does not improve pasture quality directly, except where CP% is below 7 per cent, but by producing more dry matter may allow better opportunity for selection or increased stocking rate.

However, this was not found by Santhirasegaram, Diez, Peterson, and Trigueros (1970):

Treatment	kg milk cow^{-1} day^{-1}	kg milk ha^{-1}
1. *Hyparrhenia rufa* No nitrogen	5.53	2476
2. *H. rufa* + 50 kg nitrogen after each grazing	5.32	2069
3. *Digitaria decumbens* + 50 nitrogen	4.76	2136
4. *Brachiaria decumbens* + 50 nitrogen	5.08	1750

These data demonstrate differences in nutritive value of grasses as has been shown by Stobbs (1973*c*). Cows grazing three-week regrowths of *Setaria anceps* cv. Splendida and *D. decumbens* produced more milk than cows grazing kikuyu (*P. clandestinum*).

Stage of regrowth of pastures can affect milk yield. Hamilton *et al.* (1970) found that milk yields of three-weeks regrowth of setaria gave 7.5 kg cow^{-1} day^{-1} while five-weeks regrowth gave 6.3 kg cow^{-1} day^{-1}, although there was no difference between three- and five-week regrowths of Rhodes grass (*C. gayana*). In a second trial with Rhodes grass (with a DMD% of 60.3 per cent) and Setaria (DMD, 60 per cent) milk yield declined from 9.5 kg to 5.0 kg cow^{-1} day^{-1} over eight weeks.

Although dry matter production per hectare may not be a major limitation in tropical grass pastures fertilized with nitrogen, increasing stocking rate reduces the feed on offer and amount available for selection which is particularly important in tropical swards. Colman and Kaiser (1974) studied the effects of stocking rate on milk production of kikuyu pastures fertilized at 336 kg N ha^{-1} a^{-1} (Table 5.16).

These data again show that milk yields per cow are limited by the nutritive value of tropical grasses. Colman and Holder (1968) conclude that this is mainly a limitation of low intake of digestible energy, since the feeding of sorghum grain supplements increased milk production linearly with energy intake.

The data of Colman and Kaiser (1974) also demonstrate that advantage can be taken of the high yielding ability of tropical grasses with nitrogen, by accepting a relatively small penalty in milk yield per cow, stocking rates can be increased to give marked increases in production per hectare.

Since feeding value of tropical grasses for milk production is low, conserved fodders made from these species will also be low as shown by the data of Lucci and Boin (1972):

Table 5.16. Effect of stocking rate on milk yield of kikuyu pastures fertilized at 336 kg N ha^{-1} a^{-1} (Colman and Kaiser 1974)

	Stocking rate (cows ha^{-1})	Milk yield (FCM kg cow^{-1})	Milk yield (kg ha^{-1})	Mean lactation (days)	Mean milk yield (kg cow^{-1} day^{-1})
1967–8 (rainfall (1050 mm)	2.47	2467	6093	284	8.68
	3.29	2312	7606	281	8.23
	4.94	2068	10216	278	7.44
1968–9 (rainfall 747 mm)	2.47	1964	4851	274	7.17
	3.29	1750	5757	253	6.92
	4.94	1733	8561	253	6.85

Conserved fodder	Milk yield (kg cow^{-1} day^{-1})
1. Napier grass silage + molasses grass hay	10.3
2. Napier grass silage + soyabean hay	10.3
3. Maize silage + molasses grass hay	9.6
4. Maize silage + soyabean hay	10.1

Conclusions   It can be seen from the data discussed above that while legume based tropical pastures are adequate for beef production, they are severely limiting for high milk yields per cow. The highest milk yields recorded on a long-term basis was 12.5 kg cow^{-1} day^{-1} from Fresian cows grazing *P. maximum–G. wightii* pastures, and was higher than legumes alone or grasses alone. These pastures gave yields in the range of 5–8 kg cow^{-1} day^{-1}, and in the case of some legumes milk quality was reduced.

Since energy intake is the major limitation, milk yields can be readily increased by feeding grain supplements, in direct competition with human or non-ruminant feeding requirements. Thus if we want dairy production in the tropics, then we must accept lower production per head. However, by taking advantage of the DM yield potential of tropical pastures, we can achieve respectable milk yields per ha by correct pasture management to maintain high stocking rates compatible with maintaining an adequate legume content.

Finally, in view of the limitation of tropical pastures, one must question the expense and effort put into importation of high yielding temperate dairy breeds and cross-breeding programmes, whose potential can only be achieved by the substitution of valuable grain and by-products. Far more advantage is likely to be achieved by sound development of tropical pastures whose potential may more closely match the genetic potential for milk yield of locally adapted breeds. Also in beef production, the genetic potential for LWG of most breeds exceeds the general potential of tropical pastures, although the advantages of heterosis was demonstrated in one grazing trial (Stobbs 1969*d*), where crossbred Zebu heifers gave higher liveweight gains. However, in most tropical areas poor nutrition is the greatest limitation to animal production. Pasture improvement must be the first priority. Once adequate forage resources are provided, then further gains may be achieved through breeding and other management practices.

2. Comparison of legume-based and nitrogen fertilized tropical pastures for animal production

The factors previously discussed (Sections 5.1 and 5.2) are the basic

Table 5.17. Summary of beef animal performance on nitrogen fertilized tropical grass pastures

Location	Mean annual rainfall (mm)	Pastures and treatment	Nitrogen rate kg N ha^{-1} a^{-1}	Stocking rate (animals ha$^-$)	Liveweight gain (kg head day^{-1})	(kg ha^{-1} a^{-1})	Comments	References
Sao Paulo Brazil	1400	*Panicum maximum* (cv. Coloniao)	0	1.14	0.54	227	Mean values after 3 year trial, Zebu steers. Put-and-take grazing system. Maint. fertilizer of 100 P and 67 kg S	Mott, Quinn, Bisschoff and Da Rocha (1965)
			89	2.05	0.54	403		
Sao Paulo Brazil	1400	*P. maximum* (cv. Coloniao)	0	2.42	0.28	204	Mean values over 3 year trial. Zebu steers.	Quinn, Mott, Bisschoff, and Jones (1965)
			89	2.37	0.38	330	trial. Zebu steers.	
		Hyparrhenia rufa	0	2.67	0.25	244	Basal fertilizer — 30 kg P	
			89	2.67	0.33	319	Maint. fertilizer — 35 kg P	
		Digitaria decumbens	0	1.98	0.33	240	Put-and-take grazing	
			89	2.05	0.44	330	system	
		Melinis minutiflora	0	1.90	0.18	91		
			89	1.98	0.16	119		
		Cynodon dactylon	0	1.26	0.12	56		
			89	1.43	0.27	139		
Sao Paulo Brazil	1170	*P. maximum*	0	1.21	0.53	241	Mean values over 7 year trial. Super and gypsum applied. Put-and-take grazing system.	Mott, Bisschoff, and Frietal Quinn (1970)
			200	2.88	0.55	586		
Rio de Janeiro Brazil	1300	Pangola alone	0	2.6	0.36	349	Means over 4 year trial. Steers — ½ Fresian, ½ Zebu. Put-and-take grazing system	Aronovich, Serpa, and Ribiero (1970)
		Pangola + nitrogen	100	3.3	0.44	531		
		Pangola + *Centosema pubescens*	0	2.4	0.47	410		

Location	Altitude	Species	Fertilizer	Stocking	Gain	Yield	Notes	Reference
Lismore N.S.W. Australia	1250	*Pennisetum clandestinum* (kikuyu)	0	2.2	0.47	380	Values for 1971–2 only Angus weaners, initial weight 180 kg. Set stocked, continuous grazing. Low temperatures (11.5 °C) limit pasture growth in winter	Mears and Humphreys (1974)
				3.3	0.43	520		
				4.9	0.36	640		
			134	3.3	0.49	590		
				4.9	0.40	710		
				7.4	0.24	840		
			336	4.9	0.60	1080		
				7.4	0.42	1130		
				11.1	0.28	1140		
			672	7.4	0.49	1310		
				11.1	0.37	1480		
				16.6	0.16	950		
Lajas Valley Puerto Rico	1250	*P. maximum*	0	1.38	0.78	382	Means over 4 year trial Put-and-take grazing system	Rivera-Brenes, Colon-Torres, Gelpi, and Torres-Mas (1958)
			140	2.07	0.79	594		
			280	2.94	0.83	889		
Mountain slopes Puerto Rico	1600	*P. maximum*	350	6.42	0.56	1319	Means over 4 year trial. Put-and-take grazing system. Holstein heifers	Caro-Costas, Vicente-Chandler, and Figerella (1965)
		Digitaria decumbens	350	5.93	0.52	1124		
		Pennisetum purpureum	350	5.93	0.51	1110		
Beerwah S. Queensland	1650	(a) *D. decumbens*	168	3.7	0.51	684	Continuous grazing, set stocked Hereford steers, initially 12–14 months, 250 kg. Annual maint. – 48 kg P, 63 kg K ha^{-1}	(a) Evans (1970)
			448	7.4	0.44	1176		
		(b) *D. decumbens*	448	7.4	0.47	1277		(b) Evans (1969)
			896	7.4	0.50	1362		

(continued)

335

Table 5.17. (*Cont'd*)

Location	Mean annual rainfall (mm)	Pastures and treatment	Nitrogen rate kg N kg N ha^{-1} a^{-1}	Stocking rate (animals ha^{-1})	Liveweight gain (kg head day^{-1})	(kg ha^{-1} a^{-1})	Comments	References
Irrigated trials								
South Coast		*P. purpureum*	470	5.23	0.78	1486	Means over 1½ y trial. Put-and-take grazing system	
Puerto Rico		*P. maximum*	470	5.51	0.72	1447	Basal fertilizer — 58 kg P, 316 kg K ha^{-1}	Caro-Costas, Vicente-Chandler, and Burleigh (1961)
		D. decumbens	470	4.47	0.73	1190		
Parada N. Queensland		*D. decumbens*	112	3.95	0.83	1190	Means over 3 years trials. Rotationally grazed, set stocked. Two groups of mixed store cattle per year. Maint. fert. — 25 kg P, 63 kg K ha^{-1}	Evans and Pulsford (1970)
			336	7.41	0.75	2018		

components associated with livestock production, and operate to affect both production per head and stocking rate, and so production per hectare. These parameters are entered in Table 5.17 where beef production data for a number of trials are summarized. Only beef liveweight gains are discussed in this section, as these data are less complicated compared with dairy production where management techniques, age effects on lactation, stage of lactation, and breed differences make comparisons difficult. Data for milk yields on nitrogen fertilized tropical grasses are given on pages 328–333.

Only data from trials in areas receiving above 1000 mm mean annual rainfall are considered in Table 5.17, as in lower rainfall environments responses to nitrogen are often erratic and animal production data is liable to vary greatly from year to year, except where irrigation is applied.

In the trials reported from Brazil rates of nitrogen use were low and animal production per ha also low. Production per head seldom exceeded 0.5 kg head^{-1} day^{-1} which may reflect both limitations of pasture quality or too high stocking rates. The trial reported by Mears and Humphreys (1974) provides the most useful data on interactions of stocking rate and nitrogen rate. At all nitrogen levels, as stocking rate increased, LWG head^{-1} declined. At the same stocking rates, increasing nitrogen rate tended to increase LWG head^{-1}, e.g. at 4.9 animals ha^{-1}, LWG head^{-1} was 0.36, 0.40, and 0.60 at N0, N134, and N336, respectively. However, at similar stocking pressures (animals kg^{-1} of pasture), LWG head^{-1} remained approximately constant, e.g. N0, 3.3 a ha^{-1} LWG head^{-1} = 0.43; N134, 4.9 a ha^{-1} = 0.40; N336, 7.4 a ha^{-1} = 0.42; N672, 11.1 a ha^{-1} = 0.37 kg head^{-1} day^{-1}. Although LWG head^{-1} values are not high, high levels of LWG ha^{-1} were achieved with high nitrogen levels and high stocking rates.

In the trial of Rivera-Brenes *et al.* (1958) LWG head^{-1} increased with increasing stocking rate, suggesting that at higher nitrogen levels higher stocking rates could have been used. This sort of problem is likely to arise in 'put-and-take' grazing systems, making it difficult to assess the potential of the grazing system. The later trials in Puerto Rico (Caro-Costas *et al.* 1965) and at Beerwah (Evans 1969, 1970) achieved LWG ha^{-1} in excess of 1000 kg ha^{-1}. Again LWG head^{-1} was not particularly high, but the high levels per ha were achieved through the use of adequate levels of nitrogen and high stocking rates. At Beerwah a further increase in nitrogen above 448 kg ha^{-1} a^{-1} had little effect on LWG head^{-1} or per ha.

In the irrigated trials further increases in production were achieved, particularly in the work of Evans and Pulsford (1970). In later trials Evans and Pulsford (personal communication) achieved over 2760 kg LWG ha^{-1}, using stocking rates up to 12.4 animals ha^{-1} and 672 kg N ha^{-1}. However, these high levels of production were achieved using 2–3 groups of store animals per year, so that compensatory gain was probably involved, but production at this level does indicate the potential of nitrogen fertilized pasture system.

Nitrogen fertilizer effects on animal production were mainly achieved through increasing stocking rate. As previously discussed, nitrogen fertilizer has less effect on production per head. The data indicate that given adequate nitrogen and suitable environments nitrogen fertilized tropical grass pastures have the potential to carry a large number of animals per hectare, even though production per head does not approach the values reported for temperate pastures. Thus productivity in the tropics can be achieved in a per hectare basis, but at lower productivity on a per head basis.

Comparison of legume-based and nitrogen-fertilized pastures
In rainfall regimes above 950 mm, data in Table 5.15 (Section 5.3) and Table 5.20 indicate that with the nitrogen fertilizer system stocking rates at higher levels of nitrogen are higher than in the legume system. Also liveweight gains per head tend to be higher in the nitrogen system. The average LWG per head over all entries for the nitrogen-fertilized pastures is 0.45 kg head^{-1} day^{-1}, whereas the mean over the medium and high rainfall categories in the legume pastures was 0.35 kg head^{-1} day^{-1}. Thus the combination of higher stocking rates and slightly higher LWG head^{-1} gives the nitrogen-fertilized system higher LWG ha^{-1} at nitrogen application rates of about 300 kg N ha^{-1} a^{-1} and above.

In lower rainfall environments it is unlikely that nitrogen-fertilized grass pastures can compete with the returns from *S. humilis*-based pastures. Under conditions of extensive grazing the inclusion of a legume such as *S. humilis*, which is well adapted to the monsoonal environment, has particular advantages. Also where the environment is particularly favourable for legume growth, as in the trials of Stobbs (1969*d*) in Uganda and Younge and Plucknett (1965) in Hawaii, nitrogen-fertilized grasses would have to give large increases in production to be competitive.

Economic comparisons
General economic comparisons between the two systems are extremely

difficult as economic conditions vary widely between regions, and also change with time. This latter point is clearly demonstrated by the series of budgetary appraisals of beef production on the Wallum in southern Queensland, based on trials of Evans and Bryan (1973). Internal rates of return were calculated for nitrogen and legume systems based on beef and fertilizer prices in 1971 (Mitchell, Bryan, and Evans 1972), in 1973 (Firth, Bryan, and Evans 1974), and in 1974 (Firth, Evans, and Bryan 1975) Internal rate of return (IRR) is defined as being the rate of return that brings the accumulated annual cash balance at present worth to zero.

(a) *1971* – with beef price at A\$0.60 kg^{-1} dressed weight (Mitchell *et al.* 1972).

Treatment	LWG ha^{-1}	IRR (1)	IRR (2)
Pangola + legumes	507	2.8%	5.4%
Pangola + 168 kg N	669	4.5%	6.3%
Pangola + 448 kg N	1106	2.6%	4.2%

IRR (1) = with maintenance fertilizer 500 kg superphosphate, 250 kg KCl.
IRR (2) = with maintenance fertilizer 500 kg superphosphate, 126 kg KCl.

(b) *1973* – with beef price increased to A\$0.88 kg^{-1} dressed weight (Firth *et al.* 1974).

Treatment	LWG ha^{-1}	IRR
Pangola + legume	507	9.7%
Pangola + 168 kg N	699	11.3%
Pangola + 448 kg N	1106	12.9%

(c) *1974* – with beef price A\$0.77 kg^{-1} dressed weight (Firth *et al.* 1975).

Legume-based pastures only considered. At a stocking rate of 1.65 head ha^{-1}. IRR varied from 7.1 to 13.2 per cent at 1973 prices. At 1974 fertilizer prices maximum IRR was reduced to 10.4 per cent.

However, Firth *et al.* (1975) state that 'The sharp decline (in beef) prices which occurred during 1974 and the low prices which have prevailed have again altered the picture'. Beef prices declined even further in 1975 to levels less than A\$0.5 kg^{-1}. Furthermore fertilizer prices, particularly nitrogen, increased sharply over the same period, so that it is unlikely that any production system was profitable.

The above example illustrates the difficulty of making generalized comparisons between production systems. Nevertheless, budgetary comparisons can be made within a particular region at a given point in time. The major variables which must be taken into account include:

1. *Land value* — more land will be required to carry the same number of animals in a legume system than a nitrogen system.
2. *Animal costs* — on a given area of land, more animals will be carried in a nitrogen system, thus increasing capital investment in animals, and veterinary costs.
3. *Labour costs* — probably similar if herd sizes are similar between systems.
4. *Pasture establishment* — legume-seed cost is additional in the legume system. Also as legume content tends to decline, resowing of legume must be budgeted after 10–15 years.
5. *Fertilizer costs* — maintenance requirements of phosphorus, potassium and other elements are generally applied at similar rates in the two systems. The major difference is the cost of nitrogen in the nitrogen system.
6. *Beef price* — this is a major variable affecting the profitability of both systems.

Based on the data in Tables 5.15 and 5.20, Nuthall and Whiteman (1972) made an economic evaluation of the beef liveweight gains per hectare (at various levels of nitrogen application), necessary for such a system to produce the same profit that would be achieved with a legume-based system. This was carried out for a general Wallum property (low-fertility soils, 1500 mm rainfall, cool winter season) of 600 ha.

The expected profit from a legume-based system was estimated for three stocking rates giving 280, 390, and 500 kg LWG ha^{-1}. Using this information the number of production units that a nitrogen-based system must support to enable the same profit to be made was estimated. The 'break-even' liveweight gains were calculated for a number of price and productivity situations. For each of the three legume production situations, three beef price-levels were considered, and for each of these, three different nitrogen-fertilizer price-levels and five rates of nitrogen per hectare were budgeted (Fig. 5.11).

At 1972 prices the situation represented in graph 5 (Fig. 5.11), i.e. legume-based system giving 390 kg LWG, beef price A$26 per 100lb dressed weight, nitrogen price A$0.08 per lb (17 cents kg^{-1}) was considered most appropriate. Yields necessary to break-even at 100 lb

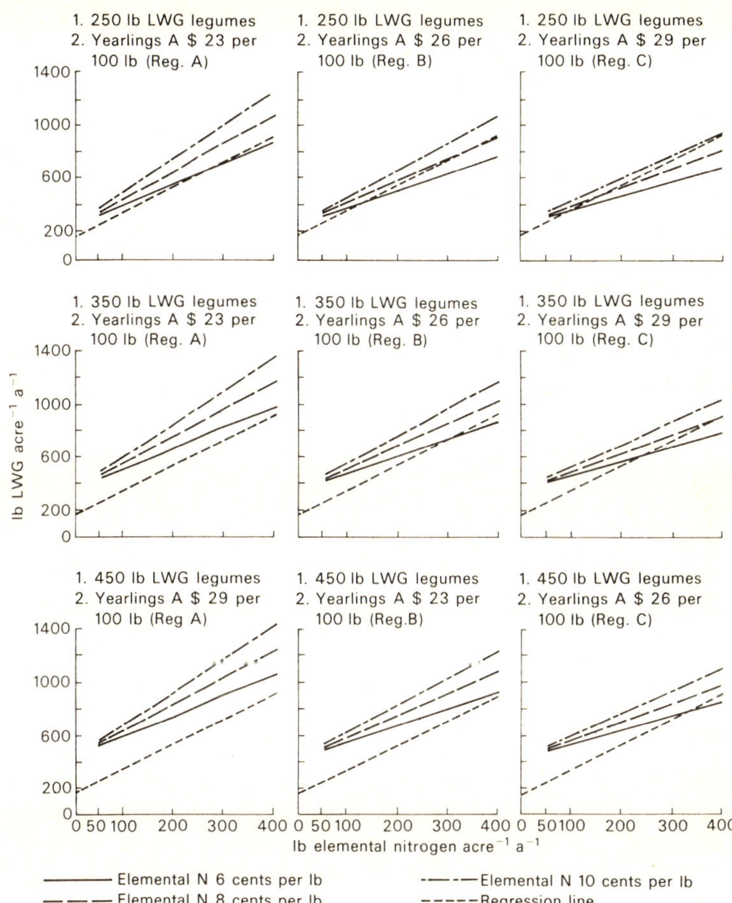

Fig. 5.11. Liveweight gain necessary for a nitrogen fertilizer system to return the same profit as a legume-based system (Nuthall and Whiteman 1972).

N acre^{-1} is 517 lb LWG acre^{-1} (580 kg ha^{-1}), while at 400 lb. N acre^{-1} 1013 lb LWG acre^{-1} (1134 kg ha^{-1}) is required. From the regression line of trial results (Table 5.15), predicted yields fall short of these levels.

In 1975 beef prices declined while nitrogen prices increased. At that stage it seemed unlikely that nitrogen systems would give better returns than legume-based systems. In a recent study of nitrogen responses of pasture grasses in a lower rainfall region (900–1000 mm) Henzell, Peake, 't Mannetje, and Stirk (1975) conclude 'At the prices paid in recent years, nitrogen fertilizers are almost certainly too expensive for commercial beef production at Narayen ('t Mannetje, unpublished data). If so, the extra nitrogen will have to be obtained from legumes, of which siratro is the most successful so far.'

Thus in Queensland while earlier studies suggested that nitrogen systems may give better returns than legume systems (Mitchell *et al.* 1972; Firth *et al.* 1974), other studies (Nuthall and Whiteman 1972) suggest that unless nitrogen price is low and beef price high nitrogen systems are less profitable. However, since these studies, nitrogen price has increased and beef price declined so that the whole picture has changed. Thus commercial producers must be flexible in their approach to take advantage of changing economic conditions.

Integration of legume-based and nitrogen-fertilized pastures

In extensive grazing situations at low stocking rates legume pastures are more widely used. Under conditions of intensive stocking, or for high value animal products, such as milk or fat lambs, use of nitrogen fertilizer may be economic. In many farming situations both legume-based and nitrogen-fertilized pastures are used for different purposes, depending on class of livestock, requirements for forage, and farm topography.

(a) Farm topography

On many farms in southern Queensland, white-clover-based pastures or nitrogen-fertilized winter forages such as rye grass (*Lolium perenne*) or oats (*Vicia sativa*) are sown on the lower farm flats which are affected by frost in winter. Summer forages such as sorghum, or *Panicum millaceum*, may also be grown. Legume-based tropical pastures are sown on the hills. During the winter season the standover tropical pastures maintain dry stock and followers, while the dairy herd is grazed on the higher quality temperate pastures and forages.

(b) Strategic nitrogen use

Even in the high rainfall environment of tropical north Queensland, growth rate of legume-based pastures declines to very low levels during the cooler dry season. At this stage pastures are most vulnerable to damage by overstocking, leading to a rapid decline in legume content. Teitzel (1969) recommends the use of nitrogen-fertilized *Digitaria decumbens* or *Brachiaria decumbens* during the dry season, sown on the lower lying areas of the property. Since these pastures are not fertilized or used over the wet season, this avoids pugging damage to swards. The legume based swards in better drained land are grazed during the wet summer period.

(c) 'Cut and cart' feeding systems

In many tropical countries with intensive land use, animals are stalled and forage is cut and hand fed. Napier grass (*Pennisetum purpureum*) is widely used. This is intensively cut and few legumes are adapted to severe regular defoliation, although some success with napier and *Glycine wightii* or *Pueraria phaseoloides* mixtures have been reported. In these intensive systems, particularly for milk production, nitrogen may be economic. The economics of nitrogen use will be enhanced if farmyard manures and urine from the stalls can be returned to forage area.

(d) Forage crops for conservation

Nitrogen may be applied to forage crops for conservation as hay or silage to increase both yield and crude protein content. Here again the economics of nitrogen compared with legumes must be closely examined.

While no clear general conclusions can be drawn on the relative economics of legumes versus nitrogen, it is now recognized in many countries that with the increase in energy and nitrogen-fertilizer costs the legume must assume increasing importance in ruminant forage production. The *Interim Report – World Food and Nutrition Study* by the USA National Academy of Science (1975) recommends as a first priority increased research into increasing the amount of biological nitrogen fixation in major food plants and in the soil.

The data presented in Sections 5.3 and 5.4 show that high levels of beef production can be attained from legume-based pastures. Further selection and breeding of tropical legumes combined with further research into grazing management and fertilizer strategies will surely increase the potential for animal production on legume-based pastures. While the emphasis in future will be on legume pastures, there is a role for nitrogen fertilizer use for special-purpose pastures, particularly for higher value products such as milk where continuity of forage supply is important.

REFERENCES

Allden, W. G. and Whittaker, I. A. M. (1970). The determinants of herbage intake by sheep: the inter-relationships of factors affecting herbage intake and availability. *Aust. J. agric. Res.* **21**, 755.

Anderson, E. W. (1967). The rotation of deferred grazing. *J. Range Mgmt* **20**, 5.

Aronovitch, S., Serpa, A., and Ribiero, H. (1970). Effect of nitrogen fertilizer and legume upon beef production of pangola grass pasture. *Proc. 11th Int. Grassl. Cong.*, p. 796.

Austin, J. D. A. (1970). Townsville stylo: looking for comparison grasses. *Turnoff* **2**, 28.

Blaxter, K. L. and Wilson, R. S. (1963). The assessment of a crop husbandry technique in terms of animal production. *Anim. Prod.* **5**, 27.

Booysen, P. de V. (1975). Economic optimisation of stocking rate and grazing management. *Proc. 6th Meeting. European Grassld. Fedn. Madrid*, p. 186.

Bryan, W. W. and Evans, T. R. (1973). Effects of soils fertilizers and stocking rates on pasture and beef production on the Wallum of south eastern Queensland. *Aust. J. exp. Agric. Anim. Husb.* **13**, 516.

Campbell, A. G. (1966). Grazed pasture parameters. I. Pasture dry matter production and availability in stocking rate and grazing management experiment with dairy cows. *J. agric. Sci., Camb.* **67**, 199.

—— (1969). Grazing interval, stocking rate and pasture production. *N.Z. Jl agric. Res.* **12**, 67.

Caro-Costas, R., Vicente-Chandler, J., and Burleigh, C. (1961). Beef production and carrying capacity of heavily fertilized, irrigated Guinea, Napier and Pangola grass pastures on the semi-arid coast of Puerto Rico. *J. Agric. Univ. P. Rico* **45**, 32.

—— —— and Figarella, J. (1965). Productivity of intensively managed pastures of five grasses on steep slopes in the humid mountains of Puerto Rico. *J. Agric. Univ. P. Rico* **49**, 99.

Chacon, E. A. (1976). The effects of sward characteristics upon grazing behaviour, intake and animal production from tropical pastures. Ph.D Thesis, University of Queensland.

Colman, R. L. (1971). Quality of pastures and forage crops for dairy production in the tropical regions of Australia. *Trop. Grassl.* **5**, 181.

—— and Holder, J. M. (1968). Effect of stocking rate on butterfat production of dairy cows grazing Kikuyu grass pastures fertilized with nitrogen. *Proc. Aust. Soc. Anim. Prod.* **7**, 129.

—— and Kaiser, A. G. (1974). The effect of stocking rate on milk production from Kikuyu grass pastures fertilized with nitrogen. *Aust. J. exp. Agric. Anim. Husb.* **14**, 155.

Conniffe, D., Browne, D., and Walshe, M. J. (1970). Experimental design for grazing trials. *J. agric. Sci., Camb.* **74**, 339.

Cowan, R. T. and O'Grady, P. (1975). Effect of presentation yield of a tropical grass-legume pasture on grazing time and milk yield of Fresian cows. *Trop. Grassl.* **10**, 213.

—— —— Moss, R. J., and Byford, I. J. R. (1974). Milk and fat yields of Jersey and Fresian cows grazing tropical grass. *Trop. Grassl.* **8**, 117.

Dale, A. B. and Holder, J. M. (1968). Milk production from tropical legume grass pastures. *Proc. Aust. Soc. Anim. Prod.* **7**, 86.

Duble, R. L., Lancaster, J. A., and Holt, E. C. (1971). Forage characteristics limiting animal performance on warm season perennial grasses. *Agron. J.* **63**, 795.

Edgley, W. H. R. and Quinlan, T. J. (1973). Managing tropical pastures on the Atherton Tableland. *Qd agric. J.* **99**, 293.

Edye, L. A., Ritson, J. B., Haydock, K. P., and Davies, J. G. (1971). Fertility and seasonal changes in liveweight of Droughtmaster cows grazing a Townsville stylo–spear grass pasture. *Aust. J. agric. Res.* **22**, 963.

Egan, A. R. (1977). Nutritional status and intake regulation in sheep. VIII. Relationships between the voluntary intake of herbage by sheep and the protein/energy ratio in the digestion products. *Aust. J. agric. Res.* **28**, 907.

Evans, J. (1970). Cattle fattening trials. *Agrost. Tech. Ann. Rep. Qld Dept. Primary Ind.*, 1970, p. 15.

—— and Pulsford, J. (1970). Agrostology. Ann. Rep. Qld. Dept Primary Ind., Brisbane.

Evans, T. R. (1969). Beef production from nitrogen fertilized Pangola grass (*Digitaria decumbens*) on the coastal lowlands of southern Queensland. *Aust. J. exp. Agric. Anim. Husb.* **9**, 282.

— (1970). Some factors affecting beef production from subtropical pastures in the coastal lowlands of southeast Queensland. *Proc. 11th Int. Grassl. Cong.*, p. 803.

and Bryan, W. W. (1973). Effects of soils, fertilizers and stocking rates on pastures and beef production on the Wallum of south eastern Queensland. II. *Aust. J. exp. Agric. Anim. Husb.* **13**, 530.

Ferguson, K. A. (1973). In *The pastoral industries of Australia: practice and technology of sheep and cattle production* (eds G. Alexander and O. B. Williams) p. 526. Sydney University Press.

Firth, J. A., Bryan, W. W., and Evans, T. R. (1974). Updated budgetary comparisons between pangola grass/legume pasture and nitrogen fertilized pangola pasture for beef production in the southern Wallum. *Trop. Grassl.* **8**, 25.

— Evans, T. R., and Bryan, W. W. (1975). Effects of soils, fertilizers and stocking rates on pastures and beef production on the Wallum of south eastern Queensland. 4. Budgetary appraisals of fertilizers and stocking rates. *Aust. J. exp. Agric. Anim. Husb.* **15**, 531.

Gillard, P. (1970). Pasture development in the dry tropics of north Queensland. *Proc. 11th Int. Grassl. Cong.*, p. 807.

Greenhalgh, J. F. D. (1970). Effects of grazing intensity on herbage production and consumption and on milk production in strip-grazed dairy cows. *Proc. 11th Int. Grassl. Cong.*, p. 856.

Grof, B. and Harding, W. A. T. (1970). Dry matter yields and animal production of Guinea grass (*Panicum maximum*) on the humid tropical coast of North Queensland. *Trop. Grassl.* **4**, 85.

Hamilton, R. I., Lambourne, L. J., Roe, R., and Minson, D. J. (1970). Quality of tropical grasses for milk production. *Proc. 11th Int. Grassl. Cong.*, p. 860.

Hayman, R. H. and Radcliffe, J. C. (1973). In *The pastoral industries of Australia: practice and technology of sheep and cattle production* (eds G. Alexander and O. B. Williams) p. 193. Sydney University Press.

Henzell, E. F., Peake, D. C., 't Mannetje, L., and Stirk, G. B. (1975). Nitrogen response of pasture grasses on duplex soils formed from granite in southern Queensland. *Aust. J. exp. Agric. Anim. Husb.* **15**, 498.

Hodgson, J., Combellas, J. J., Wade, M. H., and Booth, T. H. (1975). *Ann. Rep. (1974) Grasslds. Res. Inst. Hurley, UK*, p. 85.

Ivory, D. A., Stobbs, T. H., McLeod, M. N., and Whiteman, P. C. (1974). Effect of day and night temperatures on estimated dry matter digestibility of *Cenchrus ciliaris* and *Pennisetum clandestinum. J. Aust. Inst. Agric. Sci.* **40**, 156.

Jones, D. I. H. and Hayward, M. V. (1975). The effect of pepsin pre-treatment of herbage on the prediction of dry matter digestibility from solubility in fungal cellulose solutions. *J. Sci. Fd Agric.* **26**, 711.

Jones, R. J. (1969). A note on the *in vitro* digestibility of two tropical legumes *Phaseolus atropurpureus* and *Desmodium intortum. J. Aust. Inst. Agric. Sci.* **35**, 62.

— (1974*a*). The relation of animal and pasture production to stocking rate on legume based and nitrogen fertilized sub-tropical pastures. *Proc. Aust. Soc. Anim. Prod.* **10**, 340.

— (1974*b*). Effect of an associate grass, cutting interval and cutting height on yield and botanical composition of Siratro pastures in a subtropical environment. *Aust. J. exp. Agric. Anim. Husb.* **14**, 334.

— and Sandland, R. L. (1974). The relation between animal gain and stocking rate. Derivation of the relation from the results of grazing trials. *J. agric. Sci., Camb.* **83**, 335.

Kissock, W. J. (1966). In hill country sheep trial mob stocking proves superior. *N.Z. Jl Agric.* **113**, 28.

Laredo, M. A. and Minson, D. J. (1973). The voluntary intake, digestibility and retention time by sheep of leaf and stem fractions of five grasses. *Aust. J. agric. Res.* **24**, 875.

Lucci, C. de S. and Boin, C. (1972). Napier grass silage vs. maize silage, with hay made from molasses grass (*Melinis minutiflora*) for perennial soyabean, as roughage for dairy cows. *Dairy Sci. Abstr.* **34**, 3989.

McLeod, M. N. and Minson, D. J. (1974). The accuracy of predicting dry matter digestibility of grasses from lignin analysis by three different methods. *J. Sci. Fd Agric.* **25**, 907.

McMeekan, C. P. and Walshe, M. J. (1963). The inter-relationships of grazing method and stocking rate in the efficiency of pasture utilization by dairy cattle. *J. agric. Sci., Camb.* **61**, 147.

't Mannetje, L. (1974). Relations between pasture attributes and liveweight gains on a sub-tropical pasture. *Proc. 12th Int. Grassl. Congr.*, Moscow. Sect. Paper *Grassland utilization*, p. 386.

't Mannetje, L. (1975). Effect of daylength and temperature on introduced legumes and grasses for the tropics and sub-tropics of coastal Australia. 2. N-concentration, estimated digestibility and leafiness. *Aust. J. exp. Agric. Anim. Husb.* **15**, 256.

— Jones, R. J., and Stobbs, T. H. (1976). Pasture evaluation by grazing experiments. In *Tropical pasture research: principles and methods* (eds. N. H. Shaw and W. W. Bryan) Common. Agric. Bur. Bull. 51, pp. 194–250.

Marshall, B., Long, M. I. E., and Thonston, D. D. (1969). Nutritive value of grasses in Ankole and the Queen Elizabeth National Park, Uganda. III. *In vitro* dry matter digestibility. *Trop. Agric., Trin.* **46**, 43.

Mears, P. T. and Humphreys, L. R. (1974). Nitrogen response and stocking rate of *Pennisetum clandestinum* pastures. II. Cattle growth. *J. agric. Sci., Camb.* **83**, 469.

Milford, R. (1967). Nutritive values and chemical composition of seven tropical legumes and lucerne grown in sub-tropical south eastern Queensland. *Aust. J. exp. Agric. Anim. Husb.* **7**, 540.

— and Minson, D. J. (1966). Intake of tropical pasture species. *Proc. 11th Int. Grassl. Cong.*, Brazil, 1964, p. 814.

— — (1968). The effect of age and method of haymaking on the digestibility and voluntary intake of the forage legumes *Dolichos lablab* and *Vigna sinensis. Aust. J. exp. Agric. Anim. Husb.* **8**, 409.

Milk Marketing Board, UK (1974). *Dairy facts and figures*, 1974. Thames Ditton, Surrey.

Minson, D. J. (1966). The intake and nutritive value of fresh frozen and dried *Sorghum almum*, *Digitaria decumbens* and *Panicum maximum. J. Br. Grassld Soc.* **21**, 123.

— (1967). The voluntary intake and digestibility in sheep, of chopped and pelleted *Digitaria decumbens* (pangola grass) following a late application of fertilizer nitrogen. *Br. J. Nutr.* **21**, 587.

— (1971). The nutritive value of tropical pastures. *J. Aust. Inst. Agric. Sci.* **37**, 255.

— (1973). Effect of fertilizer nitrogen on digestibility and voluntary intake of *Chloris gayana*, *Digitaria decumbens* and *Pennisetum clandestinum. Aust. J. exp. Agric. Anim. Husb.* **13**, 53.

— (1976). Plant factors involved in digestibility and intake. Ann. Rep., CSIRO, Division of Tropical Crops and Pastures, 1975/76, p. 100.

— and McLeod, M. N. (1970). The digestibility of temperate and tropical grasses. *Proc. 11th Int. Grassl. Cong.*, p. 719.

— — (1972). The *in vitro* technique: its modifications for estimating digestibility of large numbers of tropical pasture samples. *Tech. Paper, CSIRO Div. Trop. Past. Aust.*, No. 8.

— Raymond, W. F., and Harris, C. E. (1960). Studies in the digestibility of herbage. VIII. The digestibility of S37 Cocksfoot, S23 Ryegrass and S24 Ryegrass. *J. Br. Grassld Soc.* **15**. 174.

— Stobbs, T. H., Hegarty, M. P., and Playne, M. J. (1976). Measuring nutritive value of pasture plants. In *Tropical pasture research:*

principles and methods. Bull. Comm. Bur. Past. Field Crops No. 51, p. 308.

Mitchell, T. E., Bryan, W. W., and Evans, T. R. (1972). Budgetary comparisons between Pangola grass/legume pasture and nitrogen fertilized Pangola pasture for beef production in southern Wallum. *Trop. Grassl.* **6**, 177.

Moorhouse, W. (1975). Farm management accounting service. Report No. 15 all groups 1973–74. Econ. Service Branch Qld Dept. Primary Ind.

Mott, G. O. (1960). Grazing pressure and the measurement of pasture production. *Proc. 8th Int. Grassld Congr.*, Reading, p. 606.

— Quinn, L. R. C., Bisschoff, W. V. A., and Da Rocha, G. L. (1965). Supplemental feeding of steers and nitrogen fertilization and their effect upon beef production from Guinea grass pasture. *Proc. 9th Int. Grassl. Cong.*, p. 981.

Norman, M. J. T. (1970). Relationship between liveweight gain of grazing beef steers and availability of Townsville lucerne. *Proc. 11th Int. Grassl. Cong.*, p. 829.

— and Phillips, L. J. (1970). Wet season grazing of Townsville stylo pasture at Katherine, N.T. *Aust. J. exp. Agric. Anim. Husb.* **10**, 710.

Nuthall, P. L. and Whiteman, P. C. (1972). A review and economic evaluation of beef production from legume based and nitrogen-fertilized tropical pastures. *J. Aust. Inst. Agric. Sci.* **38**, 100.

Owen, J. B. and Ridgman, W. J. (1968). The design and interpretation of experiments to study animal production from grazed pastures. *J. agric. Sci., Camb.* **71**, 327.

Owensby, C. E., Smith, E. F., and Anderson, K. L. (1973). Deferred-rotation grazing with steers in the Kansas Flint Hills. *J. Range Mgmt* **26**, 393.

Oyenuga, V. and Olubajo, F. O. (1966). Productivity and nutritive value of tropical pastures at Ibadan. *Proc. 10th Int. Grassl. Cong.*, p. 962.

Partridge, I. J. and Ranacou, E. (1974). The effects of supplemental *Leucaena leucocephala* browse on steers grazing *Dicanthium caricosum* in Fiji. *Trop. Grassl.* **8**, 107.

Payne, W. J., Laing, W. I., and Raivoka, E. N. (1951). Grazing behaviour of dairy cattle in the tropics. *Nature, Lond.* **167**, 610.

Peart, G. R. (1968). A comparison of rotational grazing and set stocking of dryland lucerne. *Proc. Aust. Soc. Anim. Prod.* **7**, 110.

Petersen, R. G., Lucas, H. L., and Mott, G. O. (1965). Relationship between rate of stocking and per animal and per acre performance on pastures. *Agron. J.* **57**, 27.

Pigden, W. J. and Greenshields, J. E. R. (1960). Interaction of design sward and management on yield and utilization of herbage in Canadian grazing trials. *Proc. 8th Int. Grassl. Cong.*, p. 594.

Playne, M. J. (1970). Differences in the nutritional value of three cuts of buffel grass for sheep and cattle. *Proc. Aust. Soc. Anim. Prod.* **8**, 511.

Playne, M. J. and Haydock, K. P. (1972). Nutritional value of Townsville stylo (*Stylosanthes humilis*) and of spear grass (*Heteropogon contortus*) – dominant pastures fed to sheep. 1. Effects of plant maturity. *Aust. J. exp. Agric. Anim. Husb.* **12**, 365.

Prajapati, M. C. (1970). Effect of different systems of grazing by cattle on *Lasiurus–Elevsine–Aristida* grasslands in arid regions of Rajasthan vis-à-vis animal production. *Ann. Arid Zone* **9**, 114.

Quinn, L. R. C., Mott, G. O., Bisschoff, W. V. A., and Frietas, L. M. M. (1970). Production of beef from winter vs summer nitrogen-fertilized colonial Guinea grass (*Panicum maximum*) pastures in Brazil. *Proc. 11th Int. Grassl. Cong.*, p. 832.

— — — and Jones, M. B. (1965). Beef production of six tropical grasses in central Brazil. *Proc. 9th Int. Grassl. Cong.*, p. 1015.

Rees, M. C. and Minson, D. J. (1976). Fertilizer calcium as a factor affecting voluntary intake, digestibility, and retention time of pangola grass (*Digitaria decumbens*) by sheep. *Br. J. Nutr.* **36**, 179.

— — and Smith, F. W. (1974). The effect of supplementary and fertilizer sulphur on voluntary intake, digestibility, retention time in the rumen and site of digestion of pangola grass in sheep. *J. agric. Sci., Camb.* **82**, 419.

Reid, R. L., Post, A. J., Olsen, F. J., and Mugerwa, J. S. (1973). Studies on the nutritional quality of grasses and legumes in Uganda. I. Application of *in vitro* digestibility techniques to species and stage of growth effects. *Trop. Agric., Trin.* **50**, 1.

Riewe, M. E. (1961). Use of the relationship of stocking rate to gain of cattle in an experimental design for grazing trials. *Agron. J.* **53**, 309.

Rivera-Brenes, L., Colon-Torres, E. N., Gelpi, F., and Torres-Mas, J. (1958). The influence of nitrogenous fertilizers on Guinea grass yield and carrying capacity in Lajas Valley. *J. Agric. Univ. P. Rico* **42**, 239.

Santhirasegaram, K., Diez, J., Peterson, N. F., and Trigueros, A. (1974). Milk production from four tropical pasture systems. *Proc. 11th Int. Grassl. Cong.*, p. 599.

Schank, S. C., Day, J. M., and de Lucas, E. N. (1977). Nitrogenase activity, nitrogen content, in vitro digestibility and yield of 30 tropical forage grasses in Brazil. *Trop. Agric., Trin.* **54**, 119.

Sharkey, M. J. and Hedding, R. R. (1964). The relationship between changes in body weight and wool production in sheep at different stocking rates on annual pastures in southern Victoria. *Proc. Aust. Soc. Anim. Prod.* **5**, 284.

Shaw, N. H. (1961). Increased beef production from Townsville lucerne (*Stylosanthes sundiaca* Taub.) in the spear grass pastures of central coastal Queensland. *Aust. J. exp. Agric. Anim. Husb.* **1**, 73.

— (1970). The choice of stocking rate treatments as influenced by the expression of stocking rate. *Proc. 11th Int. Grassl. Cong. Aust.*, p. 909.

— and 't Mannetje, L. (1970). Studies on a spear grass pasture in central coastal Queensland – the effect of fertilizer, stocking rate and oversowing *Stylosanthes humilis* on beef production and botanical composition. *Trop. Grassl.* **4**, 43.

Shaw, N. H. and Norman, M. J. T. (1970). In *Australian grasslands* (ed. R. M. Moore) p. 120. ANU Press, Canberra.

Smetham, M. L. (1973). Grazing management. In *Pastures and pasture plants* (ed. R. H. M. Langer) pp. 177–222. Reed, Wellington.

Stobbs, T. H. (1969*a*). The effect of grazing management upon pasture productivity in Uganda. I. Stocking rate. *Trop. Agric., Trin.* **46**, 187.

— (1969*b*). The effect of grazing management upon pasture productivity in Uganda. II. Grazing frequency. *Trop. Agric., Trin.* **46**, 195.

— (1969*c*). The use of liveweight gain trials for pasture evaluation in the tropics. III. The measurement of large pasture differences. *J. Br. Grassl. Soc.* **24**, 177.

— (1969*d*). The use of liveweight gain trials for pasture evaluation in the tropics. V. Type of stock. *J. Br. Grassl. Soc.* **24**, 345.

— (1970). Automatic measurement of grazing time by dairy cows on tropical grass and legume pastures. *Trop. Grassl.* **4**, 237.

— (1971*a*). Quality of pasture and forage crops for dairy production in the tropical regions of Australia. I. Review of the literature. *Trop. Grasslds* **5**, 159.

— (1971*b*). Production and composition of milk from cows grazing Siratro (*Phaseolus atropurpureus*) and greenleaf Desmodium (*Desmodium intortum*). *Aust. J. exp. Agric. Anim. Husb.* **11**, 268.

— (1973*a*). The effect of plant structure on the intake of tropical pastures. I. Variation in the bite size of grazing cattle. *Aust. J. agric. Res.* **24**, 804.

— (1973*b*). The effect of plant structure on the intake of tropical pasture. II. Differences in sward structure, nutritive value, and bite size of animals grazing *Setaria anceps* and *Chloris gayana* at various stages of growth. *Aust. J. agric. Res.* **24**, 821.

— (1973*c*). Ann. Rep. (1972/73) CSIRO Div. Trop. Past., p. 72.

— (1974*a*). Proc. Conf. on *Beef production in developing countries*, Edinburgh, 1974.

— (1974*b*). Rate of biting of Jersey cows as influenced by the yield and maturity of pasture swards. *Trop. Grassl.* **8**, 81.

— (1975*a*). Sward structure and grazing behaviour. In *Refresher course. Management of improved tropical pastures*, pp. 39–55. Aust. Inst. Agric. Sci., Qld Branch.

— (1975*b*). The effect of plant structure on the intake of tropical pasture. III. Influence of fertilizer nitrogen on the size of bite harvested by Jersey cows grazing *Setaria anceps* cv. Kazungula swards. *Aust. J. agric. Res.* **26**, 997.

— (1975*c*). Factors limiting the nutritional value of grazed tropical pastures for beef and milk production. *Trop. Grassl.* **9**, 141.

— (1977). Seasonal changes in the preference by cattle for *Macroptilium atropurpureum* cv. Siratro. *Trop. Grassl.* **11**, 87.

— and Cowper, L. J. (1972). Automatic measurement of the jaw movements of dairy cows during grazing and rumination. *Trop. Grassl.* **6**, 107.

Stobbs, T. H. and Fraser, J. S. (1971). Composition and processing quality of milk produced from cows grazing some tropical pasture species. *Aust. J. Dairy Technol.* **26**, 100.

— and Sandland, R. L. (1972). The use of a latin square change-over design with dairy cows to detect differences in the quality of tropical pastures. *Aust. J. exp. Agric. Anim. Husb.* **12**, 463.

Tayler, J. C., Aston, K. (1975). Ann. Rep. (1974). Grassld Res. Inst. Hurley, UK, p. 76.

Teitzel, J. K. (1969). Pastures for the wet tropical coast. *Qd agric. J.* **95**, 304.

Theurer, C. B. (1970). Determination of botanical and chemical composition of the grazing animals diet. *Proc. Natn. Conf. Forage Quality Evaluation and Utilization*, Lincoln, Nebraska (ed. E. Barnes) p. J1.

Thornton, R. F. and Minson, D. J. (1973). The relationship between apparent retention time in the rumen, voluntary intake, and apparent digestibility of legume and grass diets in sheep. *Aust. J. agric. Res.* **24**, 889.

Tilley, J. M. A. and Terry, R. A. (1963). A two stage technique for the *in vitro* digestion of forage crops. *J. Br. Grassl. Soc.* **18**, 104.

Vallis, I. and Jones, R. J. (1973). Net mineralisation of nitrogen in leaves and leaf litter of *Desmodium intortum* and *Phaseolus atropurpureus* mixed in soil. *Soil Biol. Biochem.* **5**, 391.

Walker, B. (1974). Stocking rate relationships with tropical pastures. *Proc. Aust. Conf. Tropical Pastures*, Townsville, Vol. 2, pp. 7–31.

— and Scott, D. S. (1968). Grazing experiments at Ukiriguru, Tanzania. 1. Comparisons of rotational and continuous grazing systems on natural pastures of hardpan soils. *E. Afr. agric. For. J.* **34**, 224.

Wheeler, J. L. (1962). Experimentation in grazing management. *Herb. Abstr.* **32**, 1.

Whiteman, P. C. (1969). The effects of close grazing and cutting on the yield, persistence and nitrogen content of four tropical legumes with Rhodes grass at Samford, south eastern Queensland. *Aust. J. exp. Agric. Anim. Husb.* **9**, 287.

Willoughby, W. M. (1958). A relationship between pasture availability and animal production. *Proc. Aust. Soc. Anim. Prod.* **2**, 42.

— (1970). Grassland management. In *Australian grasslands* (ed. R. M. Moore) pp. 392–7. ANU Press, Canberra.

Wilson, J. R. (1973). The influence of aerial environment, nitrogen supply, and ontogenetical changes on the chemical composition and digestibility of *Panicum maximum* Jacq. var. *trichoglume* Eyles. *Aust. J. agric. Res.* **24**, 543.

— (1976). Variation in leaf characteristics with level of insertion on a grass tiller. I. Development rate, chemical composition and dry matter digestibility. *Aust. J. agric. Res.* **27**, 343.

— and Ford, C. W. (1971). Temperature influences on the growth, digestibility and carbohydrate composition of two tropical grasses,

Panicum maximum var. *trichoglume* and *Setaria sphacelata* and two cultivars of the temperate grass *Lolium perenne. Aust. J. agric. Res.* **22**, 563.

— — (1973). Temperature influences on the *in vitro* digestibility and soluble carbohydrate accumulation of tropical and temperate grasses. *Aust. J. agric. Res.* **24**, 187.

Yates, J. J., Edye, L. A., Davies, J. G., and Haydock, K. P. (1964). Animal production from a *Sorghum almum* pasture in south east Queensland. *Aust. J. exp. Agric. Anim. Husb.* **4**, 326.

Younge, O. R. and Plucknett, D. L. (1965). Beef production with heavy phosphorus fertilization in infertile wetlands of Hawaii. *Proc. 9th Int. Grassl. Cong.*, p. 959.

6

Applications to pasture research and development

In the previous chapters some of the basic concepts and principles relating to pasture science have been discussed, and the main factors affecting pasture growth, yield, quality, and animal production from pastures were analysed. Pasture agronomy is essentially a practical science and this chapter is concerned with the translation of principles into field pasture development projects.

First some important aspects of developing a regional pasture research programme will be outlined, which may then allow decisions to be made on the level or intensity of pasture improvement appropriate to a particular region. The characteristics of the main pasture production systems are described and the effects of the application of increasing levels of technology on animal production are discussed.

6.1 DEVELOPING A PASTURE RESEARCH PROGRAMME

In most of the tropical countries of the world there has been little emphasis on sown pasture development. The value of tropical pastures in developing marginal areas not yet committed to, or unsuitable for, crop production, or when integrated into tropical crop production systems, have been recognized only recently. In previously undeveloped areas the major limitations to pasture growth will have to be defined, and even in existing cropping areas the limitations of climate, soils, and management may be only poorly understood.

A pasture research programme in a country is best organized on a *regional basis*, a region being defined as an area with a similar climate, and recurring patterns of soils and vegetation which usually impose similar forms of land use. The regional approach has advantages over organizations based on administrative districts, boundaries of which may be drawn on political or administrative criteria. Research findings from carefully chosen representative sites can be applied throughout the region once the main limitations to the development of sown pastures are defined. The development of a regionally organized pasture research programme in Queensland is described by Davies and Shaw (1964), where field research is conducted by CSIRO at a number of representative sites throughout the State, but scientific staff are based at two central laboratory complexes. The advantages of this type of organization are discussed by Hutton and Henzell (1976). They suggest that concentration of pasture research personnel at a central laboratory usually leads to efficiency in research. Also laboratories need to be close to important cities, not only to provide the cultural and educational amenities demanded by first-class scientists, but also to provide the commercial and technological back-up required by any reasonably sized laboratory. Thus it is better to provide adequate budgets for travel to regional research sites and maintain a larger central facility than to try and keep good staff in small regional centres.

The definition of research objectives for a pasture research programme is more important than the physical structure of the organization. It is the definition of objectives and the perception of the major limitations to increasing pasture production that will define the staff and facilities required. The development of a national or regional pasture research programme needs a great deal of survey information so that the major regions, their pasture resources, and the limitations to livestock development can be defined. The process usually develops in a series of well defined stages as suggested in Table 6.1.

This scheme shows the basic agronomic pathway evident in most country programmes, but associated with this may be a wide range of specialist programmes. These specialist projects aim to solve particular problems. Some examples might be specific problems associated with *Rhizobium* inoculation and nodulation; with specific mineral deficiencies in plants or animals; with pasture plant breeding and selection; with pasture seed production; with nutritive value of pasture species and so on. The interactions of these various disciplines and the roles they play in an overall pasture research project are well illustrated in Shaw and Bryan (1976).

Methodology of pasture development programmes

Regardless of the direction the pasture research programme finally takes, and of the major limitations to pasture production, nearly all programmes will require similar analyses, at least in the survey stages. Recommendations on development options, suitable pasture species, stocking rates, grazing management, and potential productivity can only be made after a thorough analysis of climate, topography, hydrology, soils, and ecology of the region under study.

Table 6.1. Stages in the development of a pasture research and development
programme

Survey Stage
    (i)   Define the regions within which a similar range of climate, soils, vegetation, and land use is found

    (ii)   Within regions identify those with major livestock populations

    (iii)   Define regional pasture and forage resources, and areas with potential for future development

    (iv)   Identify limitations to livestock production in existing husbandry systems and with existing forage resources

Allocate priorities for research
    (i)   Decide appropriate levels of intensification for particular regions or land units

The pasture research and development programme
  Assessment
      (i)   Assessment of soil fertility on major soil types

      (ii)   Evaluation of potentially useful grass and legume species through a plant introduction programme

  Regional testing
      (iii)   Regional testing of adapted species in grass–legume mixtures in plots with grazing

  Grazing assessment
      (iv)   Selected mixtures grazed at a range of stocking rates to determine pasture persistence, optimum stocking rate and levels of animal production

  Management studies
      (v)   Longer term studies on best mixtures to determine optimum grazing management, fertilizer management, and limitations

Application
  Extension
      (i)   Extension to farmers – field days, demonstration trials, co-operative trials with landowners

      (ii)   Develop pilot farms or commercial ranches

  Monitoring
      (i)   Identify field problems for further research

(a) *Climate analysis*

The main feature of tropical climates and the effects of climatic elements on pasture growth have been discussed in Chapter 1. Within a limited region we are concerned with a much more detailed analysis of the climatic constraints on pasture production. It is realized that in many areas, particularly in undeveloped new areas, weather records may be very limited, and require extrapolation of data collected at widely dispersed recording stations. All available data should be collated to compute the following:

(i) Rainfall: total annual

Distribution (monthly or weekly totals)
Intensity of rainfall (mm h^{-1} or mm per wet day)
Evaporation (monthly or weekly total)

Given data on soil moisture storage characteristics, water balance models can be constructed. On the basis of long-term records, probabilities of droughts, waterlogging, flooding, and lengths of growing seasons can be predicted.

(ii) Temperature: mean annual

Mean monthly or weekly temperature
Daily maximum and minimum temperatures
Grass minimum temperature
Soil temperature (5 cm)

From the long-term records, probabilities of frosts, heat waves, and limiting temperatures can be computed and used to modify predicted lengths of growing season.

(iii) Radiation: mean daily radiation receipt

Annual distribution of radiation receipt
Annual curve of photoperiod

Periods of low radiation receipt may be related to lower rates of pasture growth, while an analysis of the day-length changes is essential in predicting flowering responses in pasture species.

(iv) Relative humidity and wind
Primary effects of these parameters are integrated into the evaporation component. Computation of annual patterns of relative humidity may allow prediction of disease incidence in pasture plants and stress effects on grazing animals. Occurrence of high wind is not as important in pastures as in crop production.

Climatic factors are the primary determinant of pasture production and are the least manageable of the environmental resources. Climate will determine the limits within which pasture improvement is possible, the species which might be used, areas from which plant introductions might be sought, and the overall productivity of the system. Thus thorough analysis of the climatic factors is essential in any planning or economic assessment of development options. More detailed discussions of agricultural climatic analyses can be found in Slatyer (1960), Chang (1968), Fitzpatrick and Nix (1970), and Moore and Russell (1976).

(b) *Land evaluation*

The other factors affecting tropical pasture production, or the suitability of land for pasture improvement are land form or geomorphology, hydrology, soils, and terrestrial ecology. This chapter does not propose a detailed discussion of land-use evaluation which is covered in many publications including Stewart (1968), Brinkman and Smyth (1973), Vink (1975), and Isbell and McCown (1976).

The data relating to land resources can be accumulated through integrated survey teams. The use of combined mapping of geomorphology, soils, and vegetation will reveal patterns of association and allow prediction of the characteristics of similar land units. The presence or absence of certain plant species are often indicators of particular soil conditions such as nutrient deficiencies, impeded drainage, poor water status, regular burning, and so on. The type and density of tree-cover will also have important effects on the cost of clearing and developing improved pastures, while the form and density of ground flora is a major determinant of suitability for different types of grazing animal.

The term 'ecological survey' rather than only 'vegetation survey' has been used. In many areas the desirability of introducing domestic grazing animals will be influenced not only by the vegetation resources but also by effects on wild animal populations. Competition for available food occurs when domestic livestock are introduced to areas already occupied by feeding complexes of wildlife (FAO 1974). In East Africa, this competition becomes acute in the dry season owing to the convergence of diets of all grazing species. Ranchers are also concerned by disease transmission from wildlife to domestic species and the problems of predation and damage to fence lines and watering points. Thus development of commercial ranching may lead to a marked decline in wildlife populations in many areas and this requires surveys of animal as well as vegetation resources. Indeed, surveys in Kenya of proposed new ranching areas indicated the primary importance of wildlife resources, greatly modifying the plans for range development for domestic livestock (FAO 1973). However, in many areas in Australia the establishment of improved pastures with an increase in the number of stock watering points has lead to marked increases in kangaroo populations (Newsome 1971). Also cattle grazing on mature forage improves the availability of young green shoots to the kangaroos (Low and Low 1975).

The soil survey is an inventory of the soil resources of the region and allows prediction of soil properties and nutrient status. The soil map of the region shows the distribution and relative proportion of each soil unit. From this the dominant soil types can be assessed and development priorities allocated. This information is essential in locating experimental sites so that these are representative of the main areas under study.

In allocating priorities for research and development the social, political, economic, and infrastructural factors must also be taken into account. While a particular region may be considered to have excellent resources for pasture improvement it may be relegated to a lower priority because of a lack of roads, or distance from markets, or because of more pressing social problems elsewhere. On the basis of the data collected in the survey stage many of the problems and limitations of the environment will be defined and the pasture-research programme will be developed as outlined in Table 6.1.

6.2 PASTURE AND ANIMAL PRODUCTION SYSTEMS AND OPTIONS FOR PASTURE IMPROVEMENT

Intensification of grassland production is an evolutionary process, the level of intensification reached depending upon the physical and economic resources of the region. Historically in the major livestock-producing countries — the United States, Canada, Australia, New Zealand — the grazing industries were established on the basis of very extensive grazing on large individual land holdings (Stoddart, Smith, and Box 1975; Shaw and Norman 1970). With time, property sizes tend to decline, and stocking rates increase as improvement to watering facilities, pasture improvement, and timber clearing take effect. These effects are seen very clearly in the example of the development of the Brigalow (*Acacia harpophylla*) forests in Australia. In their natural state the brigalow forests have a carrying capacity of only 1 steer to about 20 ha (Coaldrake 1970). With clearing and sown pasture development, stocking rates can be improved to 1 steer to 1 ha, with higher levels of production per head and a marked increase in production per hectare.

In the traditional nomadic grazing lands of the Middle East and North Africa, stocking rates have also increased dramatically in the past 70 years. Pearse (1970) suggests that this has been due to increasing numbers of herds associated with the increasing human population; reduction in area of rangeland owing to increasing, cultivated areas; removal of plants from rangeland for fuel; increased sedentary (rather than nomadic) grazing; improvements in stock water availability; and improved veterinary and supplementary feeding practices. Since pasture improvement and development of forage resources has not proceeded in parallel with the increase in livestock numbers, this has lead to serious deterioration of the rangelands.

When planning pasture-improvement projects, decisions must be made on the appropriate technology to give optimum levels of animal production consistent with maintenance of, or improvement of, the forage resources. The potential for increased production depends upon the physical resources of the area, but the limits to which these can be developed depends upon economics. This is basically a function of input costs to achieve a given level of output of animal product per unit area. Obviously in areas of low rainfall, where pasture productivity is low, stocking rates will be low, and output per hectare also low. Development costs for such areas must also be low on a per hectare basis. However, current levels of production are not always a good indicator of potential productivity, as shown in the brigalow forest-clearing example above, or in the case of application of micro-elements and phosphorus to heath-lands in southern Australia, where carrying capacity is changed from almost zero to 6 to 12 sheep ha^{-1} (Newman 1970). Thus a great deal of pasture research may be required to indicate the true potential of grazing lands for improvement, and continuing research is required to ensure that these levels can be achieved commercially.

The sequence of increasing intensity of systems of pasture and animal production is shown in Table 6.2. Each step in level of intensity also generally represents an increase in stocking rate, usually an increase in production per unit area, and an increasing input of capital, fertilizers, and energy per unit area. In this final section the main features of the major systems of ruminant animal production will be described, and management options which could be applied to pasture improvement in these systems will be discussed.

Nomadic grazing
This form of grassland utilization involving the regular migration of

Table 6.2. Levels of intensification of pasture and animal protection

Major production system	Pasture or grazing management characteristics
Extensive rangeland grazing	*Nomadic herding*. Transhumant grazing. No pasture improvement
	Sedentary grazing. Animals herded around water points, or communal ranch blocks or on crop residues around villages
	Extensive ranching. Extensive grazing on range, with or without fencing, animals not herded
Managed rangeland	*Ranching*. Fencing, increased watering points, timber clearing, woody weed control, animal supplementation
Improved rangeland	*Oversown range*. Introduction of improved legumes or grasses, after burning or cultivation
	Fertilized oversown range. As above with inputs of fertilizer for the oversown species
Cultivated sown pastures	Replacement of native species with *sown pastures*, with fertilizers, cultivation, and grazing management (i) Sown grass–legume mixtures (ii) Sown pure grass plus fertilizer nitrogen (iii) Sown pure grass plus fertilizer nitrogen plus irrigation (iv) Sown annual special purpose pastures (e.g. annual rye grass plus nitrogen; subterranean clover pastures)
Fodder crops	Annually sown fodder crops, grazed
Zero grazing systems	Forages or fodder crops, cut, carted, and fed to animals in stalls or enclosures
Fodder conservation	Forages or fodder crops, ensiled or made into hay before feeding
Lot fattening	Feeding cereal grains, concentrates and supplements with cut or conserved forages

herds and flocks to seasonal grazing areas is found through the arid and semi-arid lands of North Africa, East and West Africa, the Mediterranean basin, southern Asia and Asia Minor into Northern India and parts of China. More than 15 million peope are dependent on pastoral nomadism, grazing their herds over some 67 million km^2, which is approximately twice the cultivated area in the world (Grigg 1974). It is a very important land use.

While this has been an extremely fruitful field of research for the sociologists and geographers, who have detailed the social systems, migratory paths, and family economies of most of the nomadic cultures (see Johnson 1969; Monod 1975) there is little data on pasture development or improvement in these systems. Most observers of the nomadic system agree that forage resources are declining, as any 'equilibrium' which may have existed between nomadic herds and forage resources in the past have been destroyed. As Swift (1975) observes 'Pastoralist strategy is almost always to maximize herd size; when there were few nomads and lots of land, local overgrazing and destruction of vegetation was less important: the nomad could move on so the overall state of pasture was conserved. Now there are many nomads and limited pasture, so the same strategy leads to disaster.' But nomads are behaving as in the past but in changed circumstances.

Animals are almost totally reliant on natural forages and browse, although in some areas seasonal grazing may extend into crop stubble. There are no inputs for pasture improvement, and usually very little control over stocking rates, as land is grazed on a communal basis with traditional rights of use. In times of drought, pressure on grazing and water resources is intense, leading to the marked deterioration in range productivity as noted by Pearse (1970).

A fundamental problem in most nomadic grazing systems is the dependence of the nomads on milk. As Brown (1971) points out, this ecologically unwise dependence upon milk in areas suitable only for the production of meat makes subsistence more, not less, precarious. It also means that large numbers of breeding female stock must be maintained to provide sufficient milk for a family. Brown (1971) estimates that the livestock requirement per person to provide a reasonable pastoral standard of living is 5-6 cattle, or 25–30 goats, or 2.5–3 milch camels. Supporting this number of animals puts heavy pressures on the sparse grazing resources of arid and semi-arid communities.

Although water for livestock is a major limitation in most nomadic environments, the improvement of water supplies does not necessarily lead to long-term increases in animal production. Where animals are dependent on surface-water, and thus grazing is confined to the wet season, pasture degredation occurs less rapidly. The major areas of devastation occur around the permanent dry-season water-points. Swift (1975) observed in Mali that vegetation is devastated in a radius of 20–35 km around permanent waters owing to increasing human and animal populations concentrating in the dry season. What measures can be undertaken to improve pasture resources in nomadic and transhumant grazing areas? Without control of stocking rate, and time and place of grazing, the potential for improvement is minimal, but this is the most difficult management practice to apply to communally grazed lands. If a degree of grazing control can be implemented then a number of practices can be applied to improve the grazing resource. A major FAO study in Kordofan province, Southern Sudan, in nomadic grazing areas receiving rainfalls of 280 to 570 mm per annum demonstrated the following options (Skerman 1966).

(i)   Exclosure of animals by fencing from degraded areas leads to quite rapid (2–3 years) improvement in perennial and annual grasses and legumes, and a marked increase in carrying capacity when compared with the open range as shown in Table 6.3. The increase in carrying capacity depended on the state of previous degredation, and it can be seen in the case of sites 2, 3, 5, and 6 that carrying capacity is dramatically increased by a few years' protection from grazing. Subsequent managed grazing of these areas would lead to further improvement.

Table 6.3. Effect of exclosure of animals for two to three years on estimated carrying capacity compared with adjacent unprotected open range in Kordofan, Sudan (from Skerman 1966)

No.	Site name	Mean annual rainfall (mm)	Stocking rate (ha per cattle beast)	
			Exclosure	Open range
1	Umm Higlig	408	12.0	28.0
2	Jebel Dago	408	4.4	49.0
3	Mazrub	284	6.0	17.4
4	Mahbub	300	3.2	4.4
5	Khuwei	375	8.1	32.4
6	En Nahud	415	4.4	32.4
7	Abu Laota	360	4.4	6.9

(ii) Installation of new water-points should only be undertaken where facilities can be fenced and managed, and access is based on allocation of grazing rights to control grazing pressure.

(iii) Pasture improvement can be undertaken directly in conjunction with the natural re-establishment, or planting, of gum arabic (*Acacia senegal*). The acacia provides for sap collection and browse for camels and goats, while the production of undersown grasses such as *Panicum maximum* var. *trichoglume* (green panic) or *Cenchrus ciliaris* (buffel grass) is little reduced in association with the trees. This again requires exclosure of stock during establishment.

(iv) Establishment of browse trees and shrubs. The Forestry Department enclosed some 1000 ha around the town of El Obeid, which led to a steady increase in forage and browse resources. These areas after 4–5 years could then be used for controlled rotational grazing.

The planting of browse species has been recommended in southern Sudan and Chad (Tubiana and Tubiana 1975); the species *Bauhinia rufescena*, *Cadaba farinosa*, *Acacia mellifera*, *Balanites aegyptica*, *Maerua crassifolia*, *Grewia tenax*, and *Zizyphus spina-christi* seem to be the most widely adapted and readily eaten. Protection during establishment by fencing or thorn enclosures is necessary.

(v) Fire breaks. Regular uncontrolled burning causes a loss of the current years production as well as a reduction in forage production in the subsequent year, and can also lead to the destruction of valuable browse species. An occasional controlled burn is useful to eradicate some weed species and to release nutrients held in standing dead biomass (Tothill 1971*a*). Recognizing the importance of controlling wild fire, the Range Management Service in Sudan is developing a system of fire-breaks throughout the range areas.

Not all nomadic grazing is in arid and semi-arid areas. The people of Karamoja district north-east Uganda carried out transhumant grazing in a rainfall environment of 500 to 1000 mm, albeit an unreliable rainfall with regular droughts. Here, apart from the constraints of the nomadic social system, there are many more options for pasture improvement through to the cultivation and planting of sown pasture mixtures. Thus, the two major factors determining the potential for improving grazing resources in the traditional nomadic areas are the climate/soil factor and the extent to which stocking rate and grazing patterns can be controlled.

Sedentary grazing
Patterns of sedentary grazing, where animals are herded, can arise in many ways. In some nomadic societies women and children may be left near water-points during the wet season to grow a crop. Small animals, goats, and sheep, may be herded in the vicinity of the permanent water, while the main herd is driven to distant grazing lands.

In village communities throughout Africa and Asia all classes of livestock, including draught animals, are grazed on common lands, roadsides, field verges, and forest margins around the villages. Crop stubbles may form a major part of the diet at certain times of the year. In India it was estimated in 1965 that dry fodders, mainly cereal straw, provided 49 per cent of dry matter and 55 per cent of energy eaten (Oram 1975).

Forage supply under sedentary village grazing systems is limited, firstly by the radius from the village over which animals can be herded out and back each day, secondly by the high and uncontrolled grazing pressures on communal grazing areas, by the low quality of many cereal straws, and by the fact that animals are usually coralled at night, which limits their grazing time.

This grazing system is probably the most important system in the tropics in terms of numbers of animals and people dependent on those animals. Improvements to this grazing system by improving pastures in community grazing areas, by the use of forage crops and undersown legumes in the arable areas, and more efficient utilization of stubbles and by-products will have major effects on livestock production in the tropics.

Extensive ranching

In this system animals are not herded but allowed free range on land under the control of a single manager, who is either a private owner or who holds the land on a leasehold basis. This open-range grazing on natural pastures and browse is found in North and South America, Australia, New Zealand, and parts of East and Southern Africa.

Property sizes may be large, being a function of productivity of the land and remoteness from urban centres. In Northern Australia many properties exceed 1000 km^2. On these large properties subdivision fencing is limited and stock distribution is mainly determined by availability of water.

At the first level of ranch development, animals are dependent on natural pastures and improvement in productivity is pursued through animal management, by:

control of stocking rate by off-take and selling policies;
improving stock distribution and hence pasture utilization by increasing the number of water-points and by fencing;
control of mating which allows control of weaning;
introduction of improved sires and crossbreeding;

testing and vaccination for pleuropneumonia, brucellosis, tuberculosis, and other diseases; and

supplementation during the dry season with non-protein nitrogen and minerals as required.

At this stage of development few resources are invested in direct pasture management, apart from the control of burning, through prevention of wild fires and managed burning as required.

Managed rangeland
The term rangeland usually denotes those areas of natural pastures grazed on an extensive basis in regions which are marginal for regular and sustained arable cropping. However, areas which were rangelands such as the prairies of North America and the pampas of South America have been converted to croplands through improved technology and mechanization, or in North Africa and Turkey through sheer pressure of population on the available arable lands. The major regions of managed ranchlands are found in South America, North America, parts of east and southern Africa, Australia, and New Zealand, and encompass a wide range of vegetation formations, from open monsoonal woodlands in northern Australia, through the shrub woodlands of North America, to the open grasslands and savannahs of South America and East Africa. The main rangeland grazing areas of the world are described in Stoddart *et al.* (1975).

The basic feature is still a major dependence on natural forages, but with increasing inputs for vegetation management to increase pasture productivity. The aim of rangeland management through both livestock and vegetation management is to increase the productivity of desirable grazing species and to limit the ingress or expansion of weed species. A weed species can be variously defined but includes species not eaten by grazing animals, species which are toxic, or species which produce little digestible nutrient or are poorly utilized but prevent the growth of more productive species. The main management options to meet these objectives and to improve natural grassland productivity are discussed below.

Clearing of trees
In the woodland, scrub, and scrub–savannah grasslands a major improvement in herbage production may be achieved by clearing existing tree stands. Indeed, tree clearing is essential in many parts of East and Central Africa to control tse-tse fly (*Glossinia* sp) before cattle ranching can be undertaken.

Where tree densities are high, there is a curvilinear relationship between herbage production and tree density as shown in Fig. 6.1

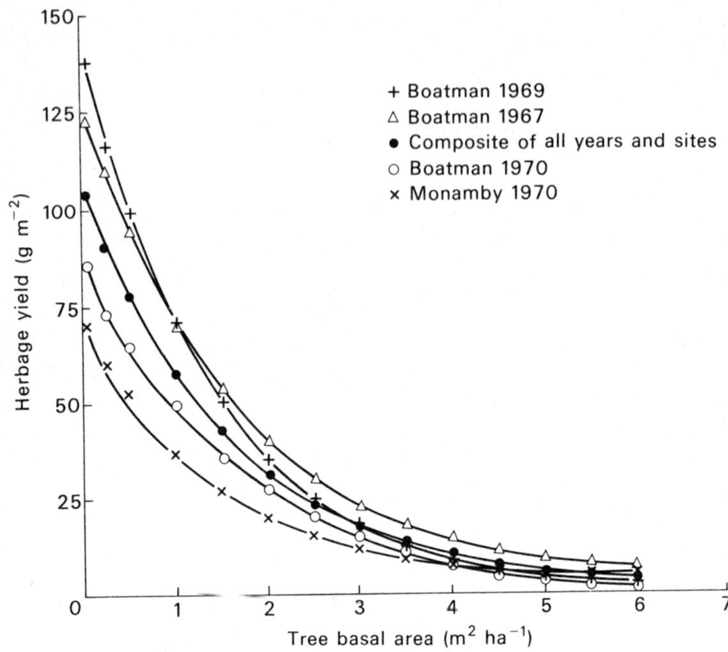

Fig. 6.1. Relation between herbage yield (g m⁻²) and basal area (m² ha⁻¹) of mulga (*Acacia aneura*) in south-west Queensland (Beale 1975).

(Beale 1973). A similar relationship was shown with clearing of *Eucalyptus populnea* by Walker, Moore, and Robertson (1972). Both studies show that the major effect is achieved with full clearing, or conversely that leaving even a few trees can significantly reduce herbage production. In semi-arid environments competition for soil water is an important factor in reducing herbage yield. Thinning and clearing *Acacia aneura* forest has been shown to reduce the total water use by the community Table 6.4 (Pressland 1976).

Table 6.4. Effect of thinning of *Acacia aneura* at Charleville, Queensland, on total evapotranspiration over 22 months (Pressland 1976)

Density (trees ha⁻¹)	0	40	167	640	4000
Et (mm)	814	835	847	870	852

The apparent lower water use at the highest tree density, where there was a complete tree canopy, was due to interception of up to 13 per cent of rainfall which was then evaporated from the *Acacia* foliage. This study also shows that even at 40 trees ha^{-1} evapotranspiration was significantly higher than with complete clearing.

However, there is some evidence from more open bush savannah communities that complete clearing may not always result in higher pasture yields. It has been noted that grass yields are sometimes higher under shade of bushes than in the open, owing in the case of mesquite, to higher soil nitrogen, lower bulk density, and higher soluble salts in the soil beneath the trees (Tiedemann 1970). Even so, clearing of mesquite leads to higher levels of herbage production as shown by Martin and Cable (1974) in studies of management of grass-shrub ranges near Tuscon, Arizona, as follows:

Year	Mesquite cleared (kg ha^{-1} total grass)	Mesquite infested (kg ha^{-1} total grass)
1959 (wet year)	437	308
1963 (wet year)	347	235
1960 (dry year)	134	90
1962 (dry year)	50	34
10 year mean	240	166

Studies in the mixed veld and *Acacia–Combretum* tree savannah in Botswana suggest that on the heavier textured soils, because herbage yields were higher under the trees and shrubs than in the open, partial charring gave higher herbage yields and more grazing days than full clearing gave higher herbage yields and more grazing days than full the sand veldt soils clearing gave increased herbage yields and improved botanical composition.

There are many methods available for timber clearing including ring barking, hand felling, felling and treatment of the cut stump with herbicide, spraying basal stems with herbicide, mechanical pushing with tractors, or pulling with chains between tractors, as previously discussed in Section 4.1. Perhaps the neatest solution to removal of trees and scrub was reported by Skovlin (1971). On one ranch in Kenya trees were cut and 'grubbed' out, and the larger material converted to charcoal, under contract, by local cultivators. The contractors made between $10 and $20 per hectare from the sale of charcoal, while the rancher doubled his carrying capacity within a year at no cost. Increased animal production after timber removal has been documented by other workers. Gillard (1970) reported that clearing of *Eucalyptus* open forest above a *S. humilis* oversown *Heteropogon* native pastures more than doubled live-weight gains over a four-year period from a mean of 17 kg ha^{-1} to a maximum 48 kg ha^{-1}. Again in *Eucalyptus* open forest, Tothill (1974) reported an increase in liveweight gain from 20 kg ha^{-1} a^{-1} to 43 kg ha^{-1} a^{-1} following clearing. Provided clearing is undertaken judiciously with due regard to the needs of fodder trees for drought reserves, shade for livestock, stream-bank protection, and timber for fuel and fencing, then removal of standing woodland can lead to significant increases in animal production.

Control of weeds and woody regrowths

Once existing tree species are cleared, control of tree regrowth or ingress of herbaceous weeds is a continuing requirement. Following clearing and disturbance, shrub species such as *Acacia* and *Cassia* may regenerate from seed, and this appears to be a greater problem on light than heavy soils as shown by the data of Moore and Walker (1972) as follows:

	Light soil		Heavy soil	
	Cleared forest	Uncleared	Cleared forest	Uncleared
Total no. of shrubs after 2 years	3370	1870	890	880
After 2 years grazing with sheep	172	–	34	–

Heavy intermittent grazing with sheep drastically reduced shrub densities to a level where they had little effect on herbage production. Tothill (1971*b*) has also shown that grazing has a far greater effect in preventing *Eucalyptus* regeneration than does burning alone or the application of fertilizers. After imposing grazing with cattle, burning, and fertilizer treatments over four years the number of tree saplings larger than 1 m tall per hectare were as follows:

Ungrazed unburnt	225	Grazed and unfertilized	15
Ungrazed burnt	103	Grazed and fertilized	6

The type of grazing animal is also important. Skovlin (1971) has reported on the beneficial effects of eland and giraffe on East African ranches in controlling shrub and tree species, and goats are widely reported to control many woody weed species (Campbell, Ebersohn, and Von Broembsen 1962).

Although the grazing animals may prevent establishment of some woody species, or reduce the density of already established species, many weed problems are a direct result of overstocking. Stocking rate can have marked differential effects on weed content of pastures and can be used to alter botanical composition. A good example is shown in the grazing trials of Harrington and Pratchett (1974) at Ankole, Uganda. They compared continuous and deferred grazing on native pastures at stocking rates of 0.28. 0.41, 0.82, 1.23, and 1.65 animals ha^{-1}. At the two highest stocking rates *Brachiaria decumbens* was encouraged at the expense of *Themeda triandra* and *Hyparrhenia filipendula*. Because of the higher nutritive value of *Brachiaria*, cattle at the highest stocking rates performed far better than would be predicted from the linear stocking rate/liveweight gain relationship.

This trial also showed that removal by hand digging of the unpalatable grass species *Cymbopogon afronadus* increased gains per hectare by over 40 per cent at a stocking rate 1.65 animals ha^{-1}. At the optimum stocking rate of 1.23 animals ha^{-1} it was calculated that the costs of clearing this weed would be recouped in one year with a steer fattening enterprise, or in two years with a breeding herd.

Burning

The control of wildfires and the use of prescribed burning are closely related tools widely employed in the management of natural pastures. Tothill (1971*a*) has listed the main uses of fire in pasture management as follows:

1. to remove accumulated rank low quality fibrous material to increase availability of new seasons growth, and to remove patchiness in unevenly grazed pastures, or to allow for oversowing of legumes;
2. to stimulate tillering and leaf production at times of the year when it might not otherwise occur;
3. to control regrowth of trees, shrubs and other seeds, and to clean up fallen timber;
4. to attract animals to areas that might otherwise be left ungrazed;
5. to remove accumulated fuel and to establish fire breaks to control wildfires; and
6. to control diseases and pests, such as ticks.

The role of burning in improving accessibility of new growth to grazing animals, or in stimulating out of season production does not appear to have been rigorously tested in terms of increasing animal production. Almost all pasture studies reported that burning reduces the total amount of growth over the ensuing season when compared with unburnt or mown treatments (Tothill 1971*a*). This is confirmed by Norman (1969) who found in a monsoonal environment favouring the accumulation of dead material that burning about every five years appeared to be more favourable for pasture production than annual or biennial burning or no burning at all.

This recommendation fits in well with management if the main use of fire is seen as a method of controlling bush regrowth. There is ample evidence of carrying capacity of pastures being doubled, trebled, or even further increased by removal of bush (Ivens 1972). In this role in East Africa burning is recommended every 3–4 years (Ivens 1972) or every 6–7 years (Heady 1972), depending on the species and their rate of growth. There is a real cost involved in the use of fire. Areas to be burnt must be rested from grazing for 6 to 12 months to build up sufficient fuel to carry a fire hot enough to control woody plants. After burning the area should not be restocked for a further 6 to 12 months to allow sufficient time for the useful plants to become well established before grazing. This means that, under managed ranch conditions where a rotational burning programme is used, approximately 25 per cent of the total grazing area is not being grazed at any one time.

The value of fire in controlling pests and diseases is not well documented. Tothill (1971*a*) suggests that burning has little effect on tick control, while Heady (1972) reports that fire has been used in tse-tse fly (*Glossinia* sp) control programmes to reduce bush, but seldom does it reduce fly populations unless vegetation is changed. Generally the influence is temporary. An exception to these generally negative reports is provided by Wright (1974), who recorded that following a major burn in Oregon, USA, that deer numbers increased. The deer were healthy and free of liver fluke and lung worm which had plagued the deer herds before the fire. It was found that the fire had eliminated the dry-land snail which is the intermediate host for liver fluke and certain lung worms.

Burning should be looked upon as a tool for managing botanical composition and maintaining overall pasture productivity and should only be undertaken at the intervals required for these purposes. With the development of supplementary feeding with protein or non-protein nitrogen during the dry season, the standing forage represents a resource which is more valuable and more readily utilized than it was formerly.

Further development of natural pastures by oversowing legumes, or replacement with sown pastures leads to the necessity to protect the areas from burning.

Oversown rangeland

This level of development recognizes that the native pastures have limitations in total production or seasonality of production, or of nutritive value, and suggests that these may be overcome by introducing other species into the existing assemblage of plants. Because of their value in nitrogen fixation and in nutritive value, legumes have been the main species oversown. Little success has been achieved in oversowing grasses into existing grass swards. Establishment and persistence of oversown species usually requires some change in management, either use of fertilizers, use of cultural practices such as fire or limited cultivation, control of grazing and changes in grazing management. However, as Humphreys (1977) points out, oversowing is more likely to be adopted by farmers if modifications to existing practices are minimal. Factors associated with the establishment of oversown legumes have been discussed in section 4.1, and some data on levels of animal production from oversown pastures were dicussed in Table 5.15. Where legume content of native pastures is low and pasture and animal production limited by protein content and total intake, introduction of a legume by oversowing can lead to marked improvements in animal production. Oversown legumes are usually more successful when fertilizer is applied, as shown in Table 6.5, where beef-cattle production is compared from experiments with native pasture, oversown pasture, or oversown fertilized pastures.

In the Australian examples, animal production was increased between two- and eight-fold by oversowing, with further improvement due to fertilizing, reflecting the generally low fertility of pastoral soils and low nutritive value of the pastures. In contrast, animal production on the pastures in the Uganda experiment was much higher, there was a lesser response to oversowing and no significant response to addition of phosphorus, potassium, and sulphur fertilizers. In the Papua New Guinea example, response to oversowing was limited by low soil fertility, and should have been much higher had fertilizer been applied. Other experiments showed significant responses to phosphorus fertilizers (Chadhokar 1977).

Since most cattle in the tropics are dependent on natural pastures, and given the demonstrated value of oversowing legumes, more research

Table 6.5. Comparison of beef cattle liveweight gains per hectare from native pastures, native pastures oversown with a legume, or oversown and fertilizer applied

Location	Natural pasture	Legume oversown	Liveweight gain (kg ha^{-1})			Time (days)	Reference
			Unimproved	Oversown	Oversown and fertilizer		
Central Queensland	*Heteropogon*	*S. humilis*	11	85	–	365	Shaw (1961)
Central Queensland	*Heteropogon*	*S. humilis*	32	103	163	365	Shaw and 't Mannetje (1970)
Katherine, NT	*Themeda, Sorghum*	*S. humilis*	–5.5	–	+150	112	Norman (1970)
Uganda	*Chloris Panicum,* } *Hyparrhenia*	*S. guianensis* } *C. pubescens*	269 245	362 339	366 373	343 343	Stobbs (1966)
Papua-New Guinea	*Imperata*	*S. guianensis*	78 78	101 112	– –	300 420	Chadhokar (1977)
Central Queensland	*Heteropogon*	siratro	43	–	115	365	Tothill (1974)
Southern Queensland	*Bothriochloa–Eragrostis*	siratro	75 92 72	– – –	91 177 143	199 264 213	Lowe, Filet, Burns, and Bowdler (1977)

is required into selecting legumes better adapted for oversowing, and in ensuring better establishment. Oversowing can be seen as a first step in improving grasslands, leading to higher fertility and eventually to fully sown pastures.

Sown pastures

The preceding chapters have concentrated on the principles and concepts of sown pasture development, and the increasing levels of input for different pasture production systems are shown in Table 6.2. Examples of increased productivity through sown pasture development are given in Tables 5.15 and 5.16.

Fodder crops

Crops may be grown on cultivated land for animal feed and harvested by grazing, cutting and hand feeding, or for forage conservation. This represents a higher cost option than permanent pastures and in the tropics is usually only justifiable where continuity of feed supply is required for a high value product such as milk. Where farmers have only arable land cultivated fodders may be grown for draught animals; for example, Berseem clover (*Trifolium alexandrinum*) is grown as a short season forage crop in Egypt, while *Crotalaria* is commonly grown for forage in India. These legumes provide a high protein forage which will assist in the utilization of lower value crop residues.

In a monsoonal environment at Katherine, Norman (1963) compared beef production from native pastures, sown *S. humilis–Cenchrus setigerus* pastures, and bullrush millet (*Pennisetum typhoides*) forage crops. Over a series of trials in which cattle were grazed on native pasture for a period at the start of the wet season and then either maintained on native pasture or transferred to sown pasture or to fodder crop *average* gains were as follows:

	Stocking rate and LWG (on treatment)		
Treatment	(beasts ha^{-1})	(kg head^{-1})	(kg ha^{-1})
Native pasture (Nov.–June)	0.25	50	12
Transferred to sown pasture (Jan. or Feb.–June)	2.47	91	225
Transferred to bullrush millet (Jan. or Feb.–June)	2.47	102	252

These data show the marked increase in production per head and per hectare when sown pastures or fodder crops are introduced into the system. Although production was higher from the fodder crop it was considered that the extra production was not achieved economically compared with the lower cost *S. humilis*-based permanent pasture. In small holder farming systems in the tropics it is also unlikely that fodder crops are a viable option for beef production when compared against the use of the same land to grow a cash crop.

However, for milk production in the tropics the intensive cultivation of fodder crops may be a viable alternative. Oram (1975) quotes results of pilot farm trials conducted by the Indian National Dairy Research Institute where 33 milking cows and two bullocks were kept on intensively irrigated small farms of only 5 ha, and produced an annual net profit of $3000. This was higher than the income obtainable from cash cropping.

Sorghum is widely used as a fodder crop in the tropics, but recent studies have shown that sorghum is usually low in sulphur and sodium (Archer and Wheeler 1978). Values in Australian sorghum forage crops for sulphur were generally below 0.13 per cent, the nitrogen/sulphur ratios above the 15:1 suggested as optimum for beef cattle, while sodium levels were generally below 0.012 per cent. Supplementation of beef cattle grazing sorghum forage with salt–sulphur blocks increased daily liveweight gain by 0.2 kg head^{-1} day^{-1} (37 per cent) (Archer and Wheeler 1978), while supplementation of dairy cattle increased milk yield by 11 per cent (Stobbs and Wheeler 1977).

While some forage legume production may be required to maintain protein levels in the diet, the data discussed in Section 5.3, suggest that the tropical legumes give only poor milk yields when fed as the main component of the diet.

In the subtropics, where it is possible to grow temperate forages such as oats (*Avena sativa*), rye grasses (*Lolium* sp) or vetches (*Vicia sativa*) during the cool season, the use of fodder crops for milk production or fattening sheep or cattle may be a viable option. These species have a high nutritive value, and grow in a period when tropical pastures are limited by low temperatures. In southern Queensland and northern New South Wales the most widely sown winter forage crops are oats and rye grass. By using a combination of planting dates, oat varieties with different maturity and rye grass a continuity of feed supply can be maintained through the cool season.

Oats has a faster initial growth rate and hence provides forage for grazing in shorter period from sowing than the rye grasses (Kemp 1974). From a comparison of a number of oats and rye grass cultivars sown at different times in south eastern Queensland, Murphy and Whiteman (unpublished data) suggest sowing saia oats (*Avena strigosa*) in early April to give early feed, at the same time planting HI rye grass (*L. perenne* × *L. multiflorum*) and Wimmera rye grass (*Lolium rigidum*) for mid-season feed and Kangaroo Valley rye grass (*L. perenne*) for late feed into the spring. This latter species has the slower early growth rate, but gave higher production in the long term as was also found by Stillman and Ostrowski (1976) in the same environment. Nitrogen applications of 30 to 60 kg N ha^{-1} after each grazing are usually required to maintain yield and protein content (Cull and Douglas 1974). They also show that it is more economical to apply nitrogen than to grow a larger area of crop to produce the required amount of feed.

Weight gains of beef cattle grazing irrigated fertilized rye grass, grown in the cool season, of 0.52 to 1.34 kg head^{-1} day^{-1} were recorded by Milles and Hall (1978). The lower values were for young *Bos indicus* weaners, and the top values for three-year-old Droughtmaster × Hereford bullocks. Over the trial period, with grazing from June to December, 230 cattle were fattened on 15.4 ha, producing dressed carcass weight gains of 1220 kg ha^{-1}. The average length of time each animal spent on ryegrass was 88 days. Values for milk production from temperate forages, including ryegrass, were given in Section 5.3, to demonstrate the much higher production potential per head compared with tropical feeds. Thus where temperate fodder crops can be integrated with tropical pastures, high levels of production per animal can be obtained to allow continuity of milk production or fattening of cattle and sheep for slaughter.

Zero grazing systems
The feeding of cut forages or 'cut and cart' systems are widely employed in intensive temperate agriculture. In most cases facilities are available for winter stall feeding so that extension to year round feeding on green or conserved forages is not difficult. The advantages claimed for zero grazing systems over field grazing are that efficiency and uniformity of utilization of forage is higher because wastage from trampling and fouling is avoided; the energy expended by the animal in grazing is reduced; pugging damage to wet soils and treading damage is avoided leading the higher forage growth rates (Smetham 1973). In economic terms less labour is required to herd stock to and from grazing and field fencing and reticulation of water to separate fields is not required. Against this, higher labour inputs may be required to cut and distribute

forage and also to remove and dispose of animal excreta, while a higher capital input may be necessary in yards, structures, and equipment for the feeding areas and for fuel costs for these operations. Furthermore, opportunity for selection by the animal is reduced when cut forage is fed (Raymond 1970). While this may impose some limitations on intake and digestibility with temperate forages, it can become a major limitation with tropical forages.

Greenhalgh, Aitken, and Reid (1972) compared beef cattle live-weight gains when zero grazed, grazed at the same stocking rate, and grazed at a variable stocking rate designed to ensure about the same rate of gain as the zero grazing group, on rye-grass-dominant pastures. The mean values for the measured pasture and animal parameters over four grazing cycles over a 115 day period are summarized in Table 6.6.

Table 6.6. Comparison of zero grazing and grazing at two stocking rates on pasture and animal production (Greenhalgh *et al.* 1972)

Grazing treatment	Stock rate (an ha^{-1})	OM intake (kg day^{-1})	OM digest (%)	LWG (kg)	
				kg day^{-1}	ha^{-1}
Zero grazing	7.8	6.54	75.0	0.98	1070
Grazed at same SR	7.8	6.18	76.5	0.78	840
Grazed at variable SR	6.7	6.87	76.8	0.90	910

These data show that the zero grazing could support a higher stocking rate which gave a higher gain per animal and a 20 per cent higher gain per hectare than grazing at the same stocking rate. However, the optimum stocking rate under grazing was lower than for zero grazing, and at this stocking rate gain per hectare was only 15 per cent higher in the zero grazing system. It then becomes an economic decision whether the higher costs involved in zero grazing are covered by this level of increased production.

In the tropics cut and cart feeding systems are widely employed because of limited grazing land, fragmentation of holdings, the lack of fences in arable crop areas, and the low opportunity cost of labour employed in cutting feed. From my own observations of smallholder cut-and-carry feeding operations there can be no doubt that animal production is usually severely limited in these for the following reasons:

(a) Operators tend to cut forage from areas that give the greatest return
for human effort. The human grass cutter is a poor simulator of the

grazing animal. Consequently, much cut forage from roadsides, waste ground, etc., tends to be mature and of low quality.

(b) The amount of forage allocated per animal is usually too small to allow for selection, or indeed for maximum intake.

(c) High producing cows, which in a grazing situation would undertake extra grazing at night (Stobbs 1970), are unable to do so, thus their productive potential is never realized.

(d) Animals are dependent entirely on feed supplied by the operator. This usually means diets are low in protein and commonly also mineral content is unbalanced.

(e) In disposing of the animal excreta, much of the urine portion tends to be lost, while the dung is returned usually to the arable crop areas, and not to the forage-producing areas, thus fertility of these sites declines.

Because of limitations of the forage supply, maintenance of reasonable levels of milk production in many areas has depended on feeding quite high levels of concentrates such as coconut meal, rice brans, or grains. The value of these products has risen sharply in the past few years, and this will demand a much greater dependence on cut forages in future. Thus a great deal of applied research on improving grass and legume forages for cut and carry feeding systems for smallholder animal production is necessary.

Fodder conservation

In many areas of the world pasture growth is highly seasonal owing to limitations of temperature or moisture. Feed requirements of a farmer's herd are relatively constant throughout the year, so the primary aim of fodder conservation is to transfer surplus forage from the peak growing period to the period of deficit. Because of the necessity to store forage for the winter season the philosophies of fodder conservation have been developed under temperate pasture conditions. These principles need to be re-examined before application to tropical pasture conditions for a number of reasons, viz:

(a) livestock are able to graze outdoors all year round;

(b) tropical forages have a lower nutritive value than most temperate forages, and thus total digestible nutrients stored per unit volume is lower and storage costs will be higher;

(c) field curing of hay during the main wet season growing period of tropical pastures can be hazardous owing to the high rainfall and frequency of rainfall events. Frequently tedded and turned *Setaria anceps* hay required 50–75 hours for curing, to reach a moisture content of 25 per cent (on a dry matter basis) (Catchpoole 1969). Rate of drying would be slower in a wet tropical environment;

(d) silage production from tropical forages is unreliable. Tropical species are more difficult to compact, and have a lower density, so that exclusion of air is more difficult (Catchpoole and Henzell 1971). Stable preservation of tropical species silage is usually associated with acetic acid production and a higher pH compared with temperate species where lactic acid is the main acid. Some species such as *M. atropurpureum* cv. Siratro and Rhodes grass (*C. gayana*) decompose after ensiling unless large amounts of molasses (7 to 8 per cent) are added (Catchpoole 1970).

While the ensiling of tropical pasture species can be difficult, silage production from sorghum and maize has been widely practised.

An important consideration is the type of production. Where a continuity of both quantity and quality of forage supply is essential as in dairy production or out-of-season lamb fattening, then fodder conservation may be necessary. For other types of production such as wool growing or maintaining a breeding herd Hutchinson (1972) has shown that in many cases the storage of energy by the build-up in body weight during the main pasture growing seasons and its subsequent breakdown to provide energy during the deficit period is more efficient than fodder conservation. Feeding of hay supplements becomes inefficient for wool production if some forage is available in the pasture. A further consideration in allowing weight loss during periods of feed deficit is that advantage can be taken of the process of 'compensatory gain'. Animals after a period of weight-loss gain weight at a faster rate on the same forage than animals maintained at a constant rate of gain (Allden (1970). There are major inefficiencies in the fodder conservation process, which penalize animal production. The first is that in reducing the total area of pasture available for grazing, by setting aside a certain proportion for fodder conservation, stocking rate is increased on the non-conserved area, so reducing production per head. Thus there is a clear interaction between stocking rate and the relative advantage of fodder conservation as shown in Table 6.7. These examples show that the advantages of fodder conservation decline at the higher stocking rates, where the need for additional fodder is greatest.

Table 6.7. The interaction of stocking rate and response to fodder conservation

Production system	Stocking rate per ha	Difference (conservation – no conservation)	Location and reference
Wool	10 wethers	+ 2 kg wool ha^{-1}	Armidale
	20 wethers	+16 kg wool ha^{-1}	(Hutchinson 1966)
	30 wethers	−15 kg wool ha^{-1}	
Weaner cattle	1.5 steers	+26 kg LWG ha^{-1}	Gayndah
	2.5 steers	+ 9 kg LWG ha^{-1}	(Scateni 1969)
Cattle (18 months)	3.7 steers	+101 kg LWH ha^{-1}	Samford
Setaria + nitrogen	5.0 steers	+34 kg LWG ha^{-1}	(Whiteman 1969)
Cattle	2.0 steers	−35 kg LWG ha^{-1}	Lawes
S. almum	3.3 steers	−26 kg LWG ha^{-1}	(Smith 1968)

These penalties are incurred because of wastages in the fodder-conservation process. In the hay-making process with tropical grasses Catchpoole (1969) recorded losses of dry matter of 10–13 per cent during curing, and about 5 per cent during storage. Losses during feeding out were not recorded, but Hutchinson (1972) found that less and less hay is consumed as the amount of live and dead herbage in the pasture increased. In silage-making with tropical species, dry-matter losses in properly sealed silos seem to average around 20 per cent, but where problems of compaction have occurred losses of up to 75 per cent are recorded (Catchpoole and Henzell 1971). Another problem with tropical silages is that intake and digestibility values are reduced after ensiling. Intake values for a range of grass silages without additives were 1.2–1.7 kg DM per 100 kg body weight, compared with values up to 2.2 kg DM per 100 kg body weight for temperate silage. Dry-matter digestibility of kikuyu grass (*P. clandestinum*) fell from 64 to 46 per cent after ensiling, and *P. dilatatum* declined from 60 to 51 per cent (Catchpoole and Henzell 1971).

Finally the costs of fodder conservation must be considered. Where the process is mechanized there is a large capital investment in specialized machinery and in storage facilities, plus the costs of fuel and labour for harvesting, storing, and feeding out. Fodder-conservation in the tropics must be considered a high cost practice which would yield less return on investment than alternative practices for increasing forage supply during deficit periods, such as planting of better species, use of fertilizers, irrigation, or supplementation with non-protein nitrogen.

Lot-fattening

This term is used to denote a production system more intensive than the 'zero-grazing' system, whereby animals are fed on concentrate feeds to ensure maximum productivity per animal. These forms of intensive animal production require high capital investment in buildings and machinery for milling, mixing, and feeding; in animals and in feedstuffs.

Lot fattening of livestock reaches its peak development in North America, where over 80 per cent of the grain produced is fed to livestock (Table 6.8) (Jasiorowski 1975). When high levels of grain and concentrates are fed with a balanced ration of minerals and vitamins production per animal can be maximized, and high conversion ratios (6.4 kg grain per kg LWG, National Research Council (1970)) can be achieved. On this basis, in terms of efficiency of human nutrition, only 21 per cent of the energy and 16 per cent of the protein fed to the animal in the grain becomes available for human consumption in the kg of liveweight gain produced. Thus feeding of grains to ruminants represents a luxury consumption of valuable resources and denies the major advantage of the ruminant, the ability to convert fibrous herbage to a high value product.

Table 6.8. The use of grain for intensive livestock production in the major economic zones (after Jasiorowski 1975)

Groups of countries	Projected grain use in 1980 (m tonnes)		% Grain used in animal feed	Amount of grain fed per kg meat
	Human food	Animal food		
Developed	58	300	84	4.95
Centrally planned	227	150	40	3.60
Developing	305	65	18	1.30

A second major problem of intensive animal feeding is in the storage, handling, and disposal of waste. In the USA approximately 600 m tonnes of solid waste, 200 m tonnes of liquid waste, and 400 m tonnes of associated wastes such as bedding and dead carcasses are produced each year from intensive animal production enterprises (Jasiorowski 1975). This is a formidable disposal problem.

In the developing world there is a place for intensive lot feeding where grazing resources are limited and population pressures are high. Lot feeding is labour-intensive, which is an advantage, and can allow for the efficient utilization of by-products such as rice bran, cannery wastes, copra meal, molasses, and fruit and vegetable processing, and household wastes.

Conclusion

This chapter has tried to demonstrate some of the management options available for pasture improvement in the developing tropics. There is now a wide range of pasture species available for sowing, and a great deal more is known about their adaptations, nutritive value, and management requirements. Furthermore some of the limitations of tropical forages for animal production are better understood, providing a basis for a more rational approach in planning livestock projects. The important requirement is now to take the principles and concepts so far developed in tropical pasture science and apply them to the many combinations of soils, climate, social, and livestock management systems in the tropics. Only when adequate nutrition is provided throughout the year for tropical livestock will the full benefits of improved animal health, parasite control, and breeding become evident.

REFERENCES

Allden, W. G. (1970). The effects of nutrient deprivation on the subsequent productivity of sheep and cattle. *Nutr. Abstr. Rev.* **40**, 1167.

Anon (1976). *An integrated programme of beef cattle and range research in Botswana, 1970–1976*. Animal Production Research Unit, Ministry of Agriculture, Gaborone, Botswana.

Archer, K. A. and Wheeler, J. L. (1978). Response by cattle grazing sorghum to salt–sulphur supplements. *Aust. J. exp. Agric. Anim. Husb.* **18**, 741.

Beale, I. F. (1973). Tree density effects on yields of herbage and tree components in south west Queensland mulga (*Acacia aneura* F. Muell.) scrub. *Trop. Grassl.* **7**, 135.

Brinkman, R. and Smyth, A. (1973). *Land evaluation for rural purposes*. Int. Inst. Land Reclamation and Improvement, Wageningen. Publ. No. 17.

Brown, L. H. (1971). The biology of pastoral man as a factor in conservation. *Biol. Conserv.* **3**, 93.

Campbell, Q. P., Ebersohn, J. P., and Von Broembsen, H. H. (1962). Browsing by goats and its effects on the vegetation. *Herb. Abstr.* **32**, 273.

Catchpoole, V. R. (1969). Preliminary studies on curing and storing Nandi Setaria hay. *Trop. Grassl.* **3**, 65.

Catchpoole, V. R. (1970). Laboratory ensilage of three tropical pasture legumes – *Phaseolus atropurpureus*, *Desmodium intortum*, and *Lotononis bainesii*. *Aust. J. exp. Agric. Anim. Husb.* **10**, 568.

— and Henzell, E. F. (1971). Silage and silage-making from tropical herbage species. *Herb. Abstr.* **41**, 213.

Chadhokar, P. A. (1977). Establishment of stylo (*Stylosanthes guianensis*) in kunai (*Imperata cylindrica*) pastures and its effect on dry matter and animal production in the Markham Valley Papua New Guinea. *Trop. Grassl.* **11**, 263.

Chang, J. (1968). *Climate and agriculture*. Aldine, Chicago.

Coaldrake, J. E. (1970). The Brigalow. In *Australian grasslands* (ed. R. M. Moore) pp. 123–40. ANU Press, Canberra.

Cull, J. K. and Douglas, N. J. (1974). Nitrogen for oats after the first grazing. *Qd Agric. J.* **100**, 10.

Davies, J. G. and Shaw, N. H. (1964). The regional approach. In *Some concepts and methods in sub-tropical pasture research*. CAB Bull 47, pp. 10–16. Hurley, Berks., England.

FAO (1973). Kenya – Rangeland surveys. Range Management Div., Min. of Agric. (FAO/AGP: SF/KEN II. Tech. Rep. 5.)

— (1974). Kenya – Range ecology, livestock production and wildlife ecology research. Range Management Div., Min. of Agric. (FAO/AGP: SF/KEN II. Tech. Rep. 4.)

Fitzpatrick, E. A. and Nix, H. A. (1970). The climatic factor in Australian grassland ecology. In *Australian grasslands* (ed. R. M. Moore) pp. 3–27. ANU Press, Canberra.

Gillard, P. (1970). Pasture development in the dry tropics of North Queensland. *Proc. 11th Int. Grassl. Cong.*, Australia, p. 807.

Greenhalgh, J. F. D., Aitken, J. W., and Reid, G. W. (1972). A note on the zero-grazing of beef cattle. *J. Br. Grassld Soc.* **27**, 173.

Griff, D. B. (1974). *The agricultural systems of the world. An evolutionary approach*. Cambridge University Press.

Harrington, G. N. and Pratchett, D. (1974). Stocking rate trials in Ankole, Uganda. I. Weight gain of Ankole steers at intermediate and heavy stocking rates under different managements. *J. agric. Sci., Camb.* **82**, 497.

Heady, H. F. (1972). *Range management in East Africa*. Govt. Printer, Nairobi, Kenya.

Humphreys, L. R. (1977). Potential of humid and subhumid rangelands. In *Potential of the world's forages for ruminant animal production*. Winrock. Int. Livestock Res. Train. Center, Arkansas, USA.

Hutchinson, K. J. (1966). A note on wool production responses to fodder conservation in pastoral systems. *J. Br. Grassl. Soc.* **21**, 303.

— (1972). Fodder conservation questioned. In *Rural research in CSIRO*, No. 76, p. 21.

Hutton, E. M. and Henzell, E. F. (1976). Planning and organising pasture research. In *Tropical pasture research, principles and methods* (ed. N. H. Shaw and W. W. Bryan) CAB Bull. 51. Hurley, Berks., England.

Isbell, R. F. and McCown, R. L. (1976). In *Tropical pasture research, principles and methods* (ed. W. H. Shaw and W. W. Bryan) CAB Bull. 51. Hurley, Berks., England.

Ivens, G. W. (1972). The problem of bush in rangeland. *Span* **15**, 23.

Jasiorowski, H. A. (1975). Intensive systems of animal production. In *Proc. 3rd World Conf. Anim. Prod.*, p. 369. Sydney University Press.

Johnson, D. L. (1969). *The nature of nomadism*. Univ. of Chicago, Dept. of Geography, Research Paper, No. 118.

Kemp, D. (1974). Comparison of oats and annual rye grass as winter forage crops. *Trop. Grassl.* **8**, 155.

Low, B. S. and Low, W. A. (1975). Feeding interactions of red kangaroo and cattle in an arid ecosystem. *Proc. 3rd World Conf. Anim. Prod*, p. 87. Sydney University Press.

Lowe, K. F., Filet, G. F., Burns, M. A., and Bowdler, T. M. (1977). Effect of sod-seeded Siratro on beef production and botanical composition of native pastures in south-eastern Queensland. *Trop. Grassl.* **11**, 223.

Martin, S. C. and Cable, D. R. (1974). Managing semidesert grass-shrub ranges. USDA Forest Service Tech. Bull. No. 1480.

Milles, A. H. and Hall, R. B. (1978). Central Burnett cattlemen make money from ryegrass. *Qd Agric. J.* **104**, 114.

Monod, T. (1975). *Pastoralism in tropical Africa*, pp. 8–186. Oxford University Press, London.

Moore, A. W. and Rusell, J. S. (1976). Climate. In *Tropical pasture research: principles and methods* (eds N. H. Shaw and W. W. Bryan) CAB Bull. 51. Hurley, Berks., England.

Moore, R. M. and Walker, J. (1972). *Eucalyptus populnea* shrub woodlands: control of regenerating trees and shrubs. *Aust. J. exp. Agric. Anim. Husb.* **12**, 437.

National Research Council (1970). Nutritional requirements of beef cattle. Nat. Acad. Sci., Washington, DC.

Newman, R. J. (1970). Dry temperate forests and heaths. In *Australian grasslands* (ed. R. M. Moore) p. 159. ANU Press, Canberra.

Newsome, E. A. (1971). Competition between wildlife and domestic livestock. *Aust. Vet. J.* **47**, 577.

Norman, M. J. T. (1963). Wet season grazing of sown pastures and fodder crops at Katherine, N.T. CSIRO Aust. Div. Land Res. Reg. Surv. Tech. Paper No. 22.

—— (1969). The effect of burning and seasonal rainfall on native pasture at Katherine, N.T. *Aust. J. exp. Agric. Anim. Husb.* **9**, 295.

—— (1970). Relationships between liveweight gain of grazing beef steers and availability of Townsville lucerne. *Proc. 11th Int. Grassl. Cong.*, Australia, p. 289.

Oram, P. A. (1975). Livestock production and integration with crops in developing countries. *Proc. 3rd World Conf. Anim. Prod.*, p. 309. Sydney University Press.

Pearse, C. K. (1970). Range deterioration in the Middle East. *Proc. 11th Int. Grassld Cong.*, Australia, p. 26.

Pressland, A. J. (1976). Effect of stand density on water use of mulga (*Acacia aneura* F. Muell) woodlands in south-western Queensland. *Aust. J. Bot.* **24**, 177.

Raymond, W. F. (1970). The utilization of grass and forage crops by cutting or grazing. *Proc. 11th Int. Grassl. Cong.*, Australia, p. A95.

Robinson, D. M. (1967). Analysis of liveweight loss in beef cattle on native pastures in the Kimberleys, Western Australia. *J. Aust. Inst. Agric. Sci.* **33**, 218.

Scateni, W. J. (1969). Sown and native pasture management. M.Agric. Sci. thesis, University of Queensland.

Shaw, N. H. (1961). Increased beef production from Townsville lucerne (*Stylosanthes sundiaca* Taub.) in the spear grass pastures of central coastal Queensland. *Aust. J. exp. Agric. Anim. Husb.* **1**, 13.

— and Bryan, W. W. (1976). *Tropical pasture research: principles and methods* CAB Bull. 51. Hurley, Berks., England.

— and 't Mannetje, L. (1970). Studies on a spear grass pasture in central coastal Queensland — the effect of fertilizer, stocking rate and oversowing *Stylosanthes humilis* on beef production and botanical composition. *Trop. Grassl.* **4**, 43.

— and Norman, M. J. T. (1970). Tropical and sub-tropical woodlands and grasslands. In *Australian grasslands* (ed. R. M. Moore) p. 112. ANU Press, Canberra.

Skerman, P. J. (1966). Land and water use survey in Kordofan Province of the Republic of the Sudan. FAO/UN Special Fund. Report FOX-SUD-A47.

Skovlin, J. M. (1971). Ranching in East Africa: a case study. *J. Range Mgmt* **24**, 263.

Slatyer, R. O. (1960). Agricultural climatology of the Katherine area, N.T. CSIRO Aust. Div. Land Res. Reg. Surv. Tech. Paper No. 13.

Smetham, M. L. (1973). Grazing management. In *Pastures and pasture plants* (ed. R. H. M. Langer) p. 179. A. H. and A. W. Reed, Wellington, New Zealand.

Smith, C. A. (1968). Annual Report, CSIRO Division of Tropical Pastures 1967–68, pp. 28–9.

Stewart, G. A. (1968). *Land evaluation*. Macmillan, Melbourne.

Stillman, S. L. and Ostrowski, H. (1976). Temperate pastures for dairying in south-east Queensland. *Qd Agric. J.* **102**, 514.

Stobbs, T. H. (1966). Beef production from Uganda pastures containing *Stylosanthes gracilis* and *Centrosema pubescens. Proc. 9th Int. Grassl. Cong.*, Brazil, p. 939.

— (1970). Automatic measurement of grazing time by dairy cows on tropical grass and legume pastures. *Trop. Grassl.* **4**, 237.

— and Wheeler, L. J. (1977). Response by lactating cows grazing sorghum to sulphur supplementation. *Trop. Agric., Trin.* **54**, 299.

Stoddart, L. A., Smith, A. D., and Box, T. W. (1975). *Range management*, 3rd edn. McGraw-Hill, New York.

Swift, J. (1975). Pastoral nomadism as a form of land use: Twareg of the Adrar N Iforus. In *Pastoral nomadism in tropical Africa* (ed. T. Monod) p. 441. Oxford University Press, London.

Tiedemann, A. R. (1970). Effect of mesquite (*Prosopis juliflora*) trees on herbaceous vegetation and soils in the desert grassland. Ph.D dissertation, Dept Watershed Manage., University of Arizona, Tuscon.

Tothill, J. C. (1971*a*). A review of fire in the management of native pastures with particular reference to north-eastern Australia. *Trop. Grassl.* **5**, 1.

— (1971*b*). Grazing, burning and fertilizing effects on the regrowth of some woody species in cleared open forests in south-east Queensland. *Trop. Grassl.* **5**, 31.

— (1974). Experiences in sod-seeding siratro into native speargrass pastures on granite soils near Mundubbera. *Trop. Grassl.* **8**, 128.

Tubiana, M. J. and Tubiana, J. (1975). Tradition et developpement au Soudan oriental: l'exemple Zaghawa. In *Pastoral nomadism in tropical Africa* (ed. T. Monod) p. 468. Oxford University Press, London.

Vink, A. P. A. (1975). *Land use in advancing agriculture*. Springer-Verlag, Berlin.

Walker, J., Moore, R. M., and Robertson, J. A. (1972). Herbage response to tree and shrub thinning in *Eucalyptus populnea* shrub woodlands. *Aust. J. agric. Res.* **23**, 405.

Whiteman, P. C. (1969). Nitrogen on intensive pastures for beef cattle. *Trop. Grassl.* **3**, 86.

Wright, H. A. (1974). Range burning. *J. Range Mgmt* **27**, 5.

Index